Springer Tracts in Nature-Inspired Computing

The book series is aimed at providing an exchange platform for researchers to summarize the latest research and developments related to nature-inspired computing in the most general sense. It includes analysis of nature-inspired algorithms and techniques, inspiration from natural and biological systems, computational mechanisms and models that imitate them in various fields, and the applications to solve real-world problems in different disciplines. The book series addresses the most recent innovations and developments in nature-inspired computation, algorithms, models and methods, implementation, tools, architectures, frameworks, structures, applications associated with bio-inspired methodologies and other relevant areas.

The book series covers the topics and fields of Nature-Inspired Computing, Bio-inspired Methods, Swarm Intelligence, Computational Intelligence, Evolutionary Computation, Nature-Inspired Algorithms, Neural Computing, Data Mining, Artificial Intelligence, Machine Learning, Theoretical Foundations and Analysis, and Multi-Agent Systems. In addition, case studies, implementation of methods and algorithms as well as applications in a diverse range of areas such as Bioinformatics, Big Data, Computer Science, Signal and Image Processing, Computer Vision, Biomedical and Health Science, Business Planning, Vehicle Routing and others are also an important part of this book series.

The series publishes monographs, edited volumes and selected proceedings.

Nilanjan Dey
Editor

Optimizing Solutions for Real-Life Problems

Springer

Editor
Nilanjan Dey
Department of Computer Science
and Engineering
Techno International New Town
Kolkata, West Bengal, India

ISSN 2524-552X ISSN 2524-5538 (electronic)
Springer Tracts in Nature-Inspired Computing
ISBN 978-981-96-3215-2 ISBN 978-981-96-3216-9 (eBook)
https://doi.org/10.1007/978-981-96-3216-9

This Springer imprint is published by the registered company Springer Nature Singapore Pte Ltd.
The registered company address is: 152 Beach Road, #21-01/04 Gateway East, Singapore 189721, Singapore

Preface

People deal with a variety of issues in their daily lives and in industry. There are numerous aspects that determine how complex a problem is. To complete various activities in the actual world, we rely heavily on our intuition. A real-world problem is converted into a mathematical description through the process of mathematical modelling. A good solution, or one that is somewhat close to the ideal, is guaranteed by mathematical modelling. In reality, the ideal solution found through mathematical modelling can be used. However, real-world situations often involve unknowns, anomalies, mysterious actions, randomness, instantaneous modifications, and other elements. Therefore, obtaining solutions to real-world problems might be challenging and open much scope for research. *This book* includes 10 chapters. For researchers, practitioners, and decision-makers, in Chap. 1, *Banerjee et al.* offered incisive advice on how to use optimization algorithms to solve issues and promote innovation in a variety of industries. This chapter covers the most modern optimization techniques for dealing with practical optimization problems. In Chap. 2, *Eakin et al.* show the potential of evolutionary approaches to shape the future of cybersecurity through a thorough analysis of existing uses and future possibilities. In Chap. 3, *Sheikhi and Dowlatshahi* present designers with the idea of differentiating between different amplifier technologies and optimizing the operating and design characteristics of each linearizer or amplifier stage. In Chap. 4, *Kamal and Alharbi* attempt to investigate how game theory and explainable artificial intelligence (XAI) may be combined to optimize software systems. In Chap. 5, *Chowdhury et al.* demonstrated how automated sentiment analysis systems can improve customer satisfaction by using a combination of classic machine learning models and a rule-based classifier to provide quick and precise insights into consumer opinions. With an emphasis on static surroundings with predefined maps and obstacles, a number of biologically inspired algorithms were evaluated in Chap. 6 by *Agarwal* in an attempt to replicate real-world AMR navigation issues. *Boopathi et al.* addressed the response analysis of the proportional integral derivative (PID)-optimized hybrid genetic firefly algorithm for frequency regulation in a two-area hybrid power system in Chap. 7. In Chap. 8, *Moradpouri and Dolatshahi* intend to describe two real-world examples from the two exploratory geochemical projects to highlight some of the most popular machine learning algorithms for the optimization of multivariate geochemical data processing. *Daw et al.* discuss the idea of the Bat algorithm and how it has been used to alter several real-world problem domains in Chap. 9. The chapter also covers a number of applications of the Bat algorithm to practical problems. In Chap. 10, *Mehta et al.* intended to assist professionals in better understanding the advantages and disadvantages of taking into account various objectives by offering a comparative analysis between multiobjective and single-objective feature selection.

The editor expresses gratitude to the excellent writers and reviewers for their contributions and accomplishments. They have worked together diligently to produce this outstanding book. The gratitude is also extended to the Springer team members for their assistance.

Kolkata, West Bengal, India Nilanjan Dey

Contents

About the Editor

Nilanjan Dey (Senior Member, IEEE) received the B.Tech., M.Tech. in information technology from West Bengal Board of Technical University and Ph.D. degree in electronics and telecommunication engineering from Jadavpur University, Kolkata, India, in 2005, 2011, and 2015, respectively. Currently, he is Associate Professor with the Techno International New Town, Kolkata and a visiting fellow of the University of Reading, UK. He has authored over 300 research articles in peer-reviewed journals and international conferences, and 40 authored books. His research interests include medical imaging and machine learning. He is also the Editor-in-Chief of International Journal of Ambient Computing and Intelligence, Associate Editor of IEEE Transactions on Technology and Society and series Co-Editor of Springer Tracts in Nature-Inspired Computing and Data-Intensive Research from Springer Nature and Advances in Ubiquitous Sensing Applications for Healthcare from Elsevier, etc.

Chapter 1
Revolutionizing Optimization in Real-Life Implementation

Archisman Banerjee, Ayan Dutta, Bitan Misra, and Sayan Chakraborty

1 Introduction

Today's competitive world demands efficient resource management to thrive. The question of how to solve these problems has emerged, and the use of optimization is the definitive answer. To perform tasks as efficiently as possible, optimization is useful in a variety of real-world situations. A few real-world instances of the optimal approach include engineering, manufacturing, production, stock maintenance, transportation, scheduling, and business solutions. Finding the optimal answer from a set of workable options is the focus of the mathematical and computer science field of optimization. Making anything operate as effectively and efficiently as possible—like a system or process—is known as optimization. Optimization challenges are classified into various types. Certain straightforward problems, including those with obvious solutions or no-choice variables, do not require formal optimization. Nevertheless, the goal is to achieve the best results most of the time, and a mathematical solution is required. Generally, most circumstances require optimization. Finding the "best" result in relation to a set of restrictions or priorities is the goal of optimization. Additionally, what-if analysis can benefit from the use of an optimization model that

A. Banerjee
Department of Electronics and Telecommunication Engineering, KIIT University, Bhubaneswar, India

A. Dutta
Aantaric Technologies Private Limited, Kolkata, India

B. Misra (✉)
Department of Computer Science and Engineering, Techno International New Town, Kolkata, India
e-mail: bitan.misra@tint.edu.in

S. Chakraborty
Department of Computer Science and Technology, JIS College of Engineering, Kalyani, West Bengal, India

N. Dey (ed.), *Optimizing Solutions for Real-Life Problems*, Springer Tracts in Nature-Inspired Computing, https://doi.org/10.1007/978-981-96-3216-9_1

is well designed, since it can highlight areas for improvement or suggest trade-off. Several academic fields use optimization to solve technical challenges [1].

The best course of action in a given scenario is ascertained using a class of algorithms called optimization algorithms. Algorithms for optimization are effective techniques for resolving challenging issues. They could completely change the way we work with data. Depending on the objective function being optimized, the optimization process entails taking a given set of parameters and determining the best possible solution that maximizes value or minimizes cost [2]. Mathematical optimization has several practical applications in a wide range of fields and industries. It offers a completely new method for approaching difficult issues and arriving at conclusions. Techniques for mathematical optimization have grown to be very valuable resources for governments, corporations, and scholars alike. They can be used for streamlining healthcare operations, improving logistics and supply chain management, and optimizing manufacturing production processes. They can also be used to maximize financial returns while minimizing risks. An introduction to popular optimization techniques and several real-world examples of their use are provided in this chapter.

1.1 Understanding the Optimization

Optimization is an action, procedure, or approach used to create anything (such as a design, system, or choice) as ideal, efficient, or effective as feasible. The term "optimize" came from the Latin word "optimum", which means "best". In its most fundamental form, optimization is the act of making something as good as possible. Therefore, the process of getting the optimal result given the restrictions can be defined as optimization. In real life, humans have to make several administrative, economic, or technological decisions at various phases of designing, building, and maintaining any system. Either maximizing the anticipated reward or minimizing the required effort is the goal of all such decisions. The process of optimizing involves identifying the conditions that produce a function's maximum or minimum value, as the effort required or the intended result may be expressed as a function of particular decision variables in any real-world situation. At this stage, a very basic example can be given to understand the requirements of optimization [2]. Let us assume that there are two points, point A and point B, as shown in Fig. 1. Now, there might be a situation of finding a path from A to B among some given paths, as shown in the figure, which will take the least amount of time under a constant speed. This is a simple problem of finding a travel path that optimizes the travel time.

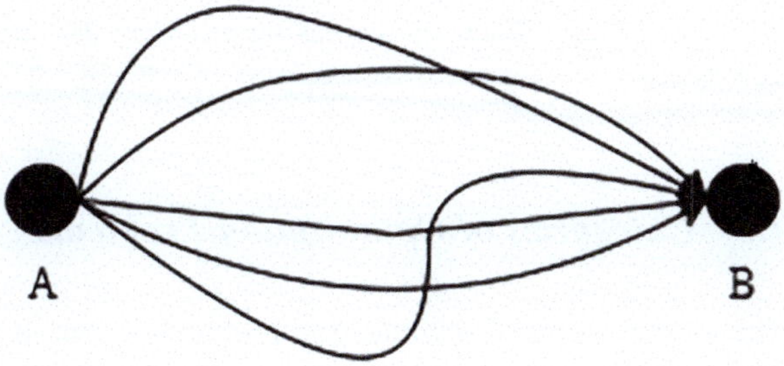

Fig. 1 Finding the shortest path

1.2 The Significance of Optimization

Optimization involves doing the most with the least number of resources. The "best of something" must be acquired through a collection of limitations or criteria that are ranked. These might include factors like output, strength, durability, dependability, efficiency, and use that need to be maximized. Thus, it is necessary to maximize these elements in any system or design. In addition, one should bear in mind the need to minimize expenses, labor, time, etc. Engineers are frequently asked with design tasks that call for them to find an efficient way to locate a design that satisfies the specified criteria while working within the various limitations. Process optimization has many advantages for businesses. The following are a few of the greatest benefits:

Greater quality of job produced: A more efficient method produces higher-quality output. A well-designed process may reduce costly mistakes and enhance the experience and outcomes for clients. Optimizing the work environment is essential, and developing a process management strategy is the best way to achieve the desired outcomes.

Boost efficiency: Naturally, an optimized process is one that is efficient. Methods to improve inconsistencies and inefficiencies are aimed at increasing production. Businesses will be able to analyze ROIs and make better decisions with an improved system.

Cutbacks on expenses: Cost reduction and waste elimination are two of the main objectives of process optimization. Organizations can save time and money by identifying inefficiencies in their current processes and putting improvements into place. They are able to function more effectively and distribute resources more wisely as a result.

Improved consistency and quality: Process standardization and variation elimination are two ways in which process optimization can increase the caliber and consistency of goods and services. Consequently, there may be an increase in client satisfaction and loyalty.

A stronger edge over competitors: Businesses that make process optimization investments can outperform their competitors. Businesses may stand out in the market and draw in new clients by increasing productivity, cutting expenses, and increasing quality.

1.3 Categories of Optimization Problems

The classification of optimization problems is based on several factors, including the kind of constraints, the characteristics of the design variables, the problem's physical structure, the type of equations involved, the variables' deterministic nature, the design variables' permissible values, the separability of the functions, and the quantity of objective functions. Figure 2 represents various categories of optimization problems.

1.3.1 Grouping Optimization Problem as Per Constraints

The constraint optimization problem can be classified into two categories:

Optimization issues with constraints: These problems have one or more restrictions or constraints.

Optimization issues without constraints: These types of optimization problems are without constraints.

1.3.2 Grouping According to the Type of Design Variables

There are two main classification categories in this categorization.

The first category involves determining a set of design parameters that will reduce or maximize a given function of these parameters under particular constraints.

The second category seeks to identify a collection of design parameters that, when subjected to a set of constraints, minimize an objective function. All of the design parameters are continuous functions of another parameter.

1.3.3 Categorization According to the Problem's Physical Structure

Based on their physical structure, optimization problems can be classified into two groups: optimal control problems and non-optimal control problems.

Optimal Control: A mathematical programming problem with multiple stages, where each step emerges from the previous stage in a predetermined way, is known as an optimum control (OC) problem. The state variables and control or design variables are the two categories of variables that characterize it.

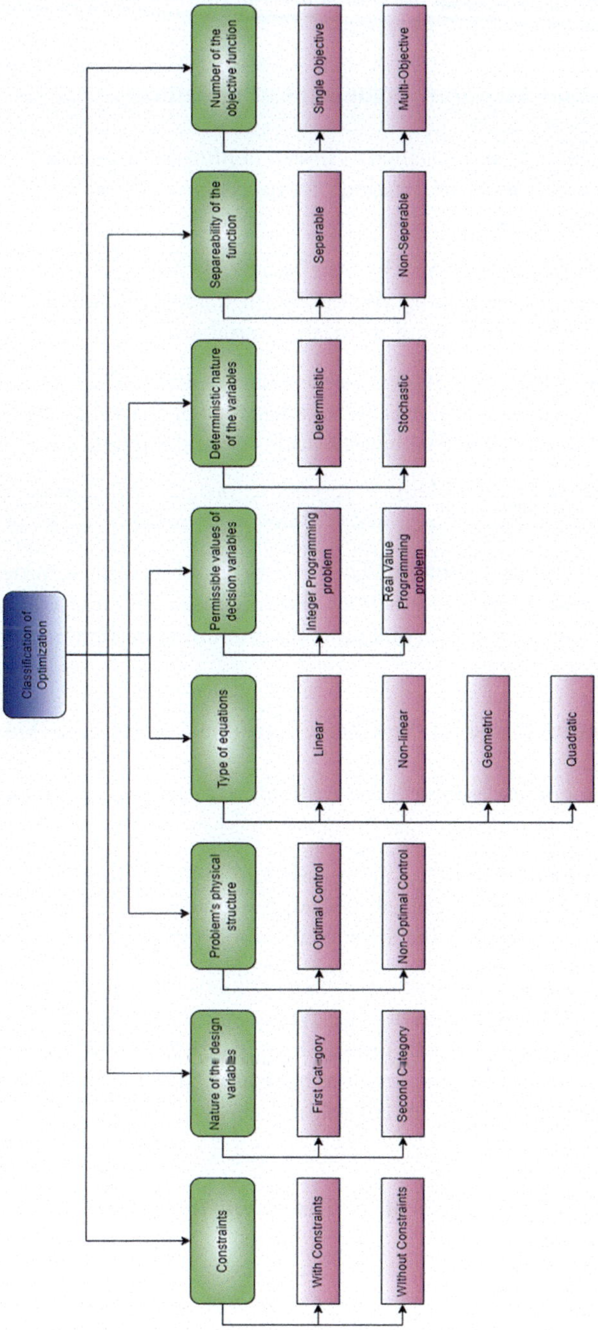

Fig. 2 Different categories of optimization problems

Non-optimal Control: The issues where we do not need to minimize their objective functions and do not need to keep track of status or control variables.

1.3.4 Categorization According to the Type of Equations

Optimization problems can be classified as linear, nonlinear, geometric, or quadratic programming problems based on the kind of expressions utilized for the objective function and the constraints.

Linear programming problem: The mathematical programming problem is referred to as a linear programming (LP) problem if the goal function and all of the restrictions are linear functions of the design variables.

Nonlinear programming problem: The problem is referred to as a nonlinear programming (NLP) problem, which is the broadest type of programming problem, if any of the relationships between the objectives and constraint functions are nonlinear.

Geometric programming problem: a geometric programming problem (GMP) where the restrictions and goal function are represented as polynomials in X.

Quadratic programming problem: The best-behaved nonlinear programming issue with a quadratic objective function, linear constraints, and concavity (for maximization problems) is called a quadratic programming problem.

1.3.5 Categorization Based on the Choice Variables' Acceptable Values

Problems can be categorized as integer and real-valued programming problems using this framework.

Integer programming problem: An optimization problem is referred to as an integer programming problem if some or all of its design variables are limited to accepting only discrete or integer values. Teachers taking time for class or nurses who are assigned to shifts in hospitals are discrete individuals.

Real-valued programming problem: This type of problem involves methodically selecting real variable values from an allowable set in order to minimize or maximize a real function. It is referred to as a real-valued programming problem when the allowable set exclusively consists of real values.

1.3.6 Classification on the Basis of the Deterministic Nature of the Variables

Optimization problems fall into one of two categories under this classification: stochastic or deterministic programming problems.

Deterministic programming problem: All of the design variables in this kind of challenge are deterministic.

Stochastic programming problem: Some or all of the parameters (design variables and/or pre-assigned parameters) in this kind of optimization problem are probabilistic (nondeterministic or stochastic).

1.3.7 Classification on the Basis of the Separability of Functions

Optimization problems can be categorized as separable or non-separable programming problems based on the separability of the objective and constraint functions.

Separable programming problems: An approximation technique for resolving certain nonlinear programming issues is separable programming. Separable functions are those that can be expressed as the sum of their sub-functions, each of which is a function of a single variable. The objective function and the constraint functions in the separable programming problem can be expressed as separable functions.

Non-separable programming problems: In contrast to the separable function, if a nonlinear function cannot be represented as a separable function, then the function is non-separable. A non-separable programming problem is one that involves optimization and occurs when either the objective function, the constraint functions, or both are non-separable.

1.3.8 Classification on the Basis of the Number of Objective Functions

Objective functions fall into one of two categories: single- or multi-objective programming challenges.

Single-objective programming problem: In this optimization problem, there is only a single objective.

Multi-objective programming problem: In this type of problem, there could be two or more objective parameters.

1.4 Mathematical Modeling for Real-Life Optimization

We can simulate and forecast the behavior of complicated systems via mathematical modeling, which uses equations and algorithms to describe real-world phenomena. We can look for the best answers to problems ranging from logistical planning to natural resource management when combined with optimization approaches. Solving real-life optimization problems involves first developing the problem statement and then defining objective functions, design constraints, and design variables.

Selecting the optimal approach to complete a certain task is the challenge of optimization. This usually means finding the maximum or minimum value of a certain parameter based on the application field. Examples of these include the least amount of money required, the maximum power that a device can generate, and the shortest time required to do a task. Numerous of these issues can be resolved by first identifying the relevant parameters and then applying mathematical methods. This is a basic example of an optimization problem in industry [3–5].

Consider a scenario in a reputed manufacturing company. The company has three distinct sorts of machinery (M1, M2, and M3) that can make three various kinds of products (P1, P2, and P3). Each machine has its own capacity to produce different products, which we call the production rate of the machine. Each machine will have a certain run-time per week, which we can call the availability of the machine. The products will also have some profit margin. Now, the company might want to increase the overall profit by the maximum usage of the machines. Here, the maximizing parameter is the profit, and the constraints are the total availability of the machine. The cost of manufacturing different products can also be considered; then, the minimizing parameter would be the production cost. The whole situation can be summarized in Table 1.

The production rate of machine M1 for P1 is R11, that for P2 is R12, and that for P3 is R13. The total number of products P1 is $\times$ 1, P2 is $\times$ 2 and P3 is $\times$ 3. Therefore, the maximizing function can be written as given in Eq. (1)

$$Profit = x_1 * F_1 + x_2 * F_2 + x_3 * F_3 \tag{1}$$

There are also constraints because the availability of machines is fixed every week, and the constraint function can be written as given in Eqs. (2), (3), (4), and (5):

$$x_1 * R_{11} + x_2 * R_{12} + x_3 * R_{13} \leq A_1 \tag{2}$$

$$x_1 * R_{21} + x_2 * R_{22} + x_3 * R_{23} \leq A_2 \tag{3}$$

$$x_1 * R_{31} + x_2 * R_{23} + x_3 * R_{33} \leq A_3 \tag{4}$$

Table 1 Example of an optimization problem in an industry

Machines	Products			
	P1	P2	P3	Availability
M1	R_{11}	R_{12}	R_{13}	A_1
M2	R_{21}	R_{22}	R_{23}	A_2
M3	R_{31}	R_{32}	R_{33}	A_3
Per unit profit	F_1	F_2	F_3	

Here, the profit function is called the objective function, the availability of the machines are design constraints, and the quantities of the products × 1, × 2, and × 3 are the variables that we want to find such that the company produces maximum profit with the given constraints.

1.4.1 Linear Programming

A straightforward technique called linear programming, or LP, uses a linear function to represent complex real-world relationships. The words "linear" and "programming" combine to form the term "linear programming". A link between numerous variables of degree one is defined as "linear". The act of choosing the optimal answer from a range of options is referred to as "programming". Relationships in the real world can be very complex. Nonetheless, these interactions can be represented via linear programming, which facilitates their analysis. Numerous industries, including manufacturing, telecommunications, energy, and transportation, use linear programming [6].

Linear programming (LP) or linear optimization is the problem of maximizing or reducing a linear function that is subject to linear restrictions. Disparities or equalities may serve as limitations. To determine the best use of resources, linear programming is crucial. For example, a delivery person has five packages to deliver from the hub (at A) to several homes (U, V, W, Y, Z) in a single day. The figure illustrates how the houses have been separated by lines. If the postman wants to figure out the quickest route that will enable him deliver the letters, then this becomes a linear programming issue. To choose the best option, which in this case is the quickest path, LP will be applied [7].

A linear programming issue consists of decision variables, an objective function, constraints, and nonnegative restrictions. A decision variable is a quantity that needs to be determined. Decision variables have a linear relationship and can be used to identify the best course of action [8]. The decision variable determines the output. It offers the conclusive solution to the problem. Finding the decision variables is the first step in solving any problem. The objective function is the following element. An issue's objective function should be expressed quantitatively, for as by maximizing profit or reducing costs. There must be a linear relationship in the function between two or more variables. This suggests that the degree of the variable is one. The limitations placed on the decision variables to restrict their value are known as constraints [9]. The nonnegative constraints provide the nonnegative value that the decision variables must always have. These limitations also apply to the resources that are at hand, including scarce personnel, supplies, equipment, etc. Redundant constraints are those that are observably present but do not impede the resolution of the researched problem. The way that the formula works is shown in Eq. (5)

$$ax + by = Z \tag{5}$$

The restrictions of the formula's operation is given in Eqs. (6) and (7)

$$fx + gy \leq h \tag{6}$$

$$cx + dy \leq e \tag{7}$$

Other non-negative limitations are x ≥ 0, y ≥ 0. The choice variables x and y in the aforementioned example dictate the output and final solution of the LP issue. The objective function, Z, is a linear function that needs to be optimized through either minimization or maximization in order to determine the solution. In this case, the specified linear function is regarded as the objective function. The collection of linear inequalities known as linear constraints must be satisfied by the objective function, which may include many variables that are subject to the requirements.

Formulating a linear programming problem with the available data is the most important step in solving it. The following are the procedures for using linear programming to solve problems:

Step 1: Identify the decision-making elements.

Step 2: Create the objective function in step two. Verify whether the function should be maximized or decreased.

Step 3: Write down the constraints.

Step 4: Verify that the choice variables are either equal to or greater than zero. (Non-detrimental constraint).

Step 5: Use simplex or graphical methods to solve the linear programming problem.

There are various approaches to solving the linear programming issue, including the graphical method, the simplex method, and the use of R, an open solver, and other tools.

Simplex Method

One common approach to solving optimization problems in linear programming is the simplex method. When dealing with situations involving two or more choice variables, the simplex method might be used. Finding the optimal solution given a maximum or minimum equation with linear constraints is known as a linear program. Finding the optimal solution given a maximum or minimum equation with linear constraints is known as a linear program. It usually consists of several limitations expressed as inequalities and functions. The solution to the inequality is typically found at a vertex, which defines a polygonal area. The vertices can be methodically examined as possible solutions via this methodology.

Through an iterative process, the technique approaches and ultimately reaches the goal function's maximum or lowest value. The approach also helps the decision maker see redundant constraints, a comprehensive resolution, several options, and an impractical problem, providing a comprehensive understanding of the business scenario.

Step 1: Using the constraints provided, formulate the linear programming problems.

Step 2: Add the slack variable to each inequality as needed to transform all of the provided inequalities into equations or equalities of the linear programming problems.

Step 3: In step 3, make the first simplex table by aligning the constraint equations in a row and positioning the goal function at the bottom row. The simplex table is the name given to the resulting table.

Step 4: Select the greatest negative item in the bottom row to determine the pivot column. By defining the highest coefficient in the goal function as the largest negative item in the bottom row, we may quickly increase the objective function's value.

Step 5: In step five, compute the quotients. Finding the quotient requires dividing the entries in the far right column by the entries in the first column, leaving out the bottom row. Using the least quotient, the row is identified. This step's row and the step's element will serve as the pivot elements.

Step 6: Pivot the column such that all other entries are zero.

Step 7: Close the procedure if the bottom row contains no negative items. If not, go back to step 4.

Step 8: Last, ascertain the answer linked to the ultimate simplex tableau.

Graphical Method

The term "graphical method" refers to the process of solving a system of linear equations by creating a graph. This also applies to problems with linear programming. Using graphical methods makes it simple to answer any optimization programming problem involving two variables. The majority of the analysis on a two-dimensional graph can be done with these variables, which can be referred to as X and Y. This approach imposes limitations on the set of inequalities. The disparities are subsequently represented on an XY plane. The viable zone can be determined by looking at the intersecting region once all of the inequalities have been drawn on the XY graph.

Step 1: Create the LPP.

First, the decision variables, objective function, and constraints must be defined for the linear programming problem. A standard form should be used for the problem.

Each constraint line is plotted, and the graph is constructed.

The limitations of this study should be noted visually. The constraint equation requires that each inequality be handled as an equation. One variable is given an arbitrary value, whereas the other variable's value is found by solving the equation. Similarly, the variable is once more given a new, random value, and it is simple to obtain the matching value for the other variable.

Step 2: Ascertain each constraint line's valid side.

Every restriction is placed on a graph. A straight line can be used to symbolize the constraint, and the part of the graph that meets every constraint is known as the viable region. The polygon created by the line junction is known as the feasible region.

Step 3: Determine the region of the feasible solution.

Finding the best answer is the next task. The graph point that maximizes or minimizes the goal function is the optimal solution. A straight line is used to represent the objective function, and the point at which the line intersects the feasible region is the optimal solution.

Step 4: Locate the optimum points.

Step 5: Determine the coordinates of the optimal points.

Step 6: At the optimal position, assess the goal function.

Step 7: Finally, consider the answer in light of the overall issue. The largest or lowest value of the goal function that can be attained while meeting every restriction is provided by the optimal solution.

For numerous reasons, linear programming is widely used in the optimization domain. The goal of linear programming is to determine a given linear function's optimal value. Either the highest or the smallest value could be the optimum value. In this case, the given linear function is thought of as the objective function. Depending on the circumstances, the objective function may contain multiple variables. It also needs to satisfy a set of linear inequalities known as linear constraints. In operations analysis, many functional issues can be modeled as linear programming issues.

1.4.2 Integer Programming

Within the field of optimization theory, integer programming is distinguished by the need for all or part of the decision variables to have integer values. Discrete numbers that reflect binary selections, counts, or allocations are called integers. Examples of integers are 0, 1, 2, or -3. Owing to the introduction of discontinuity and nonlinearity into the solution space, this constraint increases the difficulty of solving the problem. The collection of all possible values for the choice variables that meet the constraints is known as the solution space. The ideal solution for linear programming typically resides at one of the vertices or corners of the solution space, which is a convex polygon or polyhedron. In integer programming, the ideal solution is typically found somewhere in the middle of the solution space, which is typically a collection of discrete points. Integer programs look for the best possible integer solution that minimizes or maximizes a specified value of the cost function [10]. Integer programming is distinct from linear programming since it employs restrictions of integers rather than real numbers. Although linear programming techniques are widely applicable, they are not always appropriate when handling binary (0/1) and mixed-integer decisions [11].

The need for real-world situations is the reason for the significance of integer programming. Since they represent indivisible objects or choices, optimization problem solutions frequently require integer solutions. There are numerous practical uses for integer programming, such as resource allocation, portfolio selection, and logistics and transportation scheduling [12]. The ability to simulate discrete option decision-making, which is not possible with regular linear programming models alone, is one of its main advantages over classic LP methods. By mandating integer values, integer programming ensures that the solutions are not only theoretically optimal but also realistically usable and meaningful in real-world settings. Since integer programming may produce practical and workable solutions, it is extremely beneficial in a wide range of industries and applications, even though solving problems with integer programming becomes more complex [13].

Pure Integer Programming

Every decision variable in pure integer programming is limited to integer values. When making decisions about things like the amount of items or the distribution of labor, this type is commonly used because all the components are indivisible. A pure integer programming problem may have a structure similar to that of linear programming, but it is more difficult because of integer limitations.

Mixed Integer Programming

A more general variant of mixed integer programming allows for continuous variables in addition to integer variables. Practical applications frequently employ this type because it enables the simultaneous processing of both integer and noninteger decision variables. For example, in supply chain optimization, it may be necessary to choose an integer number of things to buy (integer variables) in addition to continuous factors such as transportation prices and delays.

0–1 Integer Programming

A specific type of pure integer programming known as 0–1 integer programming limits all integer variables to only accept values of 0 or 1. This kind is frequently applied to selection problem modeling, where each variable denotes a yes-or-no choice. A project in a project selection problem, for instance, can either be selected (1) or not (0).

One effective method for resolving a variety of significant optimization issues is integer programming. It helps decision-makers identify workable and effective solutions while also considering any limitations that the challenge may present. In summary, integer programming provides a useful method for locating the best answers in a variety of fields. Its ability to deliver dependable answers quickly and effectively has led to an increase in its use over time.

1.4.3 Simulated Annealing

One optimization technique that has garnered much interest is the simulated annealing method. A probabilistic method called simulated annealing (SA) is used to determine the global optimum of a given function. It is especially helpful for optimization issues with a large search space. One of the most straightforward and well-known meta-heuristic [14] techniques is called simulated annealing (SA) [15]. Randomness is a tool used by SA in its search procedure. Because of this, the technique works well for nonlinear objective functions in situations when other local search algorithms do not. The algorithm draws inspiration from metallurgical annealing [16]. The way the annealing process operates is that the material's atoms are initially excited to a high temperature, which allows them to move around a lot. Next, the excitement is gradually reduced, which causes the atoms to settle into a new, more stable configuration. Atoms travel quickly at high temperatures, and as the temperature decreases, so does their kinetic energy. The material becomes more ductile and manageable after the annealing process, as the atoms settle into more ordered conditions [17]. In a similar vein, SA searches for an optimal solution by progressively lowering the temperature (a control parameter) from a high-energy state (the starting solution) to a state of minimal energy. Similarly, by modifying the present state in accordance with a predetermined criterion, the simulated annealing algorithm creates a new potential solution (or neighbor of the current state) to the problem under consideration at each virtual annealing temperature. The metropolis criterion must then be satisfied for the new state to be accepted, and this process is repeated until convergence [18].

By randomly perturbing the initial solution and accepting the perturbation with a specific probability, the simulated annealing process iteratively improves the present solution. As the number of iterations increases, the likelihood of adopting an inferior answer steadily diminishes from its initial high level.

1.4.4 Evolutionary Algorithm

Adaptive heuristic search algorithms, or genetic algorithms (GAs), make up the majority of evolutionary algorithms. Natural selection and genetics serve as the foundation for genetic algorithms. These are creative applications of irregular searches, which concentrate the search on regions of the solution space with higher performance with the help of historical data [19]. They are widely used to generate top-notch solutions for problems relating to search and optimization.

Because genetic algorithms mimic the process of natural selection, species that are able to adapt to changes in their surroundings have a better chance of surviving, procreating, and evolving into new generations [20]. In other words, they mimic the "survival of the fittest" among people from successive generations to address an issue. Every generation is made up of a population of individuals, and each individual stands for a potential solution or point in the search space. These benefits come in numerous forms, such as the method's flexibility, resilience to changing conditions, and general ease of use [21–25].

Evolutionary systems are based on two fundamental forces:

- Recombination and mutation, two variation operators, produce the required diversity within the population, enabling originality.
- The population's mean quality of solutions is increased via selection as a force.

1.5 Real-Life Implementation of Optimization

Optimization algorithms have widespread applications and have already been implemented in solving real-life problems. The main advantage of the optimization algorithms is that their applications are not limited to a particular class of problems, so they have found applications in diverse fields. This section describes the implementation of these algorithms in different fields for optimizing the output.

1.5.1 Optimization in Manufacturing and Production

Many studies have investigated the application of optimization in manufacturing and production. In 1996, one study proposed the design and implementation of a linear programming model [26] to optimize the distribution and production of manufacturing companies. The model was proven to be efficient for parameters such as production and storage capacity analysis and cost analysis, and the number of constraints and variables can also be managed. The model was found to be effective and saved 3 million dollars for the company. Apart from linear programming, integer programming has also been found to be efficient in the optimization of production scheduling. A mixed integer programming model [27] was devised to address the problem of outsourcing. The main objective is to reduce the cost of outsourcing and take appropriate measures so that outsourcing can be performed through some subcontracts. In addition, a constraint handling technique was included in this model. The model is found to yield an optimum solution with fast convergence rates.

Multi-objective meta-heuristic optimization techniques have been found to be efficient in such studies. Specifically, the genetic algorithm (GA) is much superior for optimization problems in the production field. In a certain study, a canonical genetic algorithm approach and a multi-objective genetic algorithm approach were proposed [28]. The MOGA outperformed other techniques at that time and was found to be effective for such types of optimization problems. Therefore, the modified version of the GA is more effective than the GA is; thus, different variants of the GA have been implemented in further studies.

Meta-heuristic algorithms are found to be appropriate for space allocation in manufacturing cells. Enhanced simulated annealing algorithms play a major role in reconfigurable manufacturing [29]. Here, four variants of SA are proposed, and the variants are better than simple SA. In addition to SA, another nature-inspired meta-heuristic algorithm, particle swarm optimization, has been used in production scheduling. In [30], a hybrid version of PSO was used along with SA for scheduling and planning problems. The PSO guides global search, whereas the SA helps in local search. Another study reported that in automated manufacturing systems, the task scheduling problem for larger range implementation has been solved by using different meta-heuristic algorithms, such as the GA, SA, memetic algorithm (MA), and PSO algorithm [31]. Thus, problems in manufacturing and production can be addressed by a group of optimization algorithms, so only one algorithm cannot be declared to yield better results. Different algorithms are found to be effective for different subclasses of problems:

However, overall, meta-heuristic algorithms undoubtedly perform better in most cases. Although there are several reinforcement learning-based optimization approaches, until now, they have not provided as good a result as meta-heuristics have; however, research has still been conducted regarding such algorithms, and whether they can perform better than meta-heuristics is a matter of further research.

1.5.2 Optimization in Logistics and Supply Chain Management

The logistics and supply chain management also require optimization algorithms to be implemented for complex business problems. There are several problems in supply chain management, such as delivery problems, demand prediction, and shipping costs, which require suitable optimization techniques to address such problems. Although extensive studies have been performed on optimizing such problems via optimization algorithms, the application of linear programming optimization in these fields has been limited until now.

One study revealed that fuzzy linear programming has been used to optimize the supply chain model and minimize the cost of shipping and transportation [32]. However, both linear and integer programming models can be implemented but are not as effective; thus, research in this direction has become slightly narrower, and in this context, meta-heuristic algorithms play a major role. Although the modeling of logistics systems was performed through optimization or heuristics, in recent years, a meta-heuristic approach has been found to obtain near-optimization solutions, which normal optimization problems cannot obtain.

The supply chain network (SCN) design problem is a topic of great interest in supply chain management. A previous study proposed a GA for obtaining Pareto-optimal solutions for such design problems [33]. In this study, experimental verification is also carried out by taking data from a plastic-producing company in Turkey, and the results are satisfactory. In addition to the GA, the SA algorithm has also been implemented along with combinatorial optimization [34] in logistic supply management. This algorithm aims to solve optimized designs that help corporations plan for effective overall management at a lower cost. Here, extended SA is used for an efficient solution to this issue. Another popular meta-heuristic algorithm is PSO, which also has applications in these fields. In supply chain management, one of the main problems is uncertainty, which leads to a reduction in total cost [35].

Therefore, the PSO algorithm has been implemented for the optimization of three echelon supply networks, and good results are obtained in this case. Another classic optimization problem in supply chain management is the assignment of components and obtaining a solution that is able to deliver more orders at the proper time [36]. For such problems, both the GA and ant colony optimization (ACO) have been applied, and both algorithms perform well. However, ACO provides detailed information but at higher computational expenses.

Thus, although all the meta-heuristic methods perform well, some variants of the GA and SA have been found to yield very good results in most problems, but they are still not suitable for all types of problems in this field. In addition to meta-heuristic methods, several machine learning-based techniques have been used for such problems, and further research is also ongoing in this field. However, until now, meta-heuristic-based hybrid approaches have been found to yield faster solutions with good convergence rates and better near-optimal solutions, thus increasing the scope of research in this field.

1.5.3 Optimization in Health Care Operations

Optimization algorithms have been associated with the healthcare sector in recent years. There are certain problems in the healthcare sector, such as a reduction in treatment cost, organ allocation, capacity planning, etc., that require the implementation of optimization algorithms. Research has already been conducted, and extensive research is still ongoing regarding the application of newly evolving optimization algorithms. In this field, linear and integer programming are initially found to play a role.

Operating rooms (ORs) have always been an integral part of the healthcare sector. Therefore, it is necessary to enhance the efficiency of the ORs. To make the ORs effective, scheduling should be considered. Proper scheduling helps in proper resource allocation during operations and enables proper scheduling as well as planning. One of the most important strategies in OR is the block strategy, which cannot utilize the capacity of ORs. This is where linear and integer programming models find applications. To address such a problem, a mixed integer programming model [37] has been proposed along with a constrained programming (CP) model. However, CP is much more efficient, as it takes less computational time than integer programming. There are few studies related to smart healthcare systems that make use of linear and integer programming, but in recent years, meta-heuristic algorithms have been known to perform very well in such resource allocation problems.

Another type of scheduling problem is the nurse scheduling problem (NSP), which aims to obtain good performance that will satisfy the requirements of employees in healthcare institutions. This problem can also be considered a constrained optimization problem that may have two types of constraints: soft and hard. In [38], a GA and SA were used to solve such problems. However, SA yields more promising results than does GA; thus, SA outperforms the other methods in terms of optimizing such a class of problems. Another emerging topic in this field is location–allocation planning. An organized method is needed to address this problem, as its demand is increasing continuously [39]. The capacitated maximal covering location problem (CMCLP) has been used for this purpose. A modified version of the PSO algorithm is used here. The results show that the modified PSO yields much better results and provides good locations, thus maximizing the demand. Apart from these algorithms, other meta-heuristic approaches can also perform well, and this field is open for further research.

Although there may be studies on machine learning-based algorithms, in medical science, the algorithms used are of two main types: diagnostic and predictive. Therefore, meta-heuristic algorithms such as the GA, PSO, and SA yield optimized results in most cases, as they are best suited for detecting the optimal solution within a neighborhood and then converging on the exact solution. However, for pattern detection or classification problems, ML algorithms will be of immense use. There are certain challenges in the healthcare sector that need to be addressed by using optimization algorithms. One such challenge includes special mention here.

Medical workflows should be optimized, and medical employees need to access proper information regarding the operation of systems. Often, a patient's health report and different data are not organized; thus, this creates a barrier for proper workflow. Thus, in this context, the meta-heuristic algorithm definitely plays an important role. In addition, several other challenges, such as the efficient utilization of expensive resources in health care and expense-related issues, can be solved effectively via optimization algorithms.

1.5.4 Optimization in Financial Management

Optimization algorithms play a vital role in the field of financial management. In today's world, some of the major issues in financial management, such as cash flow management, precision planning, and financial analysis, require optimization algorithms. Thus, financial optimization can be defined as the method of allocating financial resources in an efficient manner such that net profit is maximized and expenditure is minimized. To date, several studies have been conducted on optimizing resources.

Although only a limited number of studies have used linear and integer programming models in financial management, one study suggests a mixed linear-integer programming model for a portfolio problem [40]. In this paper, it is shown that this is an NP-hard problem. Therefore, the proposed mixed linear integer optimization algorithm helps solve some practical problems, although not all problems can be solved by this model. The results obtained are quite good, and when some Italian data are applied, they over-perform, i.e., they provide detailed information about the data. In addition, this linear programming model has been implemented in this field for the last twenty years, but meta-heuristics have also had a dominant impact in this field.

Several studies related to the application of meta-heuristics in financial management exist. A GA- and SA-based mixed algorithm has been proposed for optimizing portfolio investment [41]. In this research, the GA provides the main guiding body for the algorithm, and the SA will mutate the GA, i.e., it is similar to a series–parallel structure. Thus, the solution obtained is enhanced compared with the simple use of the GA, and this hybrid algorithm is much more effective than the GA. This algorithm has achieved good efficiency in decision-making in portfolio investment. However, in another work [42], both the GA and PSO have been used in portfolio selection problems. The main aim of this work is to maximize the return while minimizing risk. Additionally, four different portfolios from the Tehran Stock Market are used. The result implies that the GA has better performance than SA and yields better average returns in a comparatively smaller period of time. On the other hand, the PSO-based portfolio performs well during the test phase. Thus, as a whole, both of these meta-heuristics are found to handle the portfolio selection problem.

There may be several other optimization techniques, such as stochastic optimization and deep learning-based techniques, that may yield good results in financial management, but meta-heuristic algorithms perform well in these types of problems and are currently used in research. Although in some cases, ML-based algorithms may perform much better, especially in financial risk management, in some subclasses of problems, further research is needed to conclude that ML algorithms are more efficient than meta-heuristics are. In financial management, many challenges can be solved only by optimization algorithms. Some of the challenges include the following:

Real-time data: A financial system should have access to all real-time data so that efficient financial planning can be performed. This requires efficient optimization algorithms.

Complex operations: Complex operations can often slow down the financial report of a company; thus, instead of manual intervention, optimization techniques can be used.

There are several other challenges, such as expenditure management and cost control, which need to be addressed, and studies are continuously ongoing to address such issues through the use of proper optimization techniques.

1.5.5 Optimization in Energy Management Systems

Energy management systems have several areas that require extensive use of optimization algorithms. For example, optimization is required for energy management in power grids, smart homes, power plants, etc. Energy management optimization refers to the process by which energy is efficiently utilized. Therefore, this basically involves different methods and techniques that can reduce energy waste and optimize energy usage via maximum energy utilization. Considerable research has been conducted in this field. There are several studies that involve mixed linear integer programming models in optimizing energy management systems.

A microgrid is an energy system that consists of energy resources, different power electronics-based devices, and loads, which are connected in a stabilized manner. These systems need to be optimized for proper energy distribution. In [43], a hybrid PV-wind-battery-connected microgrid system was designed, and a mixed integer linear programming model was used for this system. This model essentially attempts to solve the scheduling issue with the objective of reducing the cost of operation and enhancing 24-h ahead data prediction. This model has been verified experimentally in a microgrid system at Aalborg University. In addition to this work, other significant studies have used this mixed linear-integer programming model. However, following the latest trends in other fields, meta-heuristics have also been applied in energy management.

As discussed in other fields, the GA, SA, and PSO also play important roles in energy optimization. A new algorithm, GA/SA, has been developed [44] for large-scale energy management, which is very difficult for classical optimization algorithms. The GA has been modified by orthogonal crossover and effective crowding and then combined with SA to prevent early convergence problems. The result obtained by using this algorithm is better, and it converges faster and converges to the global optimum. Therefore, it performs far better than SA and GA and has been implemented to solve 167 problems.

Another important application of optimization in energy management is optimized scheduling of renewable energy in such systems [45]. This makes use of the hybrid GA and PSO algorithm. In this study, an energy hub consists of distributed generation (DG) units in addition to energy storage systems. This work aims to reduce the cost, peak-to-peak value, and emission of carbon; thus, the algorithms help in proper scheduling. This technique of combining the GA and PSO results in better convergence of the solution and a good-quality solution for such optimal scheduling problems in energy management systems.

In this era of ML, there may be several ML-based techniques, such as clustering, for efficient energy allocation; however, meta-heuristic algorithms compete with ML and have been widely implemented in this field. One of the most important challenges faced in this field that can be eradicated by using optimization algorithms include the following:

Unnecessary energy wastage is one of the most common problems faced in today's time. This leads to inefficient energy consumption and thus improper utilization of resources. Optimization algorithms have the potential to prevent such problems.

This is currently the most important optimization problem for saving the world from energy scarcity problems in the near future.

1.5.6 Optimization in Transportation and Routing

One of the most important fields of application of optimization algorithms is transportation and routing. Optimization algorithms can be used in this field for optimizing vehicle routing, flows in transport networks, etc. Many research papers mention linear and integer programming in transportation problems. There are many studies on optimal routing problems, and one such study is the solution of the routing problem for a school bus [46]. The study was carried out to transport students at a school in Central Ankara, Turkey. In this problem, distance is considered a constraint, and a linear integer model is also used. The formulation leads to a savings of 28.6% of the travel cost compared with the normal case. Thus, the model succeeded in reducing the cost to a good amount.

However, linear and integer programming can be applied for such problems, but they alone are not sufficient; thus, meta-heuristics are also needed. In the literature, much research is available in this field, which requires meta-heuristic algorithms. In routing, one of the significant NP-hard problems is vehicle routing via a time window [47]. To solve such a problem, a multi-trip approach is used. There have been different algorithms that have yielded good results, but since there is always scope in the improvement of results by using meta-heuristics, a mixed GA and SA algorithm is used. The GA performs a global search, whereas the SA performs a local search. Four different combinations of the GA–SA are used. The results show that GA–SA1 and GA–SA4 yield the best fitness values, but GA–SA4 can yield solutions faster than GA–SA1. In previous studies, hybrid GA and PSO have been implemented to solve the capacitated vehicle routing issue [48]. This model makes use of the idea of the best particle solution, i.e., the global solution in PSO, and then makes use of mutation and crossover of the GA. This also ensures that some feasible solutions are generated. This proposed model performs far better than the GA and PSO. This model has also been used in a waste collection system in Indonesia.

In addition, there are different methods, mostly the GA and PSO, which are used together to obtain optimal results in routing problems. Thus, the GA and PSO perform well in many types of problems. In this field, ML-based approaches do well, but not much research using ML algorithms exists compared with meta-heuristics. Certain challenges exist in this field that need to be optimized. Some of these are mentioned here.

Inefficient management of routing: If the routing cannot be managed, then it will affect delivery. Therefore, optimization algorithms are of immense help in this regard, as they can optimize the routing path and thus lead to proper delivery.

Increased cost of transportation: The transportation cost needs to be optimized, which is associated with inefficient routing, as untimely delivery affects the cost of transportation.

In addition, there are several other problems, such as the inability to utilize delivery capacity and the inability to track routes. Thus, it can be inferred that in transportation, the main challenge, i.e., the routing problem, needs to be addressed so that many of the linked problems can be automatically solved.

1.5.7 Optimization in Data Analysis and Machine Learning

Currently, one of the dominating fields is ML and data analysis. For proper and efficient analysis of data, different optimization algorithms are needed. In this field, mixed linear-integer programming plays a role. In machine learning and data analysis, feature selection is one of the most important tools for dimension reduction, as the complexity also decreases. Unlike traditional methods that aim to achieve better classification performance, factors such as price and complexity for feature acquisition should be considered. In [49], the support vector machine also addresses the cost of the features by preserving classification performance. A mixed linear integer programming model has been developed that can perform well on the basis of predictive performance. The algorithm used here also reduces the complexity.

A meta-heuristic algorithm, namely, the GA, plays a vital role in feature selection for big data analysis [50] and in data mining. The hybrid GA uses a weighted mechanism that classifies features into different groups, such as strong, weak, stable, and unstable. The algorithm produces new solutions by assigning more priority for stronger features than for weaker ones. It also focuses on unstable features.

In data analysis, one type of learning is online feature selection (OFS), which allows classifiers to have a small number of features. Hybrid firefly optimization and simulated annealing were used to obtain optimal features. This algorithm chooses the data from a dataset that is derived by decomposing the big data. The classifiers used are support vector machine (SVM) and logistic regression. In an experiment, 66 million data samples are taken, and on the basis of the performance matrices, the model can be said to be efficient.

Like these algorithms, other meta-heuristic algorithms, such as ant colony optimization, can also be used for data analysis as well as in neural networks, but the GA, PSO, and SA perform well in most big data problems. There are several key challenges that are faced. Optimization techniques are required to address such problems. One of the key challenges includes the following: the amount of data is growing in an un-accelerated manner; thus, for data analysis, there is a need for the optimization of these data. Thus, optimizing the growth and organization of data will make data analysis easier; otherwise, it will be difficult to address big data. Therefore, in the above section, the application of optimization algorithms in different fields is studied. From the above section, it can be concluded that optimization algorithms need to be implemented increasingly because, for the majority of types of problems, they perform well, and these algorithms are problem-independent and easy to handle.

1.6 Optimization Concerns and Their Approaches

Currently, although there has been widespread use of optimization, some factors that are key concerns for optimization still exist. Some of the concerns and approaches to address them are as follows:

Scalability: If the size of the problem or data increases, then the computational complexity and computational effort to solve the problem also increase. This is one of the greatest concerns in today's world, as a large amount of data exist. This problem can be approached via heuristics and meta-heuristic algorithms. Meta-heuristic algorithms such as the GA, SA, and PSO are found to yield good solutions within a considerable amount of time.

Constraints handling: Another concern of these optimizations is to address the constraints so that, bounded by these constraints, the problem yields an optimal solution. Several solutions exist for this problem. One of the best-known solutions is to use a penalty function to prevent constraint violation. Therefore, some penalty functions may be added to the optimization algorithm. There is also another popular method called the Lagrangian method for such constraint violation problems in optimization.

Convergence: One of the major concerns for optimization is ensuring convergence to an optimized solution. This convergence rate plays a pivotal role in any optimization process. This can also be addressed by implementing hybrid meta-heuristic algorithms. In addition to these algorithms, several stochastic optimization-based techniques can also be applied.

Robustness: Optimization should ensure that the solution does not change with changes in parameters or under any uncertainty. The algorithms should perform well under any conditions. This can be handled by developing an optimization algorithm that can reduce the worst-case scenario or include a probabilistic approach to address any kind of unpredictability.

Real-time multi-objective optimization: Optimization should consider optimizing real-time problems with multiple objective functions at a given time.

This can be addressed by using modified evolutionary optimization algorithms such as modified version of GA that can handle multiple objective functions, as in real-time problems, generally involves more than one objective function.

Thus, these are the major concerns of the optimization algorithms, and all these can be easily dealt with by using optimization algorithms, especially nature-inspired multi-objective meta-heuristic algorithms.

1.7 Emerging Trends in Optimization

With the advancement of computing capabilities and demands, certain emerging trends in optimization have been developed and applied in different fields. Some of the trends are mentioned below:

Hybrid meta-heuristic algorithms: Hybrid meta-heuristic algorithms have been developed by combining various algorithms, and in many studies, hybrid algorithms perform much better in terms of the computational time convergence rate and optimal solution than do normal meta-heuristics. Thus, hybridization of meta-heuristics with other optimization algorithms helps to develop a better problem-solving approach.

For example, a non-dominated sorting hybrid algorithm was introduced in [51]. This algorithm has a good convergence rate and can be applied for a wide range of problems, thus making it good for different engineering-based applications. For verification, it has also been applied in the optimization of weight, strength, and filtering time.

Quantum optimization algorithms: Quantum optimization algorithms have the ability to address combinatorial optimization problems in much less computational time. Quantum optimization has been a current trend, and various researchers and scientists are engaged in this field. In a certain paper [52], quantum-classical optimizers such as the variation quantum eigen solver (VQE) and quantum approximate optimization algorithm (QAOA) have been studied, and they have several applications in ML-based problems. These classifiers have increased the scope for future work in this field.

Apart from these trends, there are several other trends in optimization, such as automated optimization and tuning of hyper-parameters. In automated optimization, the models are tuned to optimize automatically, and their hyper-parameters are also tuned. Thus, optimization algorithms are revolutionizing each and every field, and they need to be integrated into all other emerging fields of science, technology, management, and business. They have potential solutions to many unsolved issues and thus play a major role in each field. Indeed, researchers are exploring new algorithms daily, and hopefully, one day, optimization will lead to efficient utilization of resources in each complex problem, which seems to be a challenge.

1.8 Conclusions

There are many different types of optimization challenges. Finding the optimal solution among a wide range of potential solutions in a given solution space is typically the aim of optimization issues. Fields such as operations research, engineering, economics, finance, logistics, and more use optimization modeling, which is a potent tool. Techniques and applications for optimization with the examples provided in this chapter offer an overview of the principles of all frequently used optimization strategies, encompassing the variety and breadth of approaches (both old and new) and algorithms.

References

1. Weise T (2009) Global optimization algorithms-theory and application. Self-Published Thomas Weise 361:153
2. Rardin RL, Uzsoy R (2001) Experimental evaluation of heuristic optimization algorithms: A tutorial. Journal of Heuristics 7:261–304
3. Foulds LR (2012) Optimization techniques: an introduction. Springer Science & Business Media
4. Venter G (2010) Review of optimization techniques
5. Yang XS (2018) Optimization techniques and applications with examples. John Wiley & Sons

6. Dantzig GB (2002) Linear programming. Oper Res 50(1):42–47
7. Karloff H (2008) Linear programming. Springer Science & Business Media
8. Best MJ, Ritter K (1985) Linear programming. Prentice Hall, Englewood Cliffs, NY
9. Eiselt HA, Sandblom CL (2007) Linear programming and its applications. Springer Science & Business Media
10. Wolsey LA (2020) Integer programming. John Wiley & Sons
11. Williams HP (2009) Integer programming. Logic and integer programming. Springer, US, Boston, MA, pp 25–70
12. Taha HA (2014) Integer programming: theory, applications, and computations. Academic Press
13. Magee TM, Glover F (1996) Integer programming. Indus Eng New York Basel Marcel Dekker Incorp 20:123–269
14. Dey N, Ashour AS (2018) Meta-heuristic algorithms in medical image segmentation: a review. Advancements in applied metaheuristic computing, 185–203
15. Van Laarhoven PJ, Aarts EH, van Laarhoven PJ, Aarts EH (1987) Simulated annealing. Springer, Netherlands, pp 7–15
16. Aarts E, Korst J, Michiels W (2005) Simulated annealing. Search methodologies: introductory tutorials in optimization and decision support techniques, 187–210
17. Delahaye D, Chaimatanan S, Mongeau M (2019) Simulated annealing: from basics to applications. Handbook of metaheuristics, 1–35
18. Hajek B (1985) A tutorial survey of theory and applications of simulated annealing. In: 1985 24th IEEE conference on decision and control, pp 755–760. IEEE
19. Bäck T, Schwefel HP (1993) An overview of evolutionary algorithms for parameter optimization. Evol Comput 1(1):1–23
20. Vikhar PA (2016) Evolutionary algorithms: a critical review and its future prospects. In: 2016 International conference on global trends in signal processing, information computing and communication (ICGTSPICC), pp 261–265. IEEE
21. De D, Mukherjee A, Das SK, Dey N (eds) (2020) Nature inspired computing for wireless sensor networks. Springer, Singapore, pp 1–18
22. Whitley D, Rana S, Dzubera J, Mathias KE (1996) Evaluating evolutionary algorithms. Artif Intell 85(1–2):245–276
23. Dey N (Ed) (2017) Advancements in applied metaheuristic computing. IGI global
24. Yu X, Gen M (2010) Introduction to evolutionary algorithms. Springer Science & Business Media
25. Kamal S, Dey N, Nimmy SF, Ripon SH, Ali NY, Ashour AS, ... Shi F (2018) Evolutionary framework for coding area selection from cancer data. Neural Comput Appl 29:1015–1037
26. Gutierrez CJ (1996) Development and application of a linear programming model to optimize production and distribution of a manufacturing company (Doctoral dissertation, Massachusetts Institute of Technology)
27. Mokhtari H, Abadi INK, Amin-Naseri MR (2011) Production scheduling with outsourcing scenarios: a mixed integer programming and efficient solution procedure. Int J Prod Res 50(19):5372–5395. https://doi.org/10.1080/00207543.2011.627687
28. Man KF, Tang KS, Kwong S, Ip WH (2000) Genetic algorithm to production planning and scheduling problems for manufacturing systems. Production Planning Control 11(5):443–458. https://doi.org/10.1080/09537280050051942
29. Musharavati F, Hamouda ASM (2012) Enhanced simulated-annealing-based algorithms and their applications to process planning in reconfigurable manufacturing systems. Adv Eng Softw 45(1):80–90
30. Bewoor LA, Prakash VC, Sapkal SU (2018) Production scheduling optimization in foundry using hybrid Particle Swarm Optimization algorithm. Procedia Manuf 22:57–64
31. Jerald J, Asokan P, Prabaharan G et al (2005) Scheduling optimization of flexible manufacturing systems using particle swarm optimization algorithm. Int J Adv Manuf Technol 25:964–971. https://doi.org/10.1007/s00170-003-1933-2
32. Paksoy T, Pehlivan NY (2012) A fuzzy linear programming model for the optimization of multistage supply chain networks with triangular and trapezoidal membership functions. J Franklin Inst 349(1):93–109

33. Altiparmak F, Gen M, Lin L, Paksoy T (2006) A genetic algorithm approach for multiobjective optimization of supply chain networks. Comput Ind Eng 51(1):196–215
34. Yanling W, Deli Y, Guoqing Y (2010) Logistics supply chain management based on multiconstrained combinatorial optimization and extended simulated annealing. In: 2010 international conference on logistics systems and intelligent management (ICLSIM) (Vol 1, pp 188–192). IEEE
35. Kadadevaramath RS, Chen JC, Shankar BL, Rameshkumar K (2012) Application of particle swarm intelligence algorithms in supply chain network architecture optimization. Expert Syst Appl 39(11):10160–10176
36. Silva CA, Runkler TA, Sousa JM, da Costa JS (2003) Optimization of logistic processes in supply-chains using meta-heuristics. In: Portuguese conference on artificial intelligence, pp 9–23. Berlin, Heidelberg: Springer, Berlin, Heidelberg
37. Younespour M, Atighehchian A, Kianfar K, Esfahani ET (2019) Using mixed integer programming and constraint programming for operating rooms scheduling with modified block strategy. Operations Res Health Care 23:100220
38. Kundu S, Mahato M, Mahanty B, Acharyya S (2008) Comparative performance of simulated annealing and genetic algorithm in solving nurse scheduling problem. In: Proceedings of the international multiconference of engineers and computer scientists Vol 1, pp 96–100
39. ElKady SK, Abdelsalam HM (2016) A modified particle swarm optimization algorithm for solving capacitated maximal covering location problem in healthcare systems. Applications of intelligent optimization in Biology and Medicine: current trends and open problems, 117–133
40. Benati S, Rizzi R (2007) A mixed integer linear programming formulation of the optimal mean/value-at-risk portfolio problem. Eur J Oper Res 176(1):423–434
41. Wang Z, Cui D (2009) A hybrid algorithm based on genetic algorithm and simulated annealing for solving portfolio problem. In: 2009 international conference on business intelligence and financial engineering, pp 106–109. IEEE
42. Talebi A, Molaei MA, Sheikh MJ (2010) Performance investigation and comparison of two evolutionary algorithms in portfolio optimization: Genetic and particle swarm optimization. In: 2010 2nd IEEE international conference on information and financial engineering, pp 430–437. IEEE
43. Luna AC, Diaz NL, Graells M, Vasquez JC, Guerrero JM (2016) Mixed-integer-linear-programming-based energy management system for hybrid PV-wind-battery microgrids: Modelling, design, and experimental verification. IEEE Trans Power Electron 32(4):2769–2783
44. Yu H, Fang H, Yao P, Yuan Y (2000) A combined genetic algorithm/simulated annealing algorithm for large scale system energy integration. Comput Chem Eng 24(8):2023–2035
45. Toro JAR, Sampietro-Saquicela JL, Suryasa IW, Solórzano LEH, Cervantes XLG, Quiñónez BFC (2024) Optimal scheduling of renewable energy resources in energy management systems using hybrid genetic algorithm and particle swarm optimization. Acta Innovations 52:19–27
46. Bektaş T, Elmastaş S (2007) Solving school bus routing problems through integer programming. J Oper Res Soc 58(12):1599–1604
47. Ariyani AK, Mahmudy WF, Anggodo YP (2018) Hybrid genetic algorithms and simulated annealing for multitrip vehicle routing problem with time windows. Int J Electr Comput Eng (2088–8708) 8(6)
48. Kuo RJ, Zulvia FE, Suryadi K (2012) Hybrid particle swarm optimization with genetic algorithm for solving capacitated vehicle routing problem with fuzzy demand–a case study on garbage collection system. Appl Math Comput 219(5):2574–2588
49. Lee IG, Zhang Q, Yoon SW, Won D (2020) A mixed integer linear programming support vector machine for cost-effective feature selection. Knowl-Based Syst 203:106145
50. Mohammed TA, Bayat O, Uçan ON, Alhayali S (2020) Hybrid efficient genetic algorithm for big data feature selection problems. Found Sci 25(4):1009–1025

51. Chatterjee S, Sarkar S, Dey N, Ashour AS, Sen S (2018) Hybrid nondominated sorting genetic algorithm: II-neural network approach. In: Advancements in applied metaheuristic computing (pp 264–286). IGI Global
52. Choi J, Oh S, Kim J (2020) The useful quantum computing techniques for artificial intelligence engineers. In: 2020 international conference on information networking (ICOIN), pp 1–3. IEEE

Chapter 2
Evolutionary Mastery Forges the Blades of Cyber Attack and Defense

Mitchell Eakin, Mahdi Khosravy, Alyssa M. Adams, and Olaf Witkowski

1 Introduction

Evolutionary and nature-inspired techniques are methodologies derived from biological processes and natural phenomena. These approaches include genetic algorithms, artificial neural networks, and swarm intelligence, among others. They are employed to address complex optimization and problem-solving tasks, taking inspiration from the adaptability and efficiency observed in natural systems [1, 2].

If we go through the mechanism and functionality of the main evolutionary techniques, the ones the most used and discussed are Genetic Algorithms, Artificial Neural Networks, and Swarm Intelligence. Genetic Algorithms are search heuristics that mimic the process of natural selection. They work by generating a population of potential solutions and iteratively selecting, combining, and mutating them to evolve better solutions over successive generations. Artificial Neural Networks modeled after the human brain, these networks consist of interconnected nodes (neurons) that process information in layers. They excel at recognizing patterns and learning from data through training processes. Swarm Intelligence is inspired by the collective behavior of social organisms like ants, bees, and birds, these algorithms optimize solutions by simulating the decentralized, cooperative actions of these groups.

Evolutionary and nature-inspired techniques have significant applications in the field of cybersecurity, addressing various challenges through their inherent strengths like (i) Intrusion Detection Systems (IDS), (ii) Malware Detection, (iii) Spam Filtering, (iv) Cryptographic Systems, (v) Network Security Optimization, etc. Intrusion Detection Systems benefit from the adaptability of evolutionary algorithms, which can dynamically update their detection rules and parameters to identify new

M. Eakin
Kent State University, Kent, OH, USA

M. Khosravy (✉) · A. M. Adams · O. Witkowski
Cross Labs, Cross Compass Ltd., Tokyo, Japan
e-mail: mahdi@corsslabs.org

N. Dey (ed.), *Optimizing Solutions for Real-Life Problems*, Springer Tracts in Nature-Inspired Computing, https://doi.org/10.1007/978-981-96-3216-9_2

and evolving threats effectively. Neural networks and genetic algorithms assist and enhance malware detection by analyzing vast datasets to identify malicious patterns and behaviors that may not be evident through traditional methods. Nature-inspired techniques also improve spam detection by continuously learning and adapting to new spam tactics, thus maintaining high accuracy in filtering unwanted emails. Cryptographic Systems contribute to the development of robust encryption techniques and cryptographic protocols, enhancing data security and privacy. Besides all the above-mentioned techniques, swarm intelligence algorithms optimize network configurations and defense strategies, providing efficient solutions for threat mitigation and resource allocation.

The advantages of evolutionary techniques in cybersecurity are mainly in adaptability, pattern recognition, and optimization capabilities. The ability to adapt to new and unforeseen threats is a critical advantage of evolutionary and nature-inspired techniques. They can continuously evolve and refine their approaches in response to emerging cyber threats. Pattern recognition techniques are highly proficient in identifying complex patterns within large datasets, making them invaluable for detecting anomalies and potential security breaches. Finally optimization capabilities of evolutionary techniques by leveraging principles of natural selection and collective behavior, these methodologies can identify optimal strategies for cyber defense and attack mitigation. Evolutionary and nature-inspired techniques represent a cutting-edge approach to cybersecurity, offering robust and dynamic solutions for detecting, preventing, and responding to cyber threats. By emulating the efficiency and adaptability found in natural processes, these techniques provide significant advancements in securing digital environments against sophisticated and evolving cyber-attacks.

The benefits of Using Evolutionary Techniques in Cybersecurity can be listed as (i) Adaptability, (ii) Optimization, (iii) Scalability, (iv) innovation, and (v) Automation. In terms of benefits of adaptability, there are in two ways; (i) Dynamic Threat Response wherein evolutionary techniques can quickly adapt to new and evolving threats, ensuring that security measures remain effective against the latest attack vectors, and (ii) Continuous Learning which is the capability of these methods to continuously improve their detection and response strategies based on ongoing data and experience. In terms of optimization, there are two main benefits by these techniques; (i) Efficient Resource Allocation: Evolutionary algorithms optimize the allocation of security resources, such as firewall rules and intrusion detection system configurations, to maximize effectiveness. (ii) Enhanced Problem-Solving: They provide efficient solutions to complex optimization problems, such as finding the best network security configurations. The benefit of the scalability of these techniques is due to large-scale data processing. They handle and analyze vast amounts of data, making them suitable for large-scale cybersecurity applications, such as network traffic analysis and user behavior monitoring. In term of innovation, evolutionary techniques often discover innovative solutions and strategies that might not be apparent through traditional approaches, enhancing the overall robustness of cybersecurity measures, and finally, automation which is due to reduced human intervention. They automate various aspects of cybersecurity, reducing the need for constant human oversight and intervention, which helps in handling large-scale security operations efficiently.

Besides the benefits, there are challenges of using Evolutionary Techniques in cybersecurity, which are as follows; (i) computational complexity, (ii) parameter tuning, (iii) overfitting, (iv) security of algorithms, and (v) integration challenges. Computational Complexity results in two issues, the first one is high resource demand and the second one is execution time. Evolutionary algorithms can be computationally intensive, requiring significant processing power and memory, which might be a limitation for resource-constrained environments. Also, the iterative nature of these techniques can lead to longer execution times compared to traditional methods. Parameter Tuning is another challenge which has two aspects; (i) Complex configuration, and (ii) Expertise requirement. The effective implementation often requires fine-tuning numerous parameters (e.g., mutation rates, population size), which can be complex and time-consuming. Also, expertise requirement for specialized knowledge is needed to configure and optimize these algorithms correctly. Overfitting is also a challenge that is related to generalization Issues. There is a risk of overfitting to specific data patterns, making the models less effective when confronted with new or unseen threats. Besides, the security of algorithms is a challenge in itself. The Algorithms have Vulnerabilities and the evolutionary algorithms themselves can become targets for adversaries, who may attempt to manipulate the optimization process or exploit weaknesses in the algorithm. Finally, there are integration challenges that can be either compatibility Issues or Interoperability. Integrating evolutionary techniques with existing cybersecurity infrastructure can be challenging, requiring significant modifications and adjustments to current systems. Interoperability is indeed Ensuring that these techniques work seamlessly with other security tools and technologies can be complex.

2 Evolutionary Techniques in Cybersecurity

The application of evolutionary techniques in cybersecurity offers powerful tools for optimizing defense mechanisms, adapting to new threats, and efficiently managing resources. By mimicking natural processes and behaviors, these techniques provide innovative solutions to complex cybersecurity challenges. Several evolutionary techniques have been successfully applied to various cybersecurity challenges. Here are some of the most prominent ones.

2.1 Genetic Algorithms in Cybersecurity

Genetic algorithms are optimization and search techniques inspired by the process of natural selection and genetics. Developed by John Holland in the 1960s and 1970s [3], GAs solve complex problems by mimicking biological evolution. GAs have gone through improvements and different variants of GA have been presented. These algorithms work with a group of potential solutions, using operators like selection, crossover, and mutation to improve solutions over successive generations.

The key Components of Genetic Algorithms are as follows;

1. Population: A group of potential solutions to the problem.
2. Chromosomes: Encoded representations of these solutions.
3. Selection: A process that chooses individuals based on their fitness to reproduce.
4. Crossover: A genetic operator that combines parts of two parent solutions to produce offspring.
5. Mutation: A genetic operator that introduces random changes to individuals to maintain diversity.

The GA working process can be summarized in the following steps; (i) Initialization: Start with a randomly generated group of potential solutions, (ii) Evaluation: Assess the fitness of each solution in the group. (iii) Selection: Select individuals based on their fitness to act as parents for the next generation. (iv) Crossover: Combine selected parents to create new offspring. (v) Mutation: Apply random changes to some offspring to introduce variability. (vi) Replacement: Form a new group, replacing less fit individuals with new offspring. (vii) Iteration: Repeat the evaluation, selection, crossover, mutation, and replacement steps for many generations until an optimal or satisfactory solution is found.

Genetic algorithms are flexible and can be applied to a variety of problems, including electrical power industries [4], electrical vehicle industries [5], machine learning [6], and artificial intelligence [7]. Their ability to explore large search spaces and find near-optimal solutions makes them especially useful for complex and non-linear problems where traditional methods might struggle. Recent inspiration from GA is Mendelian Evolutionary Theory Optimization [8]. GAs have been used in general in four categories in cybersecurity: (i) Cryptographic Systems, (ii) Malware Detection, (iii) Intrusion Detection Systems, and (iv)Intrusion Detection Systems.

GA in Cryptographic Systems GAs have been used in work by [9] for hacking the security of data hiding. The security of their scheme depends upon a secret key (chromosome) constructed using a genetic algorithm. The key specifies offsets and patterns for data embedding. By performing a key-sensitivity analysis and exploiting image pixel correlation, it is shown in their work that the secret data can be easily revealed.

The work by [10] develops a substitution box designed for combined data paths using dual-basis tower field extensions, integrating the Golod-Shafarevich theorem into an immune genetic algorithm for optimization. The enhanced immune genetic algorithm functions by first generating a population of S-boxes using chaotic systems. These S-boxes undergo optimization through a series of operators, including anti-agent extraction and immune selection, to identify those performing exceptionally well. Criteria such as differential uniformity, non-linear degree, and strict avalanche effect are then applied to analyze the S-boxes.

GA in Malware Detection GAs help in feature selection and optimization for malware detection algorithms, improving their accuracy and efficiency. In work by

[11], an Android malware detection framework GA-StackingMD is presented, which employs Stacking to combine five different base classifiers, and uses a Genetic Algorithm to optimize the framework's hyperparameters. With the rapid development of mobile communication and network technology, intelligent terminals such as smartphones and tablets have changed people's daily work and life. However, malware such as viruses, Trojans, and ransomware has introduced threats to personal privacy and security. Android malware, with its great variety and rapid updates, poses significant challenges. Android malware detection faces problems of high feature dimension and low detection accuracy using single classification algorithms.

GA in Intrusion Detection Systems The work by [12] discusses the means of computer network attack and intrusion, network security technology, and genetic algorithms. The focus is on designing an intrusion detection model based on genetic algorithms and its application deployment. They conduct specific experiments, and the results show that the computer network security protection system based on genetic algorithms has a better defense effect than conventional Internet security systems, and provides more reliable and efficient security protection services in different security attack scenarios. The implementation steps of intrusion detection based on genetic algorithms are shown in Fig. 1. This module extracts attribute features from the current detected data flow, evaluates whether there are abnormal feature fields, and encodes data packets with abnormal features according to corresponding code rules. These data packets are then selected, crossed, and mutated to form new chromosomes. Next, the test data set of normal network connection types (including both normal and abnormal connections) is used to evaluate the fitness of the new chromosomes. Chromosomes with high fitness are selected, and the above test is repeated. These high-fitness chromosomes are then added to the function library as new rules. This completes the preprocessing of data packets. Suspicious attack alerts are created, and the alerts and new rules are sent to the linkage management module.

GA in Universal perturbation attack Another aspect of the application of GAs is in Universal Perturbation Attack on AI models. A universal perturbation attack targets AI models by applying small, carefully crafted perturbations to inputs. These perturbations are designed to deceive the model consistently across different inputs, causing misclassification or erroneous predictions. Unlike targeted attacks that focus on specific inputs, universal perturbations aim to generalize across a wide range of data, making them potent threats to the robustness and reliability of AI systems in real-world applications. The perturbation crafting is done by GA by [13] wherein GAs search for the best possible perturbation attack noise on a classification system.

2.2 *Simulated Annealing Algorithm in Cybersecurity*

Simulated Annealing (SA) is a probabilistic optimization technique inspired by the annealing process in metallurgy, where a material is heated and then slowly cooled to increase the size of its crystals and reduce defects. Developed by Kirkpatrick, Gelatt,

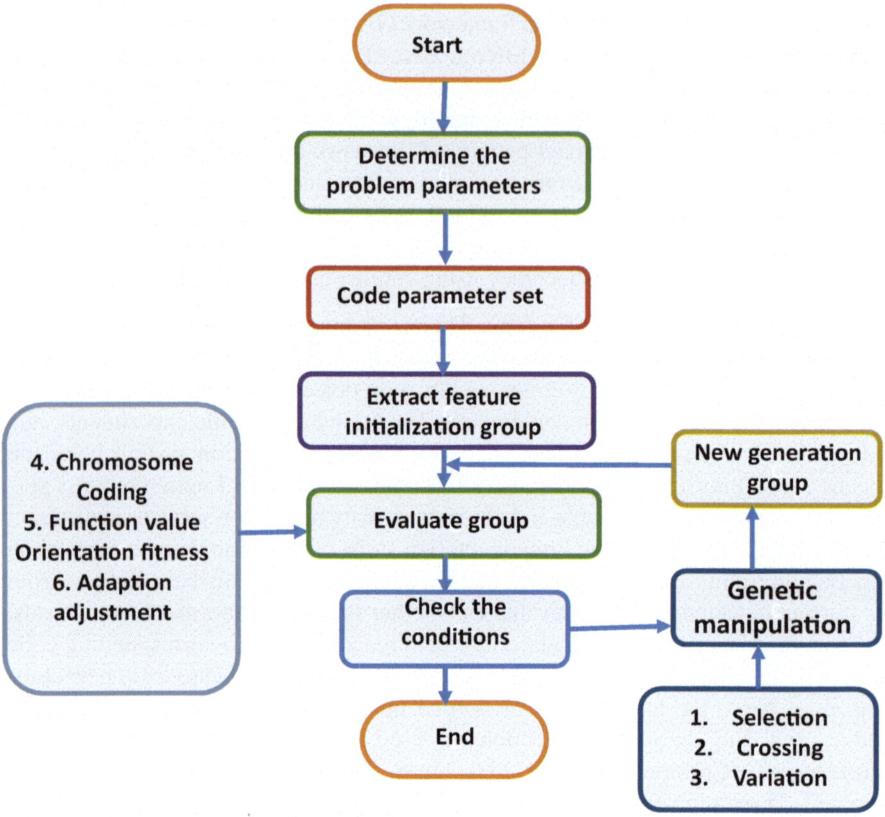

Fig. 1 Intrusion detection based on genetic algorithms

and Vecchi in 1983 [14], SA is used to find approximate solutions to optimization problems, particularly those with numerous local optima. The key components of SA are (i) Objective Function, The function to be optimized, (ii) Temperature: A control parameter that decreases over time, influencing the probability of accepting worse solutions, (iii) Cooling Schedule: The plan for decreasing the temperature, which affects the algorithm's convergence rate and solution quality, and (iv) Neighborhood: The set of potential solutions adjacent to the current solution.

The working structure of the SA is as follows: (i) Initialization: Start with an initial solution and an initial high temperature. (ii) Evaluation: Calculate the objective function value for the current solution. (iii) Perturbation: Generate a neighboring solution by making a small random change to the current solution. (iv) Acceptance Criteria: Decide whether to accept the new solution based on its objective function value and the current temperature. Accept the new solution if it is better, or with a certain probability if it is worse, to avoid being trapped in local optima. (v) Cooling: Gradually decrease the temperature according to the cooling schedule. (vi) Iteration:

Repeat the evaluation, perturbation, acceptance, and cooling steps until the system "freezes" or a stopping criterion is met.

The work [15] improves network efficiency and quickly and accurately detects abnormal network traffic, heuristic search is optimized by introducing the Simulated Annealing Algorithm (SAA). Combining the Adaptive Mutation Algorithm (AMA) with the Standard Genetic Algorithm (GA) improved by optimization strategies can effectively and quickly identify anomalies in network services.

2.3 *Ant Colony Optimization (ACO) in Cybersecurity*

Ant Colony Optimization (ACO) [16] is a nature-inspired optimization algorithm modeled after the foraging behavior of ants. Developed by Marco Dorigo in the early 1990s, ACO is based on how ants find the shortest paths between their nest and food sources. Ants communicate with each other using pheromones, which they deposit on the ground as they move. Over time, shorter paths accumulate more pheromones, attracting more ants and reinforcing the route. In computational terms, ACO is applied to solve complex optimization problems by simulating this behavior. Artificial ants traverse a graph, depositing virtual pheromones on the edges they travel. The probability of choosing a path depends on the amount of pheromone present and the path's heuristic information (e.g., distance or cost). Through iterative exploration and pheromone updating, the algorithm converges to optimal or near-optimal solutions.

ACO has been successfully applied to various problems, including the traveling salesman problem, vehicle routing, and network optimization. Its ability to find high-quality solutions in dynamic and uncertain environments makes it a powerful tool in the field of optimization.

Smart grayhole attack negatively impacts the availability and accuracy of the network by discarding data packets. The work by [17] tackles this attack by monitoring the behavior of participating nodes using a bio-inspired trust management model. This distributed trustworthiness assessment model employs the beta reputation system combined with the Ant Colony Optimization (ACO) metaheuristic. The beta reputation system primarily focuses on rating the nodes based on their successful tasks and their consumed energy, while the ACO metaheuristic maintains this reputation metric during the discovery process and computes the preference value of each traversed path to select the most secure one. The proposed model improves the traditional Dynamic Source Routing (DSR) protocol by excluding malicious nodes from participating in data packet transmission.

Another application of ACO is in Vehicular Ad Hoc Networks (VANET) security. VANET faces security threats due to its dependence on computing, control, and communication technologies. IoT-enabled VANETs face challenges such as integrity, access control, availability, privacy protection, non-repudiation, and confidentiality. Traditional cryptography-based security solutions are less efficient compared to lightweight and effective trust-based techniques. The work by [18] introduces the

Attack Resilient Position-based VANET Protocol (ARPVP) to ensure reliable communications by assessing vehicle behavior. ARPVP uses trust evaluation and artificial intelligence (AI) to detect and prevent malicious vehicles. A key feature is the use of Ant Colony Optimization (ACO) in the second phase, which forms reliable routes by avoiding attackers using a trust-based fitness function. Trust parameters such as mobility, buffer occupancy, and link quality are measured throughout the process.

ACO even contributes more to Network security as the work [19] This work aims to enhance the calculation of optimal attack paths in a target network using a Monte Carlo Graph Search algorithm integrated with ACO, referred to as ACO-MCGS. The ACO-MCGS algorithm is designed to find optimal attack paths, allowing an attacker to carry out attacks more efficiently in terms of time, energy, and cost. ACO-MCGS provides comprehensive results rapidly and addresses the issue of path loss that can occur with other methods. The ACO-MCGS operates in two main steps-selection and backpropagation: (i) Selection: In this step, the algorithm evaluates and selects potential attack paths based on specific criteria, (ii) Backpropagation: This step involves refining the selected paths to ensure optimality and efficiency. In summary, the proposed ACO-MCGS algorithm offers a robust and efficient method for calculating optimal attack paths, leveraging the strengths of both Monte Carlo Graph Search and Ant Colony Optimization.

ACO algorithm also is used in work [20] to enhance the security in sensor network systems, particularly against Sybil attacks. Sybil attacks involve a malicious node creating multiple fake identities to deceive honest nodes in the system, making the network vulnerable. The nature of ACO helps in limiting the number of attack edges efficiently and conveniently. As a result, honest nodes are more likely to accept and be accepted by other honest nodes, and the protocol ensures a high probability of rejecting Sybil nodes.

Another field in which ACO serves its role in security enhancement is social networks. Social networks connect millions of people worldwide, and data from these networks is often sold to third parties for analysis and data mining. Privacy concerns arise as user data, even when naively anonymized, can be susceptible to attacks. The walk-based attacks target social networks where an attacker creates malicious nodes linked to a target node. After a naive anonymization, the attacker tries to identify this subgraph in the anonymized data, if successful, the attack compromises the victim's identity and private information. The work [21] aims to enhance the privacy of users in social networks by using ACO to anonymize data while maintaining its utility for data mining. It aims to prevent walk-based attacks while minimizing data loss. This balance ensures that the social graph retains its value for data mining. In this regard, ACO is used in conjunction with a "user characteristics metric" to anonymize the data. ACO helps in optimizing the anonymization process, enhancing security without significantly compromising data utility. By using ACO and the user characteristics metric, the algorithm effectively mitigates walk-based attacks, thus improving user privacy while maintaining the quality of the anonymized data for analytical purposes.

ACO has been also deployed to enhance Intrusion Detection Systems (IDS) and improve network security. The work [22] aims to develop more efficient and customized IDS by using ACO to find optimal paths and solutions in complex problems, thus improving the accuracy and efficiency of detecting network intrusions. By focusing on UDP port scans and using ACO, it is contributed to the overall enhancement of network security. The specialized intrusion detection mechanism aims to provide better protection against specific types of attacks, thereby increasing the robustness of the network security infrastructure.

2.4 *Differential Evolution (DE) in Cybersecurity*

Differential Evolution (DE) is a powerful and versatile optimization algorithm inspired by the process of natural evolution. Developed by Rainer Storn and Kenneth Price in 1997 [23], DE belongs to the class of evolutionary algorithms (EAs) and is particularly known for its simplicity, efficiency, and robustness in solving complex, multidimensional optimization problems. The algorithm works by iteratively improving a population of candidate solutions with respect to a given measure of quality or fitness. The DE algorithm involves three main steps: mutation, crossover, and selection; (i) Mutation: For each individual in the population, DE creates a mutant vector by adding the weighted difference between two randomly selected population vectors to a third vector. This step introduces diversity and helps the algorithm explore the search space. (ii) Crossover: The mutant vector is then mixed with the target vector (the current individual) to form a trial vector. This combination helps in exploring new solutions while retaining useful traits from existing solutions. (iii) Selection: The trial vector is compared with the target vector, and the one with the better fitness value is selected for the next generation. This step ensures that the population gradually improves over time.

Differential Evolution's ability to optimize complex and high-dimensional problems makes it highly applicable to various cybersecurity tasks. Here are some key areas where DE has been effectively employed:

DE in Intrusion Detection Systems (IDS): DE contribution to IDS is in two aspects (i) Feature Selection wherein DE can optimize the selection of relevant features from large datasets, improving the performance and accuracy of IDS, and (ii) Parameter Tuning where DE helps in fine-tuning the parameters of the machine learning models used in IDS, enhancing their ability to detect anomalies and malicious activities.

As a feature selection by DE or IDS, the work [24] introduces a hybrid model combining ELM with adaptive differential evolution, and the goal is to enhance the performance of DDoS attack detection in cloud computing environments. Their approach optimizes the input-to-hidden layer weights and hidden layer biases of the ELM using adaptive differential evolution. This optimization helps in improving the accuracy and efficiency of the ELM-based detection model. As a result, a DDoS attack detection system is developed using the hybrid model. This system is specifically designed to address the challenges of detecting DDoS attacks within cloud computing environments.

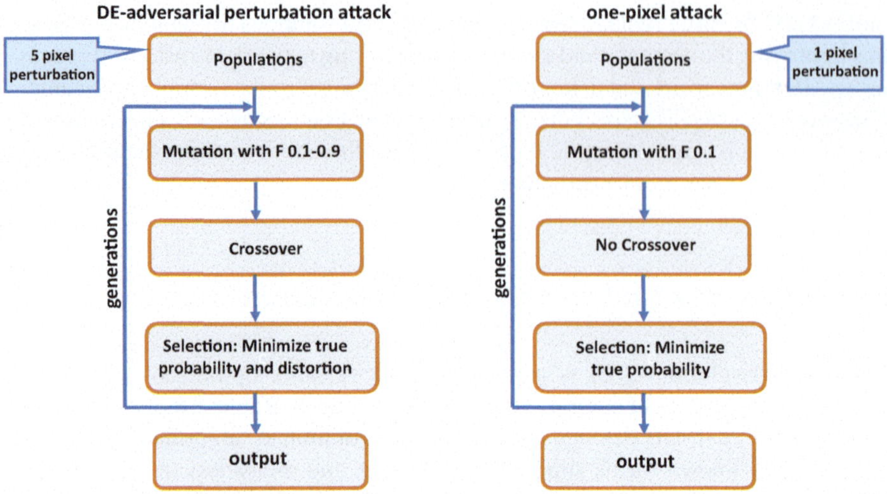

Fig. 2 Technical difference between DE-adversarial perturbation attack and one-pixel attack

Deployment of DE for IDS has been more established in work [25] for parameter tuning in practical black-box attack method against Convolutional Neural Networks (CNNs) image classifiers. Indeed the work deploys DE to optimize generated adversarial perturbations that can successfully fool CNNs. The interesting aspect of the work is it is indeed by few-pixel attack. The method is efficient, requiring modifications to only five pixels in the target image, demonstrating that even minimal changes can lead to significant misclassification. Similar to one-pixel attack [26], it highlights the vulnerability of CNNs to even small, well-crafted perturbations. It can advance the understanding of CNN vulnerabilities and demonstrate the practical application of DE in generating efficient adversarial attacks, contributing significantly to the field of cybersecurity. Fig. 2 demonstrates the technical difference between DE-adversarial perturbation attack and one-pixel attack.

DE in Cryptography: DE can be used to generate and optimize cryptographic keys, ensuring strong encryption and decryption processes. Also, DE assists in cryptanalysis by breaking cryptographic codes by efficiently searching for vulnerabilities and weaknesses in encryption algorithms. The research work [27] demonstrates that Differential Evolution (DE) is a more effective technique than GA for the cryptanalysis of the Merkle-Hellman Knapsack cipher. This is achieved by using DE's stochastic, population-based optimization capabilities, including its use of mutation, crossover, and selection operators to explore the solution space efficiently. Indeed, DE is employed as a powerful optimization strategy that iteratively evolves a population of potential solutions, refining them through its operators until the cipher is successfully broken. The study reveals that DE's approach to exploring the solution space results in a more efficient and effective cryptanalytic attack, offering improved performance metrics over traditional GA-based methods. - By establishing DE as a superior method for this cryptanalytic task, the paper suggests potential broader applications of DE in other cryptographic and cybersecurity challenges, paving the way for further research and development in this field.

DE in Network Security: In addition to optimum firewall configuration, DE plays a role in attack path analysis. DE helps in identifying and analyzing potential attack paths, allowing for better preparedness and mitigation strategies.

The research work by [28] provides a comprehensive approach to both the offensive and defensive aspects of cybersecurity in cyber-physical power systems (CPPS), making substantial contributions to the field through the integration of dynamic modeling, optimization algorithms, and advanced detection methods. The formulation of the attack design as two constrained single-objective optimization problems allows for a precise determination of both the attack location and amplitude. This dual optimization approach ensures that the attack is both strategically placed and effectively executed. Two versions of constrained differential evolution (DE) algorithms are used as solvers for these optimization problems. DE's ability to efficiently explore large search spaces and handle constraints makes it well-suited for optimizing the parameters of the dynamic false data injection attack.

The work [29] applies the Constrained Differential Evolution (CDE) to model and counteract stealthy cyber-attacks in smart grids. It introduces a novel model for stealthy sparse cyber-attacks in AC smart grids. This model uniquely considers both residual test-based detectors and interval-state-estimation-based detectors, which were not concurrently addressed in prior studies. The design of the cyber-attack model is framed as a constrained optimization problem, aiming to minimize the number of compromised meters. The constraints incorporate the characteristics of the two types of detectors, making the problem more realistic and challenging. Given the NP-hard nature of the optimization problem, CDE is proposed as the solution method. CDE is effective in navigating the complex solution space and finding optimal or near-optimal solutions under the defined constraints.

The work [30] deploys the Differential Evolution (DE) algorithm to enhance the detection and localization of Primary User Emulation Attacks (PUEA) in Cognitive Radio Networks (CRN). The study introduces an efficient localization technique using DE to identify PUEA within CRNs. By utilizing DE, the accuracy and efficiency of detecting the location of malicious users are significantly improved. Also, the DE algorithm optimizes the spectrum sensing process by accurately solving the objective function derived from the time difference of arrival values. This results in better identification and differentiation of legitimate primary users (PUs) from malicious secondary users (SUs), thus minimizing interference and improving spectrum usage.

The work [31] develops an adversarial attack algorithm for high-resolution images using differential evolution. There is a challenge for balancing efficiency and success rate in adversarial attacks on high-resolution images, which is less explored compared to low-resolution images.

DE in Malware Detection: DE can optimize the generation of malware signatures, improving the detection rate of known malware. Also, DE assists in identifying and classifying new and unknown malware based on behavior patterns, enhancing overall security measures.

DE has been employed to optimize the parameters of Support Vector Machines (SVM), transforming the classification precision into an optimization problem for the development and enhancement of an active defense strategy in malware detection [32]. This novel combination aims to enhance the accuracy of the SVM in detecting viruses based on user behavior patterns. By optimizing the SVM parameters using DE, higher classification precision compared to traditional methods such as Backpropagation Neural Networks (BPNN), standard SVM, and Genetic Algorithm-optimized SVM (GA-SVM) is achieved.

DE in Optimization of Security Protocols: DE aids in the design and optimization of secure communication protocols, ensuring that they are both secure and efficient. In addition to protocol design, DE is deployed in the optimization of the allocation of security resources, such as bandwidth and computational power, to maximize their effectiveness. The work [33] introduces a novel differential evolution attack (DEAttack) method specifically designed to generate adversarial examples for medical image segmentation models. The work shifts the focus from the more commonly attacked medical image classification models to the less studied, yet critical, medical image segmentation models. By targeting segmentation models, which are crucial in medical diagnostics and treatment planning, the study addresses a significant gap in the robustness evaluation of deep learning models in clinical settings. Unlike traditional gradient-based attack methods that require access to network structures and weights, DEAttack operates without needing such information, making it a more versatile and practical approach for real-world applications. The embedded differential evolution algorithm ensures the diversity of the optimization space, leading to more robust and varied adversarial examples that can better test the vulnerabilities of segmentation models.

2.5 *Particle Swarm Optimization (PSO) in Cybersecurity*

PSO, inspired by the social behavior of birds flocking or fish schooling, is a powerful computational method used for solving optimization problems. Introduced by Eberhart and Kennedy in 1995 [34], PSO has since found applications across various domains, including cybersecurity. Its ability to efficiently explore large search spaces and converge on optimal solutions makes it an ideal candidate for tackling the intricate challenges of cybersecurity.

In the realm of cybersecurity, PSO can be utilized for various purposes, such as enhancing intrusion detection systems, optimizing cryptographic protocols, and improving network security configurations. By leveraging the collective intelligence and adaptive learning capabilities inherent in PSO, cybersecurity professionals can develop more effective and resilient defense mechanisms.

To categorize the works based on their application of PSO in cybersecurity, we can consider two categories of 'Intrusion Detection and Attack Detection' and other PSO applications in cybersecurity. The first category contains works that directly apply PSO to cybersecurity challenges like intrusion detection, adversarial attack detection, and classification of distributed attacks. The second category includes works that, while utilizing PSO, are not directly focused on cybersecurity or are part of broader cybersecurity frameworks without a specific PSO-centric approach.

PSO is used to optimize the placement and configuration of sensors in a network for better detection of intrusions.

As an intrusion detection system of Internet of Medical Things (IoMT) environments, a PSO Deep Neural Network (PSO-DNN) has been developed and implemented [35]. The PSO-DNN is both efficient and suitable for low-computing and low-memory devices. The PSO-DNN approach possesses a high accuracy of 96% in detecting network intrusions. This performance surpasses the state-of-the-art techniques at the time, making it a significant advancement in the field of cybersecurity for IoMT. The PSO-DNN is tailored to operate within the constraints of IoMT devices, which typically have low-computing power and memory. This makes the implementation feasible and practical for real-world medical applications.

The work [36] deploys PSO in development of an adversarial attack system targeting black-box speech recognition model. The proposed system addresses the limitations of existing genetic algorithm GA-based attacks by leveraging the following key advancements: (i) Improved Efficiency in Queries: The PSO-based approach significantly reduces the number of queries required to generate successful adversarial examples. This reduction in queries lowers the risk of detection and optimizes the use of system resources. (ii) Higher Attack Success Rate: The system achieves a higher success rate in generating adversarial examples compared to GA-based methods (8% more success rate). (iii) Temporary Particle Generation: The introduction of temporary particle generation enhances query efficiency, further reducing the detection risk and preventing the wastage of system resources. This innovation ensures that the adversarial attack process is more stealthy and resource-efficient. (iv) Adaptation to Black-Box Models: The PSO-based adversarial attack system is particularly suited for black-box models, where the attacker does not have direct access to the model's parameters. This makes the approach relevant for real-world applications where speech recognition services are typically provided online and involve black-box settings. Fig. 3 shows the simple info-graphy of speech adversary.

Although IDS is considered for avoiding intrusions, it can be itself a subject of adversarial attack. The Fig. 4 illustrates various adversarial attacks on IDS. The attacker injects malicious code into the training data and attempts to extract sensitive information from it. The attacker tries to steal information from the trained IDS. As a result, the IDS predicts inaccurate classifications. WCSAN-PSO framework [37] is suggested for detecting adversarial attacks on IDS. The framework deploys a weighted conditional stepwise adversarial network (WCSAN) optimized by a PSO algorithm and uses a support vector classifier (SVC) for classification. The PSO algorithm is utilized to optimize the parameters of the generator and discriminator within the WCSAN, improving the adversarial training process. This leads to

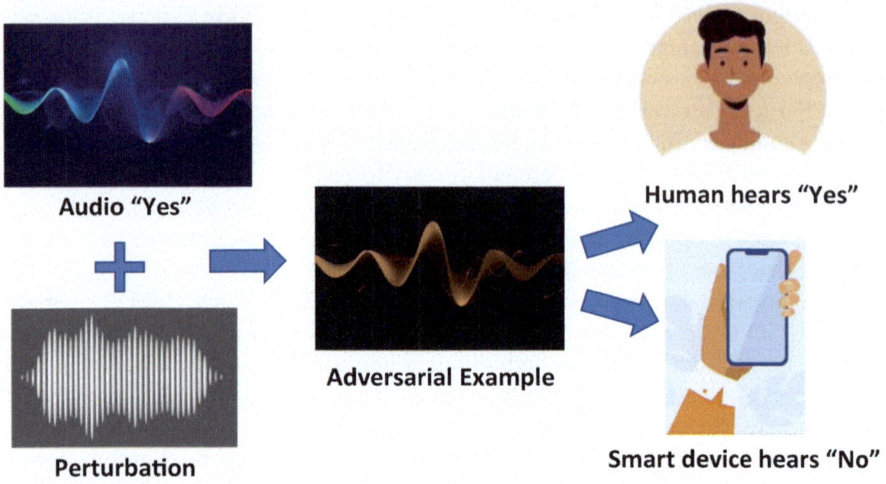

Fig. 3 Technical difference between DE-adversarial perturbation attack and one-pixel attack

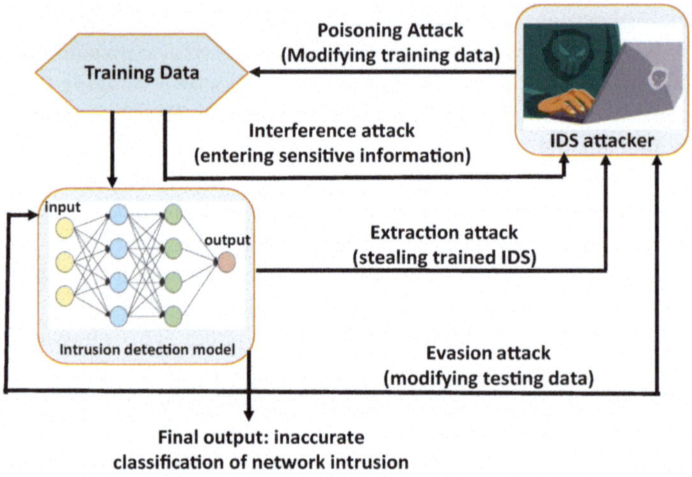

Fig. 4 Various adversarial attacks on intrusion detection system [37]

more robust IDS performance against adversarial attacks. The framework achieves improvements in detection accuracy, with reported accuracies of 99.36% for normal traffic and 98.55% for malicious traffic under adversarial conditions. This marks a notable improvement over existing methods.

The work [38] deploys PSO in combination with Radial Basis Function (RBF) to enhance the accuracy and efficiency of detecting and classifying Distributed Denial of Service (DDoS) attacks in cloud computing environments. The implementation of the detection method at the hypervisor layer allows for monitoring and securing the communication between virtual machines, which is a critical area vulnerable to DDoS attacks in cloud environments. The combination of PSO with RBF improves the performance of the detection algorithm. PSO is used to optimize the parameters of the RBF, leading to higher classification accuracy in identifying DDoS attacks. The research focuses on achieving high classification accuracy against DDoS attacks, addressing a significant challenge in cloud security. By using machine learning techniques, the method outperforms traditional detection methods and the process of identifying cloud assaults is automated, providing a scalable and efficient solution for real-time security management in cloud computing.

2.6 *Artificial Immune Systems (AIS)*

Artificial Immune Systems (AIS) [39] are a class of computational models inspired by the biological immune system. They leverage principles from immunology to solve complex problems in computer science and engineering. AIS often simulate the way biological immune systems recognize and respond to pathogens to develop algorithms for optimization, pattern recognition, and anomaly detection. These systems use mechanisms such as clonal selection, immune memory, and immune network dynamics to adapt and evolve in response to changing environments or requirements. They are particularly useful in areas like data mining, cybersecurity, and adaptive learning.

AIS in Intrusion Detection Systems (IDS): AIS-based IDS can detect unusual patterns (anomaly) that may signify potential security threats, much like the biological immune system identifies pathogens. The work [40] introduces and applies an artificial immune system approach to cybersecurity, specifically for unsupervised anomaly detection in disk images. It explores the use of artificial immune systems to address the challenge of detecting cybersecurity breaches, particularly when traditional methods are impractical due to large dataset sizes or lack of labeled training data. By using an artificial immune system, the work proposes a novel method for anomaly detection that does not rely on labeled data, making it suitable for identifying new or unknown forms of intrusion. This contrasts with machine learning algorithms that require labeled training data and are typically retrospective.

The work by [41] introduces an innovative anomaly-based Intrusion Detection System (IDS) model inspired by the human immune system. It deploys two advanced Artificial Immune System (AIS) algorithms: the Dendritic Cell Algorithm (DCA) [42] and the Clonal Selection Classifier Algorithm (CSCA) [43]. This combination uses a cascading approach, where DCA handles online detection, and CSCA is employed for classification. DCA enables real-time detection of anomalies. By incorporating segmentation techniques, the DCA's detection rate is improved, and it becomes more effective in identifying suspicious activities.

AIS in Malware Detection and Analysis: AIS leverages various approaches such as anomaly detection, behavior analysis, and predictive modeling to identify malicious activities with greater accuracy and speed. Machine learning algorithms can analyze patterns and trends in network traffic, file behaviors, and system operations to detect anomalies indicative of malware. Additionally, deep learning techniques can enhance the capability to recognize complex attack vectors and sophisticated obfuscation techniques. The integration of AIS into malware detection systems offers several advantages, including reduced false positives, improved detection rates, and the capability to handle the sheer volume and diversity of modern cyber threats. As the field continues to evolve, the synergy between artificial intelligence and cybersecurity promises to enhance our defense mechanisms and safeguard digital environments against increasingly sophisticated threats.

The work [44] introduces a framework that integrates both innate and adaptive immune functions within an artificial immune system. This dual approach enhances the system's ability to detect and respond to cyber-attacks effectively. The innate immune function is designed to detect cyber-attacks targeting known or unknown vulnerabilities. It provides immediate response by creating a new process for the server application and terminating the compromised process, thus preventing further damage. The adaptive immune function learns from the exploit code detected by the innate immune function. It enables the server application to continue operating without being terminated, even after an attack is recognized, which improves the resilience and availability of the server service.

The work [45] addresses the limitations of traditional signature-based malware detection methods through the innovative use of an artificial immune system (AIS) for smartphone malware detection, and introduces a smartphone malware detection model (SP-MDM) that leverages an artificial immune system to effectively detect such unknown threats. The integration of both static and dynamic malware analysis techniques enhances the model's ability to detect malware by considering both the static features of the malware and its behavior during execution. Malware characteristics are encoded to generate antigens, which helps in identifying and differentiating between various malware samples. Clonal Selection Algorithm enhances the detection process by selecting high-affinity detectors for further proliferation and somatic hyper-mutation. This results in the generation of more effective detector offspring, improving overall detection performance.

Another field that heavily relies on network communication is unmanned aerial systems (UAS). Implementing an artificial immune system to protect against cybersecurity threats would involve continuously adapting to and defending against evolving threats. Adaptive defense mechanisms are crucial for maintaining the security and integrity of these systems in dynamic and potentially hostile environments. The work [46] develops and applies an artificial immune system (AIS) to detect cyber-attacks on a minidrone's dynamics, with a specific focus on the following aspects: (i) zero-dynamics attack: The study designs a zero-dynamics attack using linear transfer functions to test the resilience of the minidrone's system. This attack is injected into the system to simulate potential vulnerabilities. (ii) An offline artificial immune system is created to detect these cyber-attacks. The AIS uses a discrete, binary, non-self recognition algorithm to develop detectors tailored to identify anomalies and disturbances in the minidrone's data. (iii) The AIS is employed to test and validate its effectiveness in detecting attacks, demonstrating its capability to identify disturbances and enhance the security of the minidrone system.

Another field that artificial immune system contributes to malware detection is power cybersecurity protection, wherein AIS assists to enhance the defense mechanisms of power data transmission networks. The work [47] utilizes the theory of danger to construct a power network with artificial immunity, and it incorporates the concept of detecting and responding to potential threats, enhancing the system's robustness and active defense capabilities. A novel artificial immune power cybersecurity model is proposed based on danger theory. This model includes features such as self-adaptation, self-organization, and high parallelism, aimed at improving the effectiveness and efficiency of cybersecurity measures. The established power cyber model integrates the self-adaptive and self-organizing capabilities of artificial immune systems for enhancing the resilience and defense mechanisms of power cybersecurity against evolving threats.

AIS in Fraud Detection: In fraud detection, AIS can enhance traditional methods by providing robust anomaly detection capabilities and reducing false positives. Its ability to model complex and dynamic fraud patterns makes it a valuable tool for organizations seeking to protect against evolving fraudulent schemes. Overall, the integration of AIS into fraud detection frameworks represents a forward-looking approach, harnessing the power of biological principles to safeguard financial systems and digital transactions.

The study [48] applies AIS for credit card fraud detection and compares its performance with several other methods including Neural Networks (NN), Bayesian Networks (BN), Naive Bayes (NB), and Decision Trees (DT). This comprehensive comparison helps to benchmark AIS against other popular techniques in the context of fraud detection. The research utilizes an Exhaustive Search and GA to optimize parameter sets for each method. As a result, while Bayesian Networks (BN) perform better than Neural Networks (NN), the implementation of NN in this context performed poorly. In contrast, AIS, when optimized with parameters selected by GA, shows superior performance in detecting credit card fraud.

Another work [49] introduces an AIS-based Fraud Detection Model (AFDM), which uses an improved immune system-inspired algorithm (AIRS) to enhance fraud detection efficiency. The proposed model increases the accuracy of fraud detection by up to 25% compared to the base algorithm. This improvement helps in more effectively identifying fraudulent transactions and reducing false negatives. The work achieves a significant reduction in detection system costs, up to 85%. This reduction is crucial for organizations, making fraud detection more economically feasible and scalable. The system's response time is decreased by up to 40%, leading to faster detection and response to potential fraud. This improvement enhances the overall efficiency of the fraud detection process.

3 Challenges and Future Directions in Evolutionary Techniques for Cybersecurity

3.1 Challenges

Balancing Exploration and Exploitation: One of the primary challenges of applying evolutionary algorithms to cybersecurity is finding the right balance between exploration (searching for new solutions) and exploitation (refining known solutions). Too much exploration can lead to inefficiencies, while excessive exploitation may prevent the discovery of novel and more effective defense strategies.

Real-time Adaptability: Cyber threats are constantly evolving, which means security systems must be capable of rapid adaptation. Evolutionary algorithms, although powerful, often struggle with real-time adaptability. This lag in response time can allow cyber adversaries to exploit vulnerabilities before the system has evolved to counteract them.

Integration with Existing Security Frameworks: Incorporating evolutionary techniques into established cybersecurity infrastructures can be complex. Many existing frameworks are built around traditional methods, and the seamless integration of evolutionary algorithms often requires significant re-engineering of the system.

Dynamic Nature of Cyber Threats: The constantly changing landscape of cyber threats poses a significant challenge. Evolutionary algorithms need to be continuously refined and updated to keep up with these evolving threats, which demands substantial time and computational resources.

Scalability Issues: As the size and complexity of cyber environments grow, evolutionary techniques must be able to scale accordingly. However, many current algorithms struggle with scalability, making them less effective in large-scale cyber defense applications.

3.2 Future Scope

Advances in Machine Learning and AI: The integration of machine learning and artificial intelligence with evolutionary algorithms promises to enhance their adaptability, efficiency, and overall effectiveness in cybersecurity. By leveraging AI, these algorithms can learn from vast amounts of data and improve their responses to cyber threats.

Scalability and Resilience: Future research is focusing on improving the scalability of evolutionary techniques to better address large-scale cyber environments. Enhancing the resilience of these algorithms will also ensure they can handle more complex and persistent threats.

Interdisciplinary Approaches: Combining insights from various fields such as evolutionary biology, computer science, and cybersecurity is expected to drive innovation. These interdisciplinary efforts can yield novel strategies for optimizing evolutionary algorithms and tailoring them to specific cyber threats.

Real-time Evolutionary Defense Systems: As the need for real-time security becomes more critical, developing evolutionary algorithms capable of adapting in real-time will be a key area of focus. This will enable quicker and more dynamic responses to cyber-attacks as they happen.

Innovation in Evolutionary Techniques: Ongoing innovation in the field will likely lead to the development of more robust evolutionary algorithms that can anticipate and counter emerging threats. These techniques could include hybrid models that combine evolutionary methods with other optimization approaches to enhance their efficiency and effectiveness.

By addressing these challenges and exploring the future potential of evolutionary algorithms, the cybersecurity field can significantly strengthen its defenses against increasingly sophisticated cyber adversaries.

4 Conclusion

The integration of evolutionary and nature-inspired techniques into cybersecurity represents a transformative shift in the approach to defending against and launching cyber-attacks. From Genetic Algorithms to Butterfly Optimization, these methods offer innovative solutions for enhancing security systems, improving attack detection, and optimizing defensive strategies. Throughout this article, we explored a variety of evolutionary techniques, highlighting their application in key areas such as image steganography, network security, cryptanalysis, and attack detection. Techniques like Genetic Algorithms, Ant Colony Optimization, and Particle Swarm Optimization have demonstrated significant promise in enhancing cybersecurity measures, with applications ranging from intrusion detection systems to resource allocation and attack path analysis. Artificial Immune Systems and Ant Colony optimization have further contributed to advancing defense mechanisms, providing new ways to detect and mitigate sophisticated attacks such as flooding and phishing. The incorporation of neural networks and deep learning into these evolutionary frameworks has underscored the potential for hybrid solutions to tackle complex cybersecurity challenges.

Despite these advancements, several challenges remain. Scalability issues, the security of evolutionary algorithms, and resource constraints need to be addressed to make these techniques more practical and widespread. Future directions should focus on integrating these methods with artificial intelligence and machine learning to enhance their effectiveness. Additionally, ethical and legal considerations, explainability, and benchmarking are crucial for the responsible deployment of these technologies. As we continue to innovate and refine these approaches, the future of cybersecurity will likely see a greater convergence of evolutionary strategies with advanced AI and machine learning frameworks, paving the way for more robust, adaptive, and intelligent security solutions.

References

1. Khosravy M, Gupta N, Witkowski O (2024) Frontiers in genetics algorithm theory and applications. Springer
2. Khosravy M, Gupta N, Patel N (2022) Frontiers in nature-inspired industrial optimization. Springer
3. Holland JH (1992) Genetic algorithms. Scientific American 267(1):66–73
4. Rudbari A, Khosravy M, Witkowski O (2024) Genetic algorithm applications to power management. In: Frontiers in genetics algorithm theory and applications. Springer, pp 157–175
5. Duggal R, Khosravy M, Witkowski O (2024) Application of genetic algorithms to electrical vehicle industries. In: Frontiers in genetics algorithm theory and applications. Springer, pp 179–194
6. Chandan RR, Soni S, Raj A, Veeraiah V, Dhabliya D, Pramanik S, Gupta A (2023) Genetic algorithm and machine learning. In: Advanced bioinspiration methods for healthcare standards, policies, and reform. IGI Global, pp 167–182
7. Sohail A (2023) Genetic algorithms in the fields of artificial intelligence and data sciences. Ann Data Sci 10(4):1007–1018
8. Gupta N, Khosravy M, Patel N, Dey N, Mahela OP (2020) Mendelian evolutionary theory optimization algorithm. Soft Comput 24(19):14345–14390
9. Yadav GS, Ojha A (2019) Improved security in the genetic algorithm-based image steganography scheme using hilbert space-filling curve. Imaging Sci J 67(3):148–158
10. Nandan V, Rao RGS (2023) Low-power aes s-box design using dual-basis tower field extension method for cyber security applications. Complex Intell Syst 9(3):2959–2967
11. Xie N, Qin Z, Di X (2023) Ga-stackingmd: Android malware detection method based on genetic algorithm optimized stacking. Appl Sci 13(4):2629
12. Qin S, Lao C (2023) Computer network security protection system based on genetic algorithm. In: 2023 international conference on applied intelligence and sustainable computing (ICAISC), IEEE, pp 1–8
13. Gupta N, Khosravy M, Pasquali A, Witkowski O (2024) Deep learning security breach by evolutionary universal perturbation attack (eupa). IEEE Trans Artif Intell, 1–11
14. Kirkpatrick S, Gelatt CD Jr, Vecchi MP (1983) Optimization by simulated annealing. Science 220(4598):671–680
15. Chiba Z, Abghour N, Moussaid K, El Omri A, Rida M (2019) Intelligent and improved self-adaptive anomaly based intrusion detection system for networks. Int J Commun Netw Inform Secur 11(2):312–330
16. Dorigo M, Stützle T (2019) Ant colony optimization: overview and recent advances. Springer
17. Ourouss K, Naja N, Jamali A (2021) Defending against smart grayhole attack within manets: a reputation-based ant colony optimization approach for secure route discovery in dsr protocol. Wirel Personal Commun 116:207–226
18. Maranur JR, Mathapati B (2023) Arpvp: attack resilient position-based vanet protocol using ant colony optimization. Wirel Personal Commun 128(2):1235–1258
19. Xie H, Lv K, Hu C, Sun C (2018) A monte carlo graph search algorithm with ant colony optimization for optimal attack path analysis. In: 2018 27th international conference on computer communication and networks (ICCCN), IEEE, pp 1–8
20. Zeng B, Chen B (2010) Sybilaco: ant colony optimization in defending against sybil attacks in the wireless sensor network. In: 2010 international conference on computer and communication technologies in agriculture engineering. Vol 1., IEEE, pp 357–360

21. Bhattacharya M, Roy S (2015) Prevention of walk based attack on social network graphs using ant colony optimization. In: International conference and workshop on computing and communication (IEMCON). IEEE, pp 1–5
22. Gupta A, Pandey OJ, Shukla M, Dadhich A, Ingle A, Ambhore V (2014) Intelligent perpetual echo attack detection on user datagram protocol port 7 using ant colony optimization. In: 2014 international conference on electronic systems, signal processing and computing technologies, IEEE, pp 419–424
23. Storn R, Price K (1995) Differential evolution-a simple and efficient adaptive scheme for global optimization over continuous spaces. Int Comput Sci Inst
24. Kushwah GS, Ranga V (2022) Detecting ddos attacks in cloud computing using extreme learning machine and adaptive differential evolution. Wirel Personal Commun 124(3):2613–2636
25. Su J, Vargas DV, Sakurai K (2019) Attacking convolutional neural network using differential evolution. IPSJ Trans Comput Vis Appl 11:1–16
26. Su J, Vargas DV, Sakurai K (2019) One pixel attack for fooling deep neural networks. IEEE Trans Evolut Comput 23(5):828–841
27. Sinha SN, Palit S, Molla MA, Khanra A, Kule, M (2011) A cryptanalytic attack on knapsack cipher using differential evolution algorithm. In: IEEE recent advances in intelligent computational systems. IEEE, pp 317–320
28. Lu KD, Wu ZG, Huang T (2022) Differential evolution-based three stage dynamic cyber-attack of cyber-physical power systems. IEEE/ASME Trans Mechatron 28(2):1137–1148
29. Lu KD, Wu ZG (2021) Constrained-differential-evolution-based stealthy sparse cyber-attack and countermeasure in an ac smart grid. IEEE Trans Indus Inform 18(8):5275–5285
30. Batool R, Bibi N, Muhammad N, Alhazmi S (2022) Detection of primary user emulation attack using the differential evolution algorithm in cognitive radio networks. Appl Sci 13(1):571
31. Yuan H, Li S, Sun W, Li Z, Steven X (2021) An efficient attention-based image adversarial attack algorithm with differential evolution on realistic high-resolution image. In: 2021 IEEE/ACIS 20th international fall conference on computer and information science (ICIS Fall), IEEE, pp 115–120
32. Ai J, Long Z, Gao S (2015) A novel virus detection and active defense algorithm based on svm optimized by differential evolution algorithm. In: 2015 international conference on intelligent systems research and mechatronics engineering, Atlantis Press, pp 53–56
33. Cui X, Chang S, Li C, Kong B, Tian L, Wang H, Huang P, Yang M, Wu Y, Li Z (2021) Deattack: a differential evolution based attack method for the robustness evaluation of medical image segmentation. Neurocomputing 465:38–52
34. Kennedy J, Eberhart R (1995) Particle swarm optimization. In: Proceedings of ICNN'95-international conference on neural networks. Volume 4., IEEE, pp 1942–1948
35. Chaganti R, Mourade A, Ravi V, Vemprala N, Dua A, Bhushan B (2022) A particle swarm optimization and deep learning approach for intrusion detection system in internet of medical things. Sustainability 14(19):12828
36. Mun H, Seo S, Son B, Yun J (2022) Black-box audio adversarial attack using particle swarm optimization. IEEE Access 10:23532–23544
37. Barik K, Misra S, Fernandez-Sanz L (2024) Adversarial attack detection framework based on optimized weighted conditional stepwise adversarial network. Int J Inform Secur, 1–24
38. Shukla AK, Sharma A (2023) Distributed attacks classification based on radical basis function and particle swarm optimization in hypervisor layer. In: 2023 6th international conference on information systems and computer networks (ISCON), IEEE, pp 1–4
39. Smith R, Timmis J, Stepney S, Neal M (2005) Conceptual frameworks for artificial immune systems. Int J Unconvent Comput 1(3):315–338
40. Donnachie B, Verrall J, Hopgood A, Wong P, Kennedy I (2022) Accelerating cyber-breach investigations through novel use of artificial immune system algorithms. In: International conference on innovative techniques and applications of artificial intelligence, Springer, pp 297–302

41. Enache AC, Sgârciu V (2016) Designing real-time anomaly intrusion detection through artificial immune systems. In: ECCWS2016-proceedings for the 15th european conference on cyber warfare and security, pp 333–342
42. Greensmith J (2007) The dendritic cell algorithm. PhD thesis, Citeseer
43. Brownlee J et al (2005) Clonal selection theory & clonalg-the clonal selection classification algorithm (csca). Swinburne Univ Technol 38
44. Okamoto T, Tarao M (2016) Toward an artificial immune server against cyber attacks. Artif Life Robot 21:351–356
45. Wu B, Lu T, Zheng K, Zhang D, Lin X (2014) Smartphone malware detection model based on artificial immune system. China Commun 11(13):86–92
46. Shivers M, Llanes C, Sherman M (2019) Implementation of an artificial immune system to mitigate cybersecurity threats in unmanned aerial systems. In: 2019 IEEE international conference on industrial internet (ICII), IEEE, pp 12–17
47. Zhao J, Guo S, Mu D (2020) Model design of artificial immune system in power cyber security protection. In: 2020 2nd international conference on applied machine learning (ICAML), IEEE, pp 7–11
48. Gadi MFA, Wang X, do Lago AP (2008) Credit card fraud detection with artificial immune system. In: International conference on artificial immune systems. Springer, 119–131
49. Halvaiee NS, Akbari MK (2014) A novel model for credit card fraud detection using artificial immune systems. Appl Soft Comput 24:40–49

Chapter 3
Optimization Techniques for Critical Issues in Power Amplifiers

Akram Sheikhi and Mohammad Bagher Dowlatshahi

1 Introduction to Power Amplifiers

1.1 Importance in Modern RF Systems

Power amplifiers (PAs) are essential components in radio frequency (RF) systems, particularly in communication transmitters, where they amplify RF signals to levels necessary for reliable transmission. The primary function of PAs is to boost signal power, enabling the signal to travel over long distances or through environments with significant attenuation. This makes PAs critical in a variety of wireless communication systems, including cellular networks, satellite communications, and Wi-Fi, where strong, high-quality signals are required to maintain seamless and reliable connectivity (Fig. 1).

In practical applications, PAs ensure that the transmitted signals can overcome attenuation, noise, and other forms of signal degradation that occur during propagation. For example, in cellular networks, PAs allow signals to cover larger geographic areas, ensuring reliable voice and data transmission even in challenging environments. Similarly, in satellite communications, PAs boost signals for transmission over vast distances, maintaining the integrity of the signal despite the long path between satellites and ground stations.

A key performance metric for PAs is efficiency, especially in mobile devices and infrastructure deployments. Modern PAs, as depicted in Figs. 2 and 3, are designed to maximize efficiency, reducing power consumption and extending battery life in mobile devices, [1–4]. This is critical in consumer electronics and portable devices,

A. Sheikhi (✉)
Electrical Engineering Department, Lorestan University, Khorramabad, Iran
e-mail: sheikhi.a@lu.ac.ir

M. B. Dowlatshahi
Computer Engineering Department, Lorestan University, Khorramabad, Iran
e-mail: dowlatshahi.mb@lu.ac.ir

N. Dey (ed.), *Optimizing Solutions for Real-Life Problems*, Springer Tracts in Nature-Inspired Computing, https://doi.org/10.1007/978-981-96-3216-9_3

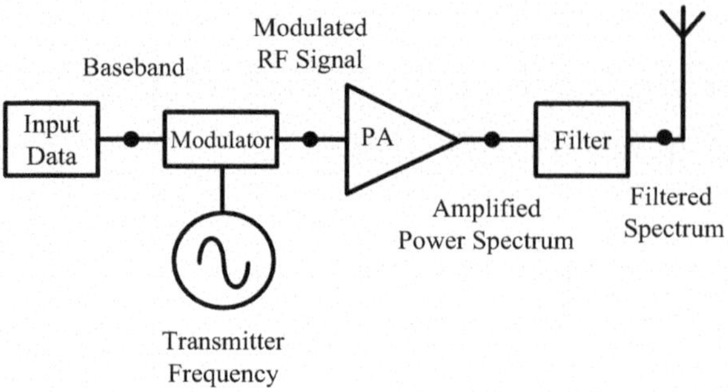

Fig. 1 The transmitter structure

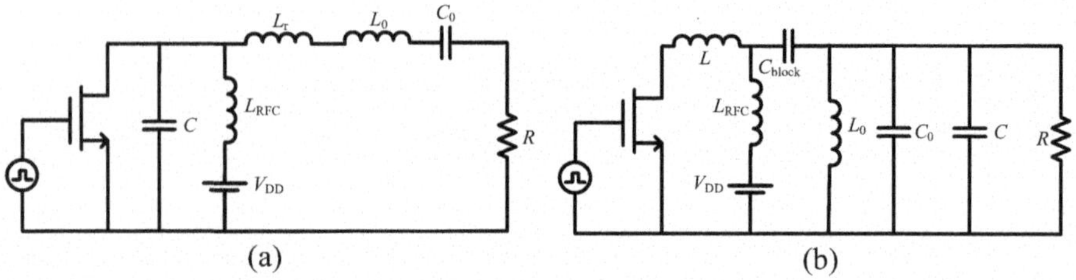

Fig. 2 **a** Class-E, and **b** Inverse Class-E

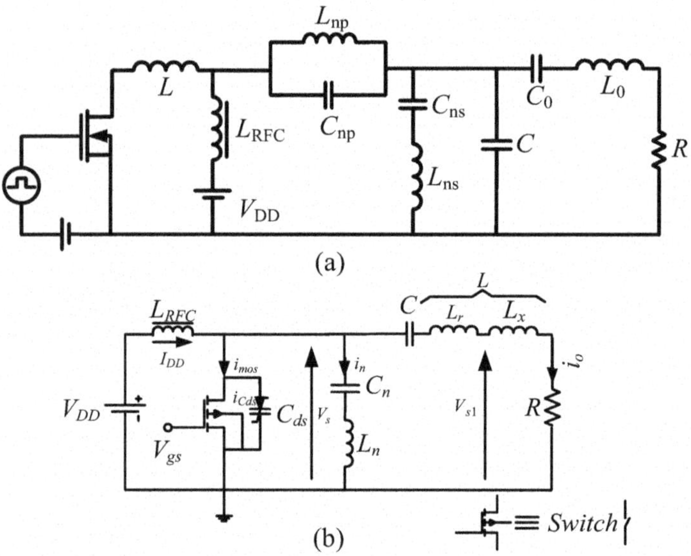

Fig. 3 **a** Class-F/E, and **b** Class-E/F

where energy efficiency is paramount for user satisfaction. In larger-scale infrastructure, such as base stations or satellite systems, efficient PAs help reduce operational costs and improve system sustainability by lowering energy demands.

In addition to efficiency, PAs must meet specific performance criteria based on the application. Telecommunications, aerospace, defense, and medical sectors each require PAs with tailored capabilities, such as high-power output, linearity, wide bandwidth, and robustness against environmental factors. For instance, defense applications may prioritize high power and ruggedness, while consumer electronics focus on miniaturization and integration.

Advancements in semiconductor technologies and PA design have led to significant improvements in PA performance, including enhanced efficiency, linearity, and reduced size. These innovations have enabled the integration of PAs into System-on-Chip (SoC) and System-in-Package (SiP) architectures, improving overall system performance while reducing manufacturing costs. As a result, modern PAs are becoming increasingly versatile and capable of meeting the evolving demands of RF systems across a wide range of industries.

1.2 Overview of Key Metrics

These key metrics–Intermodulation Distortion (IMD), receiver noise, and efficiency–provide a comprehensive framework for evaluating the performance of PAs in RF systems. Designers and engineers must balance these metrics based on specific application requirements, such as communication range, bandwidth, and power constraints. Optimizing PAs for low IMD, minimal receiver noise, and high efficiency contributes to enhanced system performance, reliability, and overall user experience in diverse RF applications.

1.2.1 IMD

IMD refers to the generation of unwanted signals at frequencies that are not present in the original input signals but are produced due to nonlinearities in the PA, Fig. 1. IMD is a critical metric because it can degrade the quality of transmitted signals, leading to interference and reduced system performance. High IMD levels can result in adjacent channel interference in communication systems, affecting signal clarity and reliability. PAs with low IMD are essential for maintaining signal integrity and minimizing interference in RF systems.

1.2.2 Receiver Noise

Receiver noise refers to the internal noise generated within the PA itself, which can degrade the signal-to-noise ratio (SNR) of received signals. This noise can originate

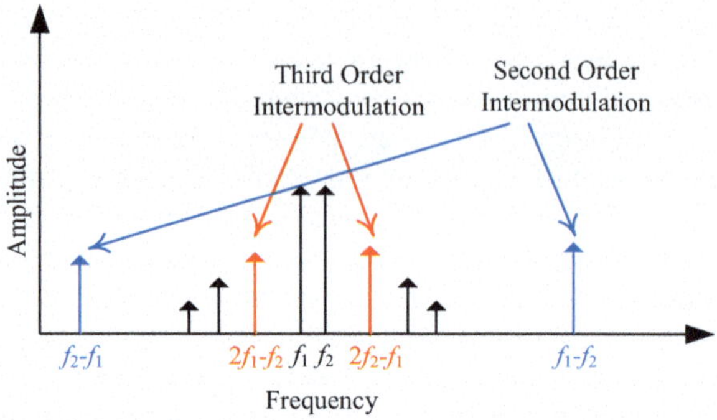

Fig. 4 Intermodulation distortion

from various sources within the amplifier circuitry, including thermal noise and shot noise. Lower receiver noise figures indicate better sensitivity and improved SNR, which are crucial for receiving weak signals with clarity and accuracy. Minimizing receiver noise is essential in applications where signal detection and sensitivity are paramount, such as in satellite communications and radio astronomy. For example, the noise figure in the receiver can be obtained as (Fig. 5)

$$NF = NF_1 + \frac{NF_2 - 1}{G_1} + \frac{NF_3 - 1}{G_1 G_2} \tag{1}$$

1.2.3 Efficiency

Efficiency in PAs refers to the ratio of output RF power to the DC power input, expressed as a percentage (depicted in Eqs. (2) and (3)). High-efficiency PAs are desirable because they minimize power consumption, reduce heat dissipation, and extend battery life in portable devices. Efficiency is particularly important in battery-operated devices, mobile communication systems, and infrastructure deployments where power consumption and operational costs are critical considerations. Advances in PA design have focused on improving efficiency while maintaining high linearity and signal fidelity, leading to more energy-efficient RF systems.

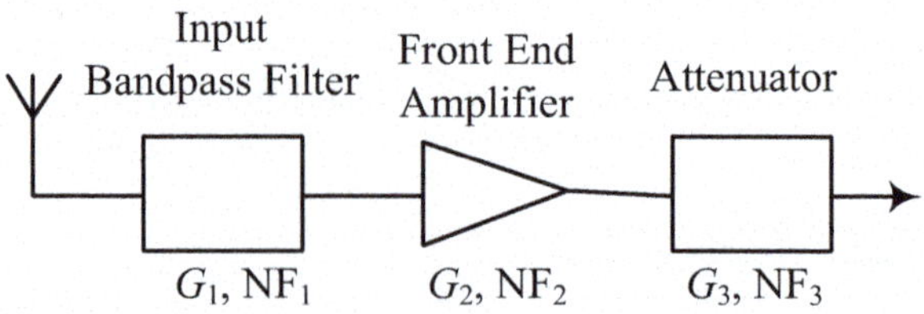

Fig. 5 Noise figure

$$\eta = \frac{P_{out}}{P_{DC}} \tag{2}$$

$$PAE = \frac{P_{out} - P_{in}}{P_{DC}} \tag{3}$$

2 Optimization Metrics for PAs

In the realm of PAs, achieving optimal performance requires careful consideration of various metrics that collectively determine their suitability and effectiveness in RF systems. One crucial metric for assessing PA performance is the Signal-to-Noise-plus-Intermodulation Ratio (SNIR). This section explores SNIR as a fundamental optimization metric for PAs in RF applications.

2.1 Signal-to-Noise-Plus-Intermodulation Ratio

The SNIR is a comprehensive metric that evaluates the combined quality of a PA's output signal by taking into account both noise and IMD. It provides a holistic view of how well a PA preserves signal integrity amidst noise and distortion, thereby influencing the overall system performance and reliability. The SNIR quantifies the ratio of the signal power to the combined power of noise and IMD, providing a measure of how well the PA amplifies the desired signal relative to unwanted distortions and noise. Linear PAs exhibit minimal IMD, preserving the fidelity of the input signal. Lower receiver noise contributes to higher SNR, thereby improving SNIR. Efficient PAs minimize power consumption, reducing thermal noise and enhancing overall SNIR. SNIR should be evaluated across the operational bandwidth to ensure consistent performance across all frequencies of interest. SNIR is critical in wireless communication systems to ensure reliable transmission and reception of signals, especially in environments prone to interference. High SNIR enables clear and reliable communication between ground stations and satellites, even under challenging atmospheric conditions. PAs with optimized SNIR contribute to improved signal coverage, reduced dropped calls, and enhanced data transmission rates. Designing PAs with low IMD and receiver noise figures. Implementing advanced linearization techniques to mitigate IMD. Enhancing PA efficiency to minimize power consumption and thermal noise. Conducting rigorous testing across various operating conditions to validate SNIR performance.

2.2 Efficiency and Power Consumption Considerations for PAs

Efficiency and power consumption are critical optimization metrics for PAs, playing pivotal roles in determining their performance, cost-effectiveness, and environmental impact. This section explores these metrics in detail and their significance in the context of PA design and operation. Factors Influencing Efficiency are class of operation, load matching, supply voltage and biasing, and thermal management. Different amplifier classes, e.g., Class A, Class AB, Class B, and Class C have varying efficiency characteristics. Classes like Class A typically have lower efficiency but better linearity, whereas Classes like Class D are highly efficient but may introduce more distortion. Proper impedance matching between the PA and the load, e.g., the antenna enhances efficiency by maximizing power transfer. Optimal supply voltage and biasing conditions ensure the PA operates within its efficient range without unnecessary power loss. Effective heat dissipation techniques, such as heatsinks or fans, prevent thermal inefficiencies that can degrade amplifier performance. Efficiency and power consumption are integral metrics in optimizing PAs for RF applications. Engineers can achieve superior performance, cost-effectiveness, and environmental stainability across diverse RF systems and applications by prioritizing high efficiency and low power consumption in PA design and implementation.

2.3 Saturated Versus Nominal Power Levels in PAs

Understanding the concepts of saturated and nominal power levels is crucial for optimizing PAs in various RF applications. This section explores these metrics and their significance in PA design and performance evaluation. Understanding and optimizing saturated and nominal power levels are essential for designing efficient and reliable PAs in RF systems. By balancing performance requirements with efficiency considerations at both saturated and nominal power, engineers can achieve optimal PA design tailored to specific application needs, ensuring high performance and longevity in operation.

3 Technological Constraints and Applications

In the realm of PAs, understanding the role of saturated power in PAs is essential for optimizing performance across telecommunication satellites, mobile handsets, and RFID systems. By balancing saturated power P_{sat} with efficiency, linearity, and application-specific requirements, engineers can design PAs that meet the demanding needs of modern communication and wireless connectivity technologies.

1. Role of Saturated Power:
   - Definition and Significance: P_{sat} denotes the maximum output power a PA can deliver while maintaining specified levels of linearity and efficiency. It serves as a critical parameter in determining the upper operational limits of a PA without introducing significant distortion or inefficiency.
2. Applications in Telecommunication Satellites:
   - Signal Transmission Requirements: Telecommunication satellites rely on PAs with high P_{sat} to ensure robust transmission of signals across vast distances in space.
   - Interference Management: PAs in satellites must handle varying signal strengths from multiple ground stations and users, necessitating high P_{sat} to maintain signal integrity amidst potential interference.
3. Impact on Mobile Handsets:
   - Power Efficiency: Mobile handsets require PAs optimized for efficiency to prolong battery life, yet capable of delivering sufficient power during transmission.
   - Dynamic Signal Handling: PAs with well-defined P_{sat} allow mobile devices to handle fluctuating signal strengths encountered in different environments (urban vs. rural) while minimizing power consumption.
4. Role in RFID Systems:
   - Range and Reliability: RFID systems benefit from PAs with sufficient P_{sat} to ensure reliable communication over varying distances between tags and readers.
   - Signal Integrity: Maintaining signal integrity is crucial in RFID applications, where PAs with high linearity at P_{sat} minimize errors and ensure accurate data transfer.
5. Technological Constraints and Challenges:
   - Heat Dissipation: High P_{sat} PAs generate substantial heat, requiring efficient thermal management to prevent performance degradation and ensure longevity.
   - Size and Integration: Miniaturization of PAs for mobile handsets and RFID tags requires innovative designs to balance P_{sat} with size constraints without compromising performance.
6. Future Directions and Innovations:
   - Advancements in Semiconductor Technologies: Continued advancements in semiconductor materials and fabrication techniques are enhancing PA's P_{sat} capabilities while reducing power consumption and size.
   - Integration with IoT: PAs are evolving to support the expanding Internet of Things (IoT) ecosystem, requiring versatile designs capable of adapting to diverse application requirements.

4 Design Challenges and Considerations

When designing PAs, engineers encounter various challenges and must consider critical factors such as antenna size constraints and the influence on Equivalent Isotropic Radiated Power (EIRP) and gain-to-temperature ratio. Additionally, strategies for selecting and optimizing input and output matching networks in amplifier technologies (Fig. 6) play a crucial role in achieving desired performance metrics, [5]. Addressing antenna size constraints, optimizing EIRP and G/T ratio, and strategically selecting amplifier technologies are crucial steps in designing efficient and high-performance PAs. By leveraging advancements in technology and innovative design approaches, engineers can overcome challenges and meet the stringent requirements of modern wireless communication systems. Equivalent Isotropic Radiated Power (EIRP) quantifies the power that an antenna transmits in a specific direction, relative to an isotropic antenna. It considers both the PA's power output and the antenna's gain. Larger antennas typically offer higher gain but are impractical in many applications due to size limitations, especially in mobile devices and compact systems. Designing PAs for small antennas requires balancing power output (and hence P_{sat}) with antenna efficiency to maximize EIRP within space and weight constraints.

5 Future Directions and Emerging Technologies

Advancements in PA technology and design methodologies are poised to revolutionize RF systems, offering potential improvements in efficiency, performance, and reliability. This section explores the emerging trends and technologies shaping the future of PAs.

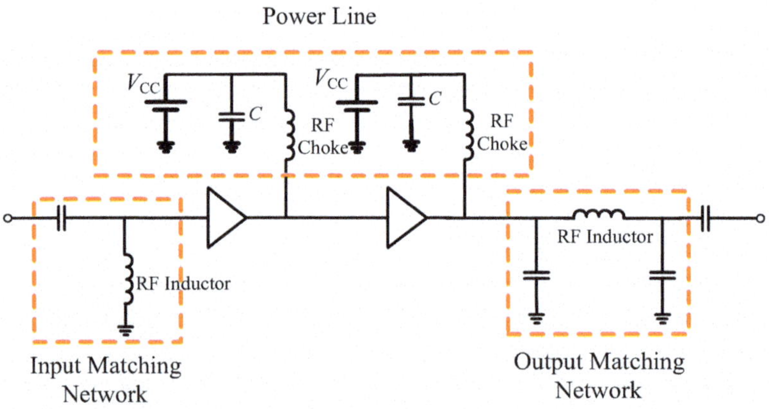

Fig. 6 Power amplifier circuit

- Advancements in PA Technology: Continued integration of Gallium Nitride (GaN) and Gallium Arsenide (GaAs) technologies for higher efficiency and power density. Adoption of multi-stage amplifier architectures for improved linearity and reduced distortion. Enhanced digital predistortion (DPD) algorithms leveraging machine learning and artificial intelligence techniques for real-time adaptive linearization. Development of PAs capable of operating across broader frequency ranges, supporting multi-band and wideband communication systems. Shrinking PA footprint through advanced packaging techniques and integration with RF front-end modules.
- Potential Improvements in Efficiency and Performance: Optimization of PA efficiency through envelope tracking, envelope elimination and restoration (EER), and efficiency-enhanced architectures. Advancements in linearization techniques such as Cartesian feedback, feedforward linearization, and adaptive DPD to mitigate nonlinear distortion. Improved thermal management strategies to handle higher power densities and enhance reliability. Increasing RF power output while maintaining efficiency and minimizing size constraints for high-power applications.
- Integration with Emerging Technologies: Adaptation of PAs to meet the stringent requirements of 5G networks, including higher data rates, lower latency, and increased network capacity. PAs are tailored for IoT devices with low power consumption, extended battery life, and robust connectivity. High-power, high-efficiency PAs for satellite communication payloads, enabling global coverage and reliable data transmission. PAs are designed for automotive radar, vehicle-to-vehicle communication (V2V), and other automotive RF systems.

5.1 *Linearization Techniques*

In wireless communication systems, the rising demand for high data rates and efficiency introduces numerous practical challenges in designing direct conversion transmitters for system-on-chip (SoC) or system-in-package (SiP) applications. Existing research highlights that the nonlinearity of PAs significantly impacts the quality of transmitted signals. This chapter delves into these issues to enhance the signal quality by reducing the distortion in the transmitter's output. The quadrature modulation-based direct conversion transmitter (DCT) is a widely adopted architecture known for its low modulation distortion, regardless of data rates. The DCT primarily includes a local oscillator (LO) to provide a pure sinusoidal carrier, a quadrature modulator for baseband I/Q modulation, and a PA to boost the transmitted signal power level, Fig. 7.

Also known as zero-IF (intermediate frequency) architecture, direct conversion executes frequency conversion between the RF signal and the baseband signal directly, without intermediate frequency stages. This architecture simplifies the design and reduces the number of components, which can lead to cost and power consumption benefits. We can classify the advantages:

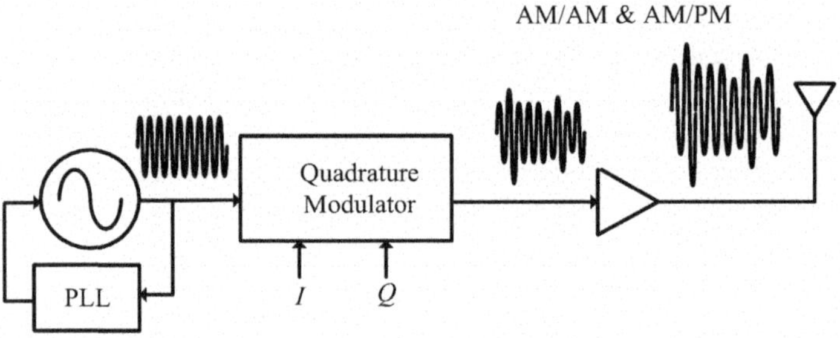

Fig. 7 Signal quality degradation owing to PA nonlinearity in a DCT

- Low Modulation Distortion: Maintains low distortion across varying data rates, ensuring high signal quality.
- Simplified Design: Eliminates the need for intermediate frequency stages, reducing complexity and component count.
- Integration: Suitable for integration into System-on-Chip (SoC) or System-in-Package (SiP) applications, enabling compact and efficient transmitter designs.

Despite its advantages, the DCT architecture faces several distortion-related challenges that need to be addressed to achieve successful deployment in complex wideband modulation applications, especially within SoC/SiP environments.

- I and Q Imbalance: Differences in gain and phase between the I and Q channels can introduce distortion.
- DC Offset: Unwanted DC components can arise from imperfections in the circuitry, leading to signal distortion.
- Nonlinearities: The PA can introduce nonlinear distortion, especially when operated near saturation to maximize efficiency.

Addressing these distortion problems is crucial for the effective implementation of DCTs in advanced communication systems. Various linearization techniques, such as predistortion and feedforward linearization, are employed to mitigate these issues and ensure high-fidelity signal transmission. Linearization techniques are essential in wireless communication systems to mitigate the nonlinear distortion introduced by PAs. These distortions can significantly degrade signal quality, reducing the efficiency and performance of communication systems. Several linearization methods are employed to address these issues, each with its advantages and challenges.

The main goal of linearization techniques is to mitigate these nonlinear effects, ensuring that the amplified signal closely matches the input signal's characteristics. Traditional methods include predistortion techniques, where the input signal is deliberately distorted to counteract PA nonlinearities, and feedback techniques, which adjust the input signal based on the PA's output to minimize distortion. Neural networks offer a data-driven approach to PA linearization:

- Data-Driven Approach: Neural networks learn the complex relationship between the input signal and the distorted output signal directly from training data. This data typically consists of pairs of input signals and their corresponding distorted outputs after amplification.
- Architecture: Different neural network architectures such as Convolutional Neural Networks (CNNs) or Recurrent Neural Networks (RNNs) can be employed based on the nature of the input-output relationship. CNNs are effective for spatial dependencies in signals, while RNNs are suitable for handling temporal dependencies.
- Training Process: During training, the neural network adjusts its internal parameters (weights and biases) to minimize the difference between the predicted and actual distorted outputs. This process requires a diverse dataset that covers various operating conditions and PA behaviors.
- Real-Time Adaptation: Once trained, the neural network can be deployed in real-time to predict the distortion characteristics of the PA based on the input signal. This prediction enables the generation of a pre-distorted signal that, when amplified by the PA, results in a more linear output signal.

Advantages of Neural Network-Based Linearization

- Flexibility: Neural networks can model highly nonlinear and dynamic behaviors of PAs.
- Adaptability: They can adapt to changes in PA characteristics and environmental conditions.
- Performance: Neural network-based approaches often achieve better linearization performance compared to traditional methods.

Challenges and Considerations

- Training Data: Acquiring high-quality training data that represents diverse operating conditions and PA behaviors is crucial for effective neural network training.
- Computational Complexity: Implementing neural network-based solutions may require significant computational resources, especially for real-time applications.
- Deployment: Ensuring the robustness and stability of neural network models in real-world PA applications requires thorough validation and testing.

Ongoing research focuses on enhancing the efficiency and effectiveness of neural network based PA linearization techniques. These techniques find applications in various wireless communication systems, including cellular networks, satellite communications, and radar systems, where maintaining high signal fidelity and spectral efficiency is critical.

5.1.1 Feedforward Linearization

A feedforward linearizer, as depicted in Fig. 8, comprises a PA that introduces nonlinear distortion, an error amplifier (EA), couplers, and delay elements (d), all operating at RF. Part of the RF input signal passes through a 180o delay line, while the rest is directed to the PA. A sample of the PA's output is then compared with the delayed input signal. If the PA is linear and the coupling is properly adjusted, these signals will match, resulting in zero signal input to the EA, which is part of the initial calibration process. Nonlinear distortion in the PA produces an RF error signal at the output of the combiner. This error signal is linearly amplified by the EA and fed back to a properly delayed PA output, ideally canceling all unwanted components, leaving only the desired amplified signal for transmission from the antenna. The primary challenge is to optimize distortion cancellation across the amplifier's dynamic range and evaluate the impact of compensation under varying conditions like aging and temperature fluctuations. The EA must be highly linear to avoid additional impairments and exhibit minimal and stable propagation delay. This technique is sensitive to parameters such as temperature and power level changes, necessitating adaptive control circuitry. Power efficiency in this setup is influenced by coupling factors and the EA's power rating and operating point. Typically, an output coupler with 8 to 10 dB coupling is used, with the EA operating at around 25% of the PA power. This setup results in an overall DC to RF efficiency of 5 to 10% for a feedforward cellular amplifier. Despite its ability to significantly improve linearity over wide bandwidths, feedforward linearization is costly and bulky due to the required tuning components, making it more suitable for multicarrier base station applications.

5.1.2 Envelope Elimination and Restoration (EER)

Initially developed for SSB and TV transmitters, EER has been adapted for multicarrier modulation schemes in next-generation wireless systems. Also known as the Kahn technique, EER combines linearization and efficiency enhancement. It extracts and amplifies the envelope of the RF signal separately, then uses this envelope to modulate the PA's DC power supply, as shown in Fig. 9. The phase information of the RF signal, with the envelope information removed, is fed into the PA. The original signal envelope is restored to the amplified RF signal, achieving optimal efficiency with classes C, D, E, and F PAs. However, EER faces challenges, such as ensuring the linearity of the envelope amplifier and maintaining phase coherency between the envelope and phase signals during restoration.

Fig. 8 Feedforward linearization

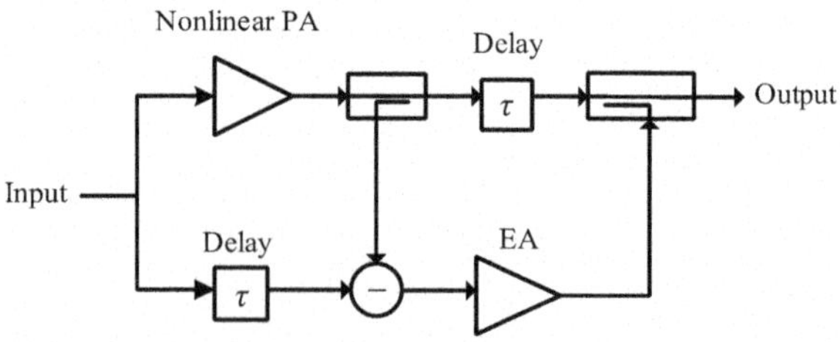

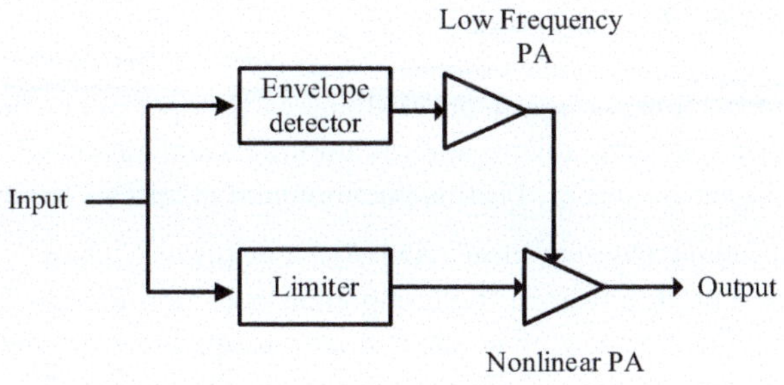

Fig. 9 Envelope elimination and restoration

5.1.3 Cartesian Feedback

As shown in Fig. 10, Cartesian feedback involves demodulating a portion of the PA output signal and feeding it back negatively at the baseband through differential amplifiers to the I/Q modulator, thus compensating for PA distortion if the loop gain is correct, as shown in Fig. 4. This method provides good linearity improvements across the entire transmission path but is inherently narrowband. Challenges include the need for an extra I/Q demodulator and stability issues dependent on loop gain, PA phase shift, time delays, and loop-filter bandwidth. Loop delays can introduce output distortion. The main advantage of Cartesian feedback is its straightforward implementation and adaptability to any modulation scheme.

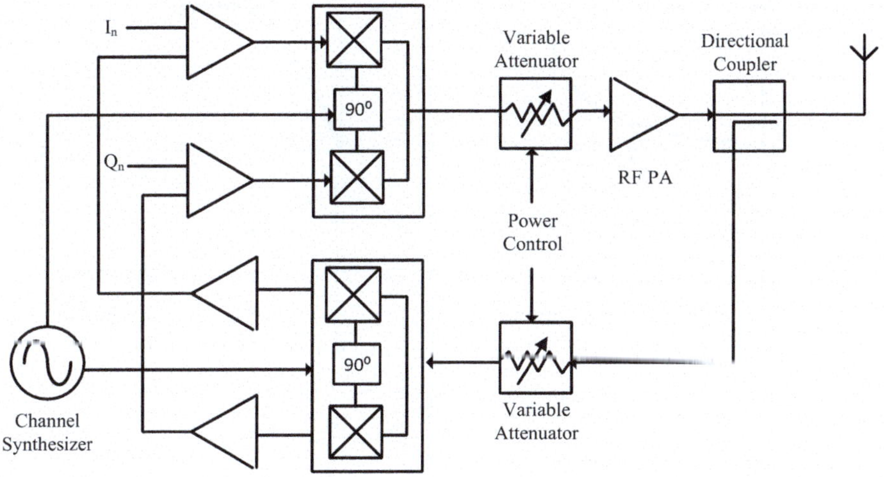

Fig. 10 Cartesian feedback

5.1.4 Predistortion

Distortion is a significant challenge in the operation of nonlinear systems, particularly affecting components such as amplifiers and mixers when driven at high signal levels. As shown in Fig. 11, this phenomenon manifests in several distinct ways:

1. Gain Compression (AM-AM Distortion): Gain compression occurs when the output power of the component does not increase linearly with the input power Pin. Instead of a linear relationship, the gain starts to compress as the input power rises. This nonlinearity results in distorted amplitude modulation (AM) characteristics of the output signal.
2. Phase Distortion (AM-PM Distortion): Phase distortion refers to changes in the phase of the output signal relative to the input signal as a function of input power. In an ideal linear system, the phase relationship remains constant. However, in nonlinear systems, especially at high-power levels, the phase shift varies nonlinearly with the input signal amplitude, leading to phase modulation (PM) distortion.
3. Unwanted Frequencies (Spurious Outputs or Spurs): Nonlinear systems generate additional frequencies in the output spectrum that are not present in the input signal. These unwanted frequencies appear as spurious outputs or spurs. When these spurs interfere with the desired signal, they degrade the Signal-to-Noise plus Distortion Ratio (SINAD) or Signal-to-Noise Ratio (SNR), impacting the overall signal quality.

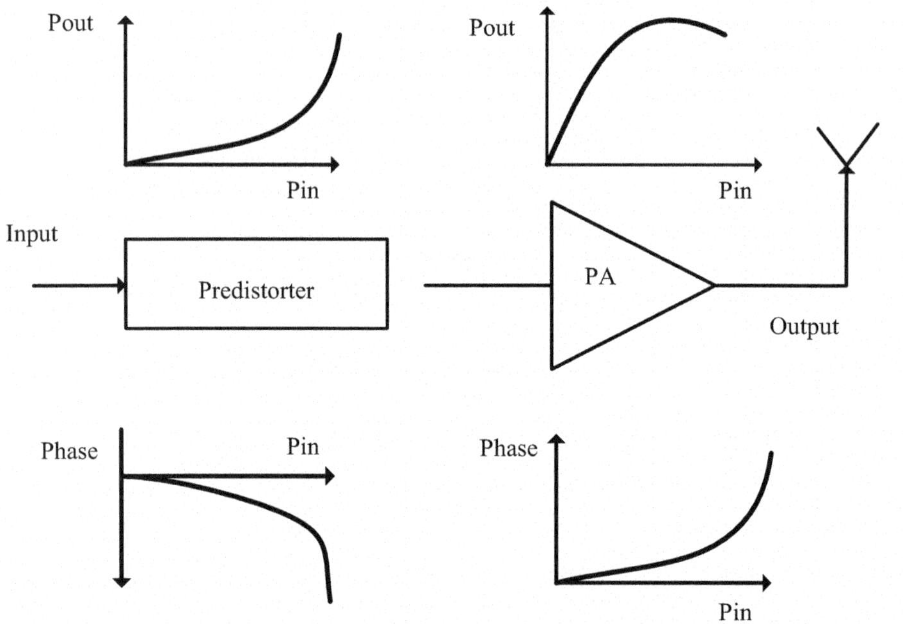

Fig. 11 Predistortion

4. IMD: IMD occurs when multiple input signals interact nonlinearly within the system, producing additional frequencies at the output. This in-band distortion manifests as undesired signals at frequencies that are sums and differences of the original input frequencies. IMD is particularly problematic in communication systems where multiple signals are simultaneously present.

Distortion limits the upper operational range of analog systems by compromising signal fidelity and reducing the effective dynamic range. High levels of distortion degrade the accuracy and reliability of signal transmission, affecting applications ranging from wireless communications to audio processing. To mitigate distortion in nonlinear systems, various techniques are employed:

- Linearization Techniques: Predistortion, feedforward, and feedback linearization methods are used to counteract nonlinearities and maintain signal integrity.
- Component Design: Improving the linearity of components through careful design and selection of materials helps minimize distortion effects.
- Operational Considerations: Proper operational tuning and signal conditioning can reduce the impact of nonlinear distortion, ensuring optimal performance across varying signal levels and frequencies.

Understanding and managing distortion in nonlinear systems are critical for achieving high-performance and reliable operation in modern communication and signal processing applications. Efforts to mitigate distortion contribute significantly to enhancing signal quality and system efficiency. P1dB, or 1 dB compression point, is a crucial metric used to quantify gain compression in amplifiers. It represents the input power level at which the amplifier's gain decreases by 1 dB from its linear small-signal gain, Fig. 12. This point is significant because it indicates the onset of nonlinear behavior in the amplifier. When an amplifier reaches its P1dB, it means that further increases in input power will cause the amplifier's output to compress, resulting in a reduction of gain. This gain compression is accompanied by the generation of severe distortion in the output signal. The distortion arises because the amplifier's nonlinearity causes the output signal to deviate from the ideal linear relationship with the input signal.

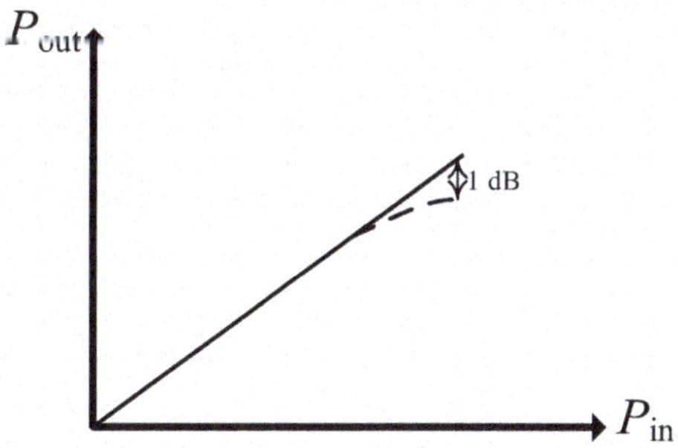

Fig. 12 1 dB compression point

The severity of distortion at the P1dB point indicates that the amplifier can no longer accurately amplify the input signal, leading to degradation in signal quality. This distortion manifests as harmonic distortion (HD) and intermodulation distortion (IMD), which introduce unwanted frequencies into the output spectrum. These distortions can significantly degrade the Signal-to-Noise Ratio (SNR) and other key performance metrics of the amplifier, impacting its suitability for applications requiring high fidelity and low distortion. In summary, P1dB serves as a straightforward measure to assess gain compression in amplifiers, and an amplifier operating at or near its P1dB point will exhibit severe distortion, making it critical to ensure operation below this threshold for maintaining signal integrity and fidelity. Harmonic distortion (HD) is a type of distortion caused by the nonlinearity of the signal path in amplifiers and other electronic systems. It manifests as additional frequency components in the output signal that are integer multiples (harmonics) of the input signal frequencies. For instance, if the input signal has a fundamental frequency f_0, the harmonic distortion would produce components at frequencies like $2f_0$, $3f_0$, $4f_0$, etc. In many cases, harmonic distortion can be mitigated or reduced effectively through filtering techniques. Filters can be designed to attenuate or remove these unwanted harmonics from the signal, thereby improving the overall signal quality. One specific type of harmonic distortion is third-order harmonic distortion, often denoted as HD3. It arises due to the cubic nonlinearity in the signal path. Unlike linear systems where distortions would ideally scale proportionally to the input signal, in nonlinear systems experiencing third-order distortion, the power of the third harmonic is proportional to the cube of the input amplitude. When plotting the output power Pout versus the input power Pin on a dBm scale, the power of third-order harmonic distortion increases at a rate three times faster than the rate at which the fundamental power increases with input power. This characteristic is sometimes referred to as the system being "well behaved" in terms of distortion behavior, as it follows a predictable cubic relationship. While harmonic distortion is relatively straightforward to mitigate through filtering, IMD poses greater challenges. IMD results from the interaction of multiple signals in a nonlinear system, producing sum and difference frequencies that are not harmonically related to the input signals. IMD can significantly degrade signal quality, especially in communication systems where multiple signals are present simultaneously. Predistortion involves creating an inverse characteristic of the PA at RF to cancel out PA nonlinearity. This technique is cost-effective, utilizing an open-loop and inherently stable circuit that consumes less power compared to other methods. However, to counteract the effects of temperature changes, aging, power level variations, and frequency shifts, adaptive predistorters are necessary. It's simplicity and wideband capabilities make predistortion an attractive option for future broadband systems, such as multicarrier Orthogonal frequency-division multiplexing (OFDM) modulation. Simulations of a predistortion linearizer provided valuable insights. The typical amplitude (AM/AM) and phase (AM/PM) characteristics of the one sample PA are illustrated in Fig. 13. The input and output power back-off (IBO, OBO) is measured relative to the saturated power (IBO = $P_{in}-P_{sat}$, etc.). Both the PA and the predistorter were modeled using a four-coefficient power series in MATLAB, with a sampling frequency set to avoid aliasing. A key reference measure is the

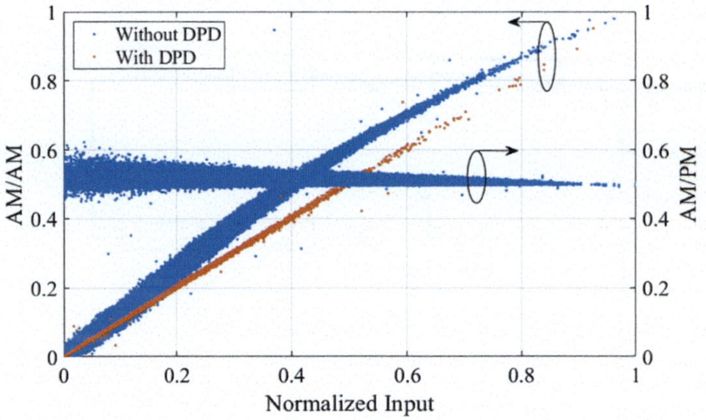

Fig. 13 PA AM-AM and AM-PM plots

linearization improvement as a function of IBO for third-, fifth-, and seventh-order intermodulation distortion, observed using two equally powered carrier input tests. Additionally, the improvement in modulation fidelity of digitally modulated signals is crucial.

AM-to-AM Response

The AM-to-AM response of a PA is a critical metric that illustrates the output power relative to the input power, providing insight into the linearity and correctability of the PA. Ideally, a linear PA exhibits a straight AM-to-AM line, where each 1dB increase in input power results in a 1dB increase in output power, indicating no distortion. However, in wireless infrastructure applications, PAs are typically operated near saturation to maximize efficiency, leading to nonlinear behavior and necessitating linearization techniques to meet spectral emission requirements while maintaining high efficiency. In practical scenarios, an unlinearized PA may show a gain ripple, which is the fluctuation in gain across its operational range. For optimal linearization performance, the gain variation should not exceed 2dB. This ensures that the PA operates close to the ideal linear response, even when approaching compression. Class AB amplifiers generally show a monotonic gain response, with gain decreasing as output power approaches saturation, a phenomenon known as gain compression. More advanced PA architectures, like Doherty amplifiers as described in [6] and illustrated in Fig. 14, exhibit more intricate AM-to-AM responses.

Achieving minimal gain ripple and understanding the impact of bias conditions are crucial for optimizing PA performance and ensuring effective linearization. Properly designed PAs with controlled AM-to-AM responses enable better spectral emission control and improved efficiency, which are essential for modern wireless communication systems.

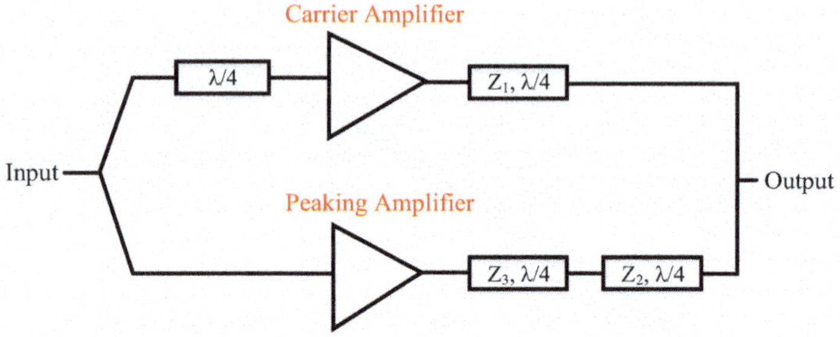

Fig. 14 The DPA architecture with a two-section peaking network

AM-to-PM Response

The phase response, or AM-to-PM response, is a key factor in designing a correctable RF amplifier using RF Power Amplifier Linearization (RFPAL). As the PA's output power nears P_{sat}, its phase begins to rotate. The degree of this rotation and the smoothness of the phase change are critical for effective linearization. For optimal results, phase rotation should remain below 15o when the gain variation is under 1 dB. However, if the gain variation exceeds 1 dB, the phase rotation (AM-to-PM) should be kept under 10o to ensure good linearization. Linearization improves significantly when the AM-to-PM curve is smooth and monotonic, without inflection points. A monotonic phase change means that the phase continuously increases or decreases without reversing direction, which simplifies the linearization process. Figure 13 illustrates an ideal smooth and monotonic AM-to-PM curve that is conducive to effective linearization. For an RF amplifier to be effectively linearized using RFPAL, it is essential to maintain minimal phase rotation and ensure the AM-to-PM response is smooth and monotonic. By adhering to these criteria, the linearization process can significantly improve the amplifier's performance, reducing distortion and enhancing signal fidelity.

Digital predistortion is a common technique for correcting nonlinear distortion in PAs, [7–12]. Traditional digital predistortion methods, including polynomial models or deep neural networks (DNNs), are typically designed for a single specific state. When the PA's operating conditions change, retraining and updating the model's coefficients become necessary, highlighting the limitations of DNN models regarding generalization. To tackle this problem, this letter introduces a novel modeling approach that creates a generalized PA model capable of operating under multiple states using DNNs. This method incorporates a set of coding vectors that represent different states to construct the generalized model. Experimental results indicate that, compared to conventional DNN models, the proposed approach effectively builds a PA model that accommodates multiple states while maintaining excellent modeling performance. Yin et al. [13] enhanced the input vector of neural networks (NNs) to adapt to multiple working states in a multiple-input-multiple-output system. In [14], a self-sensing (SS) module was developed based on signal features, resulting in the SS DPD NN, which achieves better performance in linearizing PAs under varying transmission signals compared to [13]. Wu et al. [15] introduced the uniform neural network (UNN) DPD model, which includes additional bandwidth (BW) and power networks, providing high-performance linearization of PAs. This model surpasses the SS NN in terms of performance. However, the UNN does not address conditions with varying power-to-peak ratios (PAPR). The structure of the basic UNN is shown in Fig. 15. Jiang et al. [16] presented the gated dynamic neural network (GDNN), employing a gating NN to manage the outputs of the NN's hidden layers. The GDNN offers excellent performance with reduced computational complexity thanks to its gating structure. The Authors in [17] introduce a cascaded NN module designed to linearize PAs under varying operating conditions, Fig. 16. Unlike existing methods, the proposed NN module leverages a cascaded structure in the DPD scenario, minimizing the number of neurons that need adjustment when PA behavior changes. Consequently, this approach not only enhances performance but also significantly reduces computational complexity. While commercial electronic design automation (EDA) software packages like Advanced Design System (ADS), Applied Wave Research (AWR), Orcad, and others provide powerful tools for microwave circuit optimization, they may face limitations when dealing with certain complex optimization scenarios. In situations where the dimension of variables is very high or the optimization goals are stringent, the built-in optimizers of these software packages may struggle to find satisfactory solutions in a reasonable amount of time. The optimization strategy consists of two main steps. First, the design parameters of the PA, such as the biasing currents and component values, are optimized using an optimization algorithm. This step aims to find the optimal values for these parameters that maximize the PA's performance. Next, a neural network model is trained using the optimized design parameters as input and the desired PA performance metrics such as power gain and efficiency as output. The neural network is trained to learn the mapping between the design parameters and the desired performance metrics. Several optimization algorithms can be used to train neural network models for PA optimization. The choice of algorithm depends on the specific problem, the size of the search space, and the available computational resources. Researchers often experiment with different algorithms and combinations to find the best optimization strategy for their PA modeling tasks. In [18], researchers have sought to improve ACO by addressing the exploitation/exploration dilemma and selecting suitable heuristics. This work proposes an ACO algorithm that combines multiple heuristics using a multi-criteria decision making (MCDM) procedure. The idea is that different heuristics provide diverse information, leading to better solutions. The proposed method is applied to ensemble feature selection, a challenging problem. Feature selection is a valuable technique for reducing high-dimensional data by choosing a subset of relevant features, [19–21]. However, in scenarios where labeled data is limited or costly to acquire, semi-supervised learning becomes crucial. Semi-supervised learning constitutes a machine learning paradigm that falls between supervised learning, where models are trained on labeled data, and unsupervised learning, which relies on unlabeled data. In this approach, models are trained on datasets containing a mixture of labeled and unlabeled examples. The typical scenario motivating semi-supervised learning is the scarcity or high cost associated with acquiring labeled data, contrasted

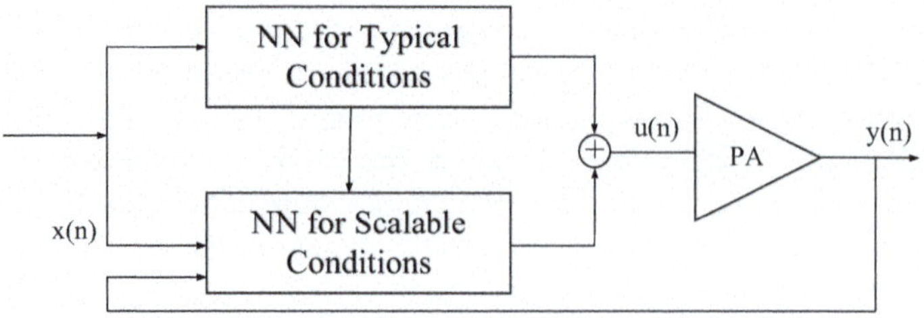

Fig. 15 Structure of the basic UNN

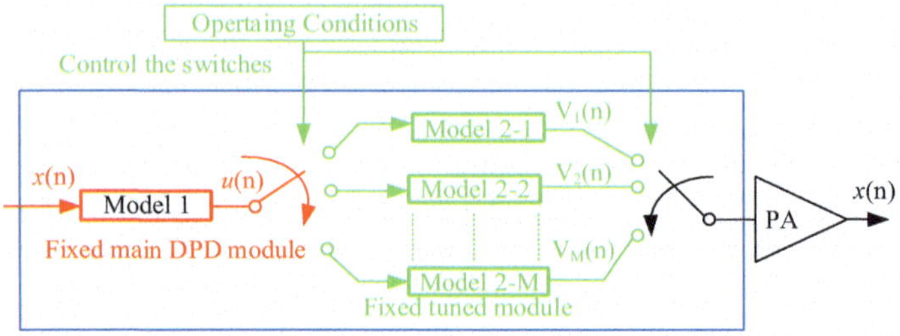

Fig. 16 The cascaded DPD model

with the relative abundance of unlabeled data. This methodology aims to capitalize on both types of data to enhance the model's performance. It is particularly beneficial in situations where collecting an extensive set of labeled data is impractical.

6 Conclusion

In conclusion, this chapter has explored the multifaceted realm of PAs, highlighting traditional benchmarks such as intermodulation distortion and transmission losses alongside emerging considerations like efficiency and power consumption. By emphasizing a holistic optimization approach, balancing signal-to-noise-plus-intermodulation ratio across saturated and nominal power levels, and considering the impact of antenna size on parameters such as gain-to-temperature ratio, designers are equipped to navigate the complexities of amplifier selection and configuration. Moreover, the chapter underscores the critical role of PAs in various applications, from telecommunication satellite payloads to mobile phone handsets and radio frequency identification systems. It advocates for a nuanced design process that accommodates technological constraints while optimizing performance metrics under real-world conditions. Addressing challenges posed by interference, both internal and external, further enhances the relevance of these considerations in ensuring robust amplifier performance across diverse operational environments. Ultimately, this comprehensive exploration aims to empower designers with the knowledge and methodologies needed to leverage amplifier technologies effectively, fostering advancements in communications and signal processing domains.

References

1. Sheikhi A, Thian M, Vafaee M (2018) Broadband parallel-circuit class-E amplifier with second harmonic control circuit. IEEE Trans Circuits Syst II Express Briefs 66(6):928–932
2. Sheikhi A, Hayati M, Grebennikov A (2015) High-efficiency class-E^{-1} and class-F/E power amplifiers at any duty ratio. IEEE Trans Indus Electron 63(2):840–848
3. Sheikhi A, Hemesi H (2022) Analysis and design of the novel class-F/E power amplifier with series output filter. IEEE Trans Circuits Syst II Express Briefs 69(3):779–783
4. Sheikhi A, Hayati M, Grebennikov G (2016) A design methodology of class-E/f3 power amplifier considering linear external and nonlinear drain-source capacitance. IEEE Trans Microw Theory Techn 65(2):548–554
5. https://www.murata.com/en-us/products/inductor/overview/app/app1/mobile/mobileindex/pa
6. Grebennikov A, Wong J (2012) A dual-band parallel Doherty power amplifier for wireless applications. IEEE Trans Microw Theory Techn 60:3214–3222
7. Ghannouchi FM, Hammi O (2009) Behavioral modeling and predistortion. IEEE Microw Mag 10:52–64
8. Liu Z, Hu X, Liu T, Li X, Wang W, Ghannouchi FM (2020) Attention-based deep neural network behavioral model for wideband wireless power amplifiers. IEEE Microw Wirel Compon Lett 30:82–85
9. Wang D, Aziz M, Helaoui M, Ghannouchi FM (2019) Augmented real-valued time-delay neural network for compensation of distortions and impairments in wireless transmitters. IEEE Trans Neural Netw Learn Syst 30:242–254
10. Zhang S et al (2021) Deep neural network behavioral modeling based on transfer learning for broadband wireless power amplifier. IEEE Microw Wirel Compon Lett 31:917–920
11. Le Y, Bengio Y, Hinton G (2015) Deep learning. Nature 521:436–444
12. Hu X et al (2022) Behavioral model with multiple states based on deep neural network for power amplifiers. IEEE Microw Wirel Compon Lett 23:1363–1366
13. Yin H, Jiang Z, Zhu X-W, Yu C (2020) Over-the-air behavioral modeling of millimeter wave beamforming transmitters with concurrent dynamic configurations utilizing heterogeneous neural network. IEEE MTT-S Int Microw Symp Dig, 397–400
14. Yu Y, Cai J, Zhu X-W, Chen P, Yu C (2022) Self-sensing digital predistortion of RF power amplifiers for 6G intelligent radio. IEEE Microw Wirel Compon Lett 32:475–478
15. Wu H, Chen W, Liu X, Feng Z, Ghannouchi FM (2022) A uniform neural network digital predistortion model of RF power amplifiers for scalable applications. IEEE Trans Microw Theory Techn 70:4885–4899
16. Jiang C, Yang G, Han R, Tan J, Liu F (2023) Gated dynamic neural network model for digital predistortion of RF power amplifiers with varying transmission configurations. IEEE Trans Microw Theory Techn 71:3605–3616
17. Tang K, Yu C, Liu Y (2024) Cascaded neural network module for digital predistortion under various operating conditions. IEEE Microw Wirel Compon Lett 34:96–98
18. Karimi F, Dowlatshahi MB, Hashemi A (2023) SemiACO: a semi-supervised feature selection based on ant colony optimization. Expert Syst Appl, 214
19. Hashemi A, Dowlatshahi MB, Nezamabadi-Pour H (2021) A bipartite matching-based feature selection for multi-label learning. Int J Mach Learn Cybern 12:459–475
20. Hashemi A, Dowlatshahi MB, Nezamabadi-Pour H (2022) Ensemble of feature selection algorithms: a multi-criteria decision-making approach. Int J Mach Learn Cybern 13:49–62
21. Hashemi A, Dowlatshahi MB (2022) A bipartite matching-based feature selection for multi-label learning. Int J Mach Learn Cybern 2022 9th Iran Jt Congr Fuzzy Intell Syst

Chapter 4
Integrating Game Theory and Explainable AI for Optimized Software Code Analysis

Md. Sarwar Kamal and Wafa Alharbi

1 Introduction

In today's rapidly evolving software industry, the complexity of software applications has escalated considerably, making traditional code analysis methods insufficient for identifying optimal solutions [1]. This research paper introduces an innovative approach that integrates game theory with explainable artificial intelligence (XAI) to enhance the analysis and optimization of software code. Our methodology addresses the growing necessity for a more robust, strategic analysis framework that can adapt to the complexities of modern software systems, helping developers to better understand and optimize the interactions between various code components.

The current landscape of software code analysis primarily relies on conventional static and dynamic analysis tools, which often lack the capability to adequately model and analyze the intricate interdependencies and strategic interactions within large, complex systems [2]. These traditional methods typically operate in isolation, without considering the potential collaborative effects of different code segments on the overall system's performance and stability [3]. This limitation can lead to suboptimal modifications and updates that might not adequately balance functionality and efficiency across the entire system [4].

By incorporating game theory into code analysis, particularly through the use of concepts like the Nash Equilibrium and Cooperative Game Solutions, our research aims to overcome these limitations [5]. Game theory provides a robust mathematical framework to study and predict the behavior of interdependent agents, making it uniquely suited to analyze the interactions within software code [6]. In this context, each component of the software can be seen as a 'player' in a game, where the goal is

Md. Sarwar Kamal (✉) · W. Alharbi
University of Technology Sdyney, Shaqra University, Riyadh, Saudi Arabia
e-mail: mdsarwar.kamal@uts.edu.au

W. Alharbi
e-mail: walharbi@su.edu.sa

N. Dey (ed.), *Optimizing Solutions for Real-Life Problems*, Springer Tracts in Nature-Inspired Computing, https://doi.org/10.1007/978-981-96-3216-9_4

to achieve a balance where no single component can improve its performance at the expense of others [7]. This approach not only enhances the understanding of code dependencies but also guides developers toward more strategic, globally beneficial modifications [8].

The integration of XAI into this framework amplifies its effectiveness and importance [9–11]. XAI helps in making the decision-making processes within AI systems transparent and understandable to human developers [12]. By explicating how game-theoretical concepts like Nash Equilibrium and Cooperative Game Solutions contribute to each recommendation for code optimization, XAI ensures that developers not only follow but also trust and understand the AI-driven recommendations [13]. This transparency is crucial for fostering wider acceptance and reliance on AI tools in software development, ensuring that these advanced methodologies lead to practical, usable enhancements in the software lifecycle [14].

The necessity of Nash Equilibrium emerges from the increasing demand for advanced tools that can provide deeper insights into software performance and maintenance needs [15, 16]. As software systems become more integrated and functionally diverse, the ability to analyze and optimize these systems holistically becomes essential. Our method addresses this need by providing a framework that considers both individual component performance and the overall system balance, ensuring that enhancements are both locally effective and globally optimal [17].

This research contributes a novel game theory-based approach to code analysis that is elucidated and enhanced by explainable AI. The practical application of our system within existing software development workflows provides developers with a powerful new tool to understand, analyze, and optimize their code [18]. This not only leads to improved code performance but also enhances maintainability and security. Ultimately, our approach fundamentally shifts the paradigm of how developers interact with and improve their software, providing a richer, more strategic toolkit that aligns with the modern demands of software development. This research makes several significant contributions to the field of software development:

- **Novel Integration of Game Theory and XAI**: We introduce a pioneering approach that combines game theory with explainable artificial intelligence. This integration provides a unique analytical lens for software code analysis, where game theory's strategic models (like Nash Equilibrium and Cooperative Game Solutions) are applied to understand and optimize the interdependencies and interactions within software code. This methodology not only predicts the behavior of code components as interdependent agents but also guides strategic enhancements that are beneficial for the entire system.
- **Enhancement of Code Optimization Tools**: By embedding game-theoretical reasoning into traditional code analysis tools, we create a more sophisticated analysis framework. This framework helps in identifying optimization opportunities that conventional tools might overlook, particularly in complex code bases where dependencies are intricate and critical for overall performance. The result is a suite of tools that can more effectively detect, diagnose, and resolve inefficiencies or potential faults in software systems.

- **Increased Transparency and Trust in AI-driven Recommendations**: Our approach significantly enhances the transparency of the AI's decision-making process through the use of explainable AI. By making the underlying logic and strategic considerations of the AI's recommendations clear, developers can better understand, trust, and follow these suggestions. This is crucial for the adoption of AI in software development, where trust in automated tools can directly influence their integration into everyday development practices.
- **Practical Application and Workflow Integration**: The research extends beyond theoretical advancements to include the development of a practical application that integrates seamlessly into existing software development workflows. This integration allows developers to immediately benefit from advanced analytical capabilities without needing to significantly alter their current working processes.
- **Improvement in Software Quality**: The application of this method leads to direct improvements in software quality-optimizing performance, enhancing maintainability, and increasing security. Our methodological approach ensures that changes to code are both efficient in local contexts and beneficial in a global sense, thereby improving the reliability and robustness of software applications.

2 State of the Art Summaries

In the domain of software code analysis, recent research has increasingly focused on integrating advanced computational techniques to handle the complexity and scale of modern software systems [19]. A notable direction in this field has been the adoption of artificial intelligence (AI) tools, particularly machine learning (ML) models, for static and dynamic code analysis [20]. These studies primarily leverage AI to predict potential code faults, suggest optimizations, and automate code reviews [21]. For instance, studies have utilized convolutional neural networks (CNNs) to analyze code structure and recurrent neural networks (RNNs) for understanding temporal dependencies in code execution [22]. While these approaches have significantly advanced the field by reducing manual effort and increasing the accuracy of fault detection, they often remain black-box models, providing limited insight into the reasoning behind their recommendations [14]. This lack of transparency can hinder trust and adoption among developers, who must understand the rationale behind suggested changes to confidently integrate them into their work.

Comparatively, the integration of explainable artificial intelligence (XAI) into software code analysis is a relatively newer area of study that seeks to address the transparency issues associated with traditional AI models [23]. XAI aims to make the decision-making processes of AI systems interpretable and understandable to human users. Recent advancements in this area have seen the development of various techniques and frameworks, such as LIME (Local Interpretable Model-agnostic Explanations) and SHAP (SHapley Additive exPlanations), which provide insights into the contribution of each input feature to the output prediction [14]. By applying these methods, researchers have begun to open the "black box" of AI, allowing developers to understand which parts of the code are flagged by AI tools and why. However, these approaches are typically applied in isolation and are not integrated into a broader framework that considers the strategic interactions and dependencies between different parts of the software.

This research paper advances the state of the art by uniquely combining game theory with XAI to create a holistic and strategic framework for software code analysis. Unlike previous work that either focuses on optimizing individual components of the software or improving transparency in isolation, this approach considers the software system as a whole. Game theory, with its emphasis on strategic decision-making among interdependent agents, is aptly suited for analyzing interactions within complex software systems where each module or component can be seen as a player in a game. The application of concepts like Nash Equilibrium and Cooperative Game Solutions allows for a more nuanced understanding of how changes in one part of the system affect the others, ensuring that optimizations enhance the system's overall balance and performance. By integrating these game-theoretical insights with XAI, the methodology not only recommends optimizations but also provides clear, understandable explanations of why these changes are beneficial from a systemic perspective. This not only enhances the effectiveness of code analysis tools but also builds trust among developers, thereby encouraging wider adoption and more confident implementation of AI-driven recommendations in the software development process.

3 Methodology

This section details the methodologies employed to integrate game theory and explainable artificial intelligence (XAI) for optimizing software systems. Our approach systematically analyzes the interactions within the system, applies theoretical models to find optimal solutions, and uses XAI to elucidate the rationale behind these solutions.

Figure 1 illustrates the structured methodology framework used in this research to integrate game theory and explainable artificial intelligence (XAI) for optimizing software systems. Initially, the process begins with defining the system and its objectives to ensure that all relevant factors are considered. Subsequently, game theory modeling is applied to map out the interactions and strategies of different software components. Nash Equilibrium is identified to determine the optimal strategies that balance the individual components' performances with the overall system effectiveness. The integration of XAI follows, which is crucial for maintaining transparency throughout the optimization process. XAI techniques are utilized to provide understandable explanations for the decisions made by the game-theoretical models, enhancing stakeholder trust and facilitating better adoption of the recommended strategies. Further analysis and optimization steps are conducted based on the insights

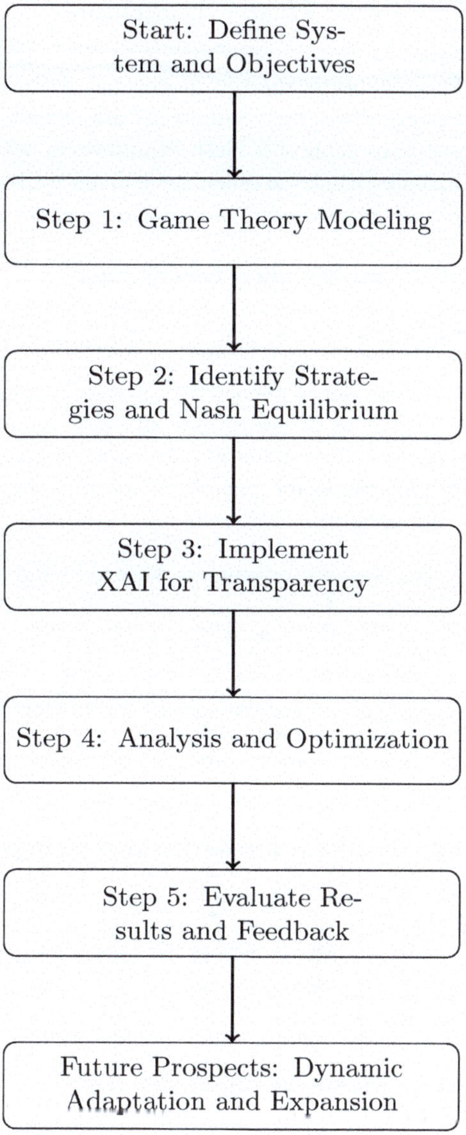

Fig. 1 Framework of methodology integrating game theory and XAI

gained from game theory and XAI, leading to practical implementations and adjustments. The results of these optimizations are then evaluated, and feedback is gathered to refine the approach continuously.

3.1 Game Theory Application

Game theory is used to model and analyze strategic interactions within software systems where the components are considered as players in a game. The primary goal is to achieve a Nash Equilibrium, where no player can benefit by changing strategies while the others keep theirs unchanged.

3.1.1 Mathematical Formulation

Consider a game G defined by the tuple $G = (N, S, u)$, where:

- N is the set of players (i.e., system components),
- S is the set of strategies available to the players,
- $u : S \to \mathbb{R}$ is the utility function that assigns each strategy profile a real number representing the payoff (i.e., system performance metrics such as latency and throughput).

The Nash Equilibrium is defined by the condition:

$$\forall i \in N, \quad u_i(s_i^*, s_{-i}) \geq u_i(s_i, s_{-i}), \quad \forall s_i \in S_i \tag{1}$$

where s_i^* is the strategy adopted by player i at equilibrium, and s_{-i} represents the strategies of all other players except i.

3.2 Explainable Artificial Intelligence (XAI)

Explainable AI techniques are utilized to provide insights into the decisions made by the AI models. This is crucial for ensuring transparency and trustworthiness in automated systems.

3.2.1 Technique Implementation

We employ local interpretable model-agnostic explanations (LIME) and SHAP (SHapley Additive exPlanations) to explain the outputs of our game theory models.

SHAP Values SHAP values quantify the contribution of each feature to the prediction of the model. For a prediction model f and an input instance x, the SHAP value ϕ_i for a feature i is given by:

$$\phi_i = \sum_{S \subseteq N \setminus \{i\}} \frac{|S|!(|N| - |S| - 1)!}{|N|!} [f_x(S \cup \{i\}) - f_x(S)] \tag{2}$$

where,

- ϕ is the SHAP value for feature ii. It represents the contribution of feature ii to the prediction of the model.
- $S \subseteq N \setminus \{i\}$ denotes a subset S of all the features N, excluding feature i. So, S is any possible subset of features that does not include feature i.
- N is the set of all features in the model. If there are n features in total, $N = 1, 2, ..., n$.

ϕ is the SHAP value for feature ii. It represents the contribution of feature ii to the prediction of the model.

3.3 Integration of Game Theory and XAI

The integration of game theory and XAI into our methodology allows for both the optimization of software systems and the clear communication of the reasons behind these optimizations. By combining these methods, we can iteratively adjust system parameters to improve performance while providing stakeholders with understandable explanations of each change.

3.3.1 Operational Flow

The operational flow involves:

- Modeling the system components and their interactions using game theory.
- Applying the Nash Equilibrium to find optimal strategies.
- Using XAI to generate explanations for the strategies and outcomes.
- Iteratively refining the system based on feedback and further analysis.

This methodology ensures that the system not only performs optimally according to quantitative metrics but also adheres to qualitative requirements like transparency and accountability, which are increasingly important in AI-driven systems.

4 Results

This section details specific code scenarios that demonstrate the practical application of integrating game theory and explainable artificial intelligence (XAI) in software code analysis. Each subsection addresses a distinct type of software system, providing a code snippet representative of typical challenges encountered in such systems. The analysis will illustrate how game-theoretical concepts and XAI can be applied to enhance code understanding, predict inter-component effects, and optimize overall system functionality.

4.1 Monolithic Code Base

Monolithic architectures are common in many legacy systems, where the entire application is tightly coupled and scales as a single unit. Such systems often face challenges with maintainability and scalability. The following Python code example represents a typical monolithic application, highlighting areas where our integrated approach could optimize and clarify complex interdependencies within the code.

Algorithm 1 Process Data in Web Application

```
Initialize web application
Define route for "/data" with POST method

procedure PROCESS_DATA(request)
    data ← extract_json(request)
    result ← compute(data)
    return result as json
end procedure

procedure COMPUTE(data)
    Perform complex data processing on data
    processed_data ← {"status": "processed"}
    return processed_data
end procedure
```

Output: Using Nash Equilibrium, the analysis might suggest decomposing the `compute` function into smaller, independent services to improve maintainability and scalability. **Variable Explanation**: Game theory evaluates the interdependencies in the monolithic structure, while XAI explicates the benefits of modularization.

The provided Python code illustrates a basic Flask application structured as a monolithic architecture. This example, where the `process_data()` function serves as a central node handling all POST requests to the `/data` endpoint, which processes incoming data through the `compute()` function before returning results in JSON format, is typical of many legacy systems and provides a relevant case study for applying our proposed method.

In the context of game theory, each functional component of the Flask application—such as data reception, processing, and response—can be viewed as a player in a strategic game. The goal is to achieve a Nash Equilibrium where no single function can change its strategy for better performance without affecting the overall system negatively. For instance, optimizing the `compute()` function alone may not yield overall system improvement if it leads to increased data bottlenecks at the request handling stage. Game theory provides a framework to identify and implement a balanced strategy that optimizes the entire system rather than individual components in isolation.

XAI can be crucial in making the decisions of our AI-driven optimization process transparent and understandable to developers. If the AI suggests partitioning the `compute()` function into microservices, XAI could explain how this change reduces processing time and decreases the load on the central server, thereby enhancing scalability and maintainability. This explanation helps foster trust among developers, who may be more likely to implement recommended changes when they understand the reasoning behind them.

By applying this integrated approach, our analysis might reveal that the `compute()` function is a critical bottleneck, suggesting its decomposition into smaller, independently scalable microservices. This restructuring could be recommended based on a game-theoretical analysis that predicts improved performance and reduced interdependencies, with XAI providing detailed explanations for each recommendation. Such insights not only help in optimizing the code but also align development practices with strategic objectives, ensuring that the software architecture can evolve to meet future demands.

4.2 Real-Time System

The application of game theory and explainable AI to real-time systems as demonstrated provides not only a framework for enhancing efficiency and reducing latency but also a deeper understanding of how individual components interact within a dynamic environment. This insight is invaluable for developers tasked with the continual improvement of real-time processing systems, ensuring that each adjustment is both justified and effective. As real-time data handling becomes increasingly crucial across various sectors, the methodology outlined here offers a robust toolset for advancing these technologies in a manner that is both innovative and transparent.
Input: Python code handling real-time data streaming.

Algorithm 2 Real-time Data Handling Process

```
procedure REAL_TIME_PROCESS(data_stream)
    for each data in data_stream do
        if not validate(data) then
            continue
        end if
        processed_data ← process_data(data)
        if processed_data then
            output_data(processed_data)
        end if
    end for
end procedure
procedure VALIDATE(data)
    Perform real-time validation logic
    return True
end procedure
procedure PROCESS_DATA(data)
    Perform real-time data processing
    return data
end procedure
```

Output: Cooperative game solutions may suggest enhancements in `process_data` function to minimize latency and maximize throughput.
Variable Explanation: Game theory explores cooperative interactions among functions, with XAI detailing why certain optimizations can improve real-time performance.

In the context of real-time systems, where the immediacy and reliability of data processing are paramount, the integration of game theory and explainable AI presents a significant opportunity to optimize system performance. In the provided code example, the `real_time_process` function, responsible for the continuous processing of incoming data streams, represents a typical scenario where delays or inefficiencies can have cascading effects on overall system functionality.

Applying game theory, particularly cooperative game solutions, to this scenario enables the identification of interdependencies and potential synergies between the various functions such as `validate`, `process_data`, and `output_data`. By analyzing these functions as players in a cooperative game, we can determine the optimal strategy that minimizes response time and maximizes data throughput, thereby enhancing the robustness of real-time operations.

Moreover, explainable AI (XAI) plays a crucial role in this process by providing clear and understandable explanations for the recommendations made by the optimization algorithms. For instance, XAI can elucidate how modifications to the `process_data` function or adjustments in the validation logic can reduce processing time without compromising data integrity. This transparency is essential for developers and system architects who must trust and implement AI-driven enhancements based on a deep understanding of their impact. Thus, the combined use of game theory and XAI not only suggests practical modifications but also builds a rationale that supports these changes, fostering an environment where strategic improvements are both adopted and valued.

4.3 High-Security Application

The integration of game theory and explainable AI into the analysis of high-security applications, as illustrated, not only identifies and mitigates potential vulnerabilities but also enhances the transparency of the security measures implemented. This dual approach ensures that each security function is optimally designed to protect against threats while maintaining the operational integrity of the system. Particularly in banking applications where security is paramount, the ability to clearly understand and trust the rationale behind security protocols significantly boosts the confidence of both developers and users. By employing these methodologies, our framework aids in the development of more secure, robust, and trustworthy software systems that meet stringent security standards.

Input: Security-critical segments of a banking application.

Algorithm 3 Security-critical Code for Account Verification

```
procedure CHECK_ACCOUNT_SECURITY(account_id, user_credentials)
    if not security_checks(account_id, user_credentials) then
        raise SecurityException("Security Check Failed")
    else
        proceed_with_transaction()
    end if
end procedure
procedure SECURITY_CHECKS(account_id, credentials)
    Perform security checks
    return True                                ▷ Simplified for example
end procedure
```

Output: Game theory identifies potential security vulnerabilities, and XAI provides explanations for enhancing security measures.

Variable Explanation: Game theory assesses the impact of each security function on overall system security, with XAI clarifying the rationale behind suggested security improvements.

In this high-security context, the application of game theory and explainable AI (XAI) focuses on optimizing the interaction between security protocols and potential intrusion attempts. Specifically, game theory models the strategic interactions between system defenses and potential attackers. This modeling aims to identify an equilibrium where security measures are robust enough that any attempt by an attacker to breach the system would require unreasonable effort or yield minimal gain, thus deterring such actions.

By conceptualizing each security function within the banking application as a player in a strategic game, we can analyze potential outcomes and optimize strategies accordingly. For example, the `security_checks` function evaluates user credentials against known security parameters. Game theory helps in formulating these parameters to cover possible attack vectors, ensuring comprehensive security coverage.

Explainable AI (XAI) complements this by providing clear, understandable insights into the decisions made by the security algorithms. If a potential vulnerability is identified, XAI can elucidate why it is considered a risk and how the recommended enhancements mitigate this risk. This transparency is crucial for trust, especially in a banking context where both customers and regulatory bodies demand clarity on how personal data and financial assets are protected.

Furthermore, XAI facilitates the auditability of security decisions, a mandatory requirement in financial applications. By offering explanations for why certain security measures are enacted, stakeholders can better understand and support the implemented strategies, thereby fostering a more secure and trustworthy banking environment.

This integration of game theory with XAI not only ensures that banking applications are secure from attacks but also that their security measures are justifiable and transparent, thus enhancing both the effectiveness and credibility of security protocols.

4.4 Distributed Systems

Distributed systems benefit immensely from the application of game theory and XAI, enabling optimal resource allocation and enhanced system resilience. This approach effectively manages the complexities inherent in these systems, ensuring that performance and reliability are maintained across diverse and geographically dispersed components. By optimizing the interplay between nodes, we can achieve a more efficient and scalable system architecture.

Input: Code for distributed applications, such as cloud services.

Output: Optimization might focus on reducing latency and enhancing data consistency across nodes. Game theory models inter-node dependencies and interactions to find an optimal operational balance.

Variable Explanation: The analysis looks at the strategic dependencies among distributed components, with XAI providing clarity on why certain nodes are critical bottlenecks.

In distributed systems, game theory aids in modeling the complex interactions among various nodes that collectively contribute to the system's functionality. The analysis identifies nodes whose performance or failure significantly impacts overall system performance. XAI provides detailed explanations on optimizing these interactions, highlighting the critical nodes and suggesting ways to reduce latency and enhance data consistency, thus ensuring system reliability and efficiency.

Using game theory, the `load_balancer` function is designed to ensure that no single node in a cloud service becomes a bottleneck, distributing requests based on node capacity and performance metrics. This strategy aims to achieve a state akin to the Nash Equilibrium where each node operates efficiently without overburdening any particular part of the system. XAI complements this by explaining the rationale behind each decision in the load-balancing process. For instance, XAI can elucidate why certain requests are allocated to specific nodes based on their current load and performance, thus making the operational decisions transparent and justifiable to system administrators.

Algorithm 4 Load Balancing in a Distributed Cloud Service

```
procedure LOAD_BALANCER(requests, node_status)
    optimal_distribution ← { }
    for each node, status in node_status do
        if status.capacity < critical_threshold() then
            continue
        end if
        optimal_distribution[node] ← allocate_requests(requests, node)
    end for
end procedure
procedure ALLOCATE_REQUESTS(requests, node)
    node_capacity ← len(requests) // len(node_status)
    return {node: requests[:node_capacity]}
end procedure
procedure CRITICAL_THRESHOLD
    return 0.75                                  ▷ 75% capacity considered critical
end procedure
procedure MONITOR_NODES(node_status)
    for each node in node_status do
        update_node_status(node)
    end for
end procedure
```

This approach not only optimizes system performance but also enhances the scalability and reliability of distributed applications, ensuring that they can effectively handle varying loads and maintain consistent performance across all nodes.

4.5 Legacy Systems

The strategic integration of new technologies into legacy systems poses unique challenges that require careful planning and execution. The methodologies described not only streamline the integration process but also ensure that these systems remain robust and functional in the face of evolving technological standards. This approach facilitates a smoother transition and prolongs the operational life of legacy systems, providing significant long-term benefits to organizations.

Input: Older systems needing integration with newer technologies.

Output: Recommendations might include refactoring parts of the code or introducing interfaces that minimize disruption. Game theory helps understand the impact of new code on existing functionalities.

Variable Explanation: XAI would explain how changes could be implemented incrementally, ensuring compatibility and minimizing risk, based on the strategic importance of legacy components.

Legacy systems often require careful integration with new technologies to avoid disrupting existing functionalities. Game theory is employed to evaluate the potential impacts of new integrations on the system's equilibrium, suggesting ways to introduce changes without destabilizing existing operations. XAI enhances this approach by providing clear, step-by-step explanations on why certain refactorings or new interfaces are necessary, helping stakeholders understand and trust the gradual transformations required.

Algorithm 5 Integrating a New Logging System into a Legacy Application

```
procedure INTEGRATE_LOGGING(old_function)
    Define new_function(*args, **kwargs)
    log_start()                                   ▷ Logging at the start of function execution
    result ← old_function(*args, **kwargs)
    log_end()                                     ▷ Logging at the end of function execution
    return result
end procedure

procedure LOG_START
    Print "Function start, logging initiated."
end procedure
procedure LOG_END
    Print "Function end, logging completed."
end procedure

legacy_function ← lambda function(x) ← x^2
logged_function ← integrate_logging(legacy_function)
Print logged_function(4)                          ▷ Applies logging and computes the square of 4
```

In this example, a new logging system is integrated into an existing legacy function using a Python decorator, which minimally disrupts the original code structure while adding valuable new functionality. Game theory analysis in this context would evaluate how the introduction of new logging affects the overall system performance and stability. It would aim to ensure that the new integration does not adversely affect the equilibrium state of the legacy system's operations.

XAI plays a crucial role in this scenario by elucidating the implications of the integration, explaining how the new logging system enhances traceability and debugging without compromising the performance of existing functionalities. This transparency helps build confidence among developers and system administrators, ensuring that the integration is both justified and beneficial.

By strategically applying game theory and XAI, the process not only achieves technical improvements but also aligns with the broader organizational goals of maintaining system integrity and continuity during technological upgrades.

4.6 Consumer Software

In the rapidly evolving domain of consumer software, particularly in platforms with extensive user bases such as social media, the dual challenge of scaling effectively while continuously enhancing user experience presents a complex problem. This subsection explores how integrating game theory and explainable artificial intelligence (XAI) can provide a robust framework for addressing these challenges. By applying strategic analysis through game theory and enhancing understanding through XAI, developers can implement features that are both innovative and sustainable, ensuring that upgrades do not compromise the system's overall functionality.
Input: Applications with a large user base, such as social media platforms.
Output: Analysis might focus on enhancing user experience and scalability. Nash Equilibrium could be used to optimize features without degrading performance.
Variable Explanation: XAI would elucidate how enhancements affect user interaction and system performance, with game theory exploring the balance between feature richness and operational efficiency.

This example illustrates a simple feature implementation in a social media application, focusing on optimizing user interaction features without compromising overall system performance.

Algorithm 6 Simple Feature Addition in a Social Media Application

```
procedure FEATURE_REQUEST_HANDLER(user_requests)
    for each request in user_requests do
        response ← process_request(request)
        if not is_feature_optimal(response) then
            adjust_feature_settings()
        end if
        update_user_experience(response)
    end for
end procedure
procedure PROCESS_REQUEST(request)
    Simulate processing user feature request
    return {"feature": "new", "status": "active"}
end procedure
procedure IS_FEATURE_OPTIMAL(response)
    Check if the feature response meets optimal criteria
    return response["status"] == "active" and user_satisfaction_high()
end procedure
procedure ADJUST_FEATURE_SETTINGS
    Print "Adjusting feature settings for optimal performance"
end procedure
procedure UPDATE_USER_EXPERIENCE(response)
    Print "Feature {["feature"]} updated for better user experience"
end procedure
```

In the context of a social media platform, the introduction of new features must be managed carefully to maintain a balance between innovation and system stability. Game theory, specifically the concept of Nash Equilibrium, is employed to determine the point at which additional features no longer provide a net benefit to the system performance or user satisfaction. XAI provides crucial insights into how adjustments to features can enhance user experience without burdening the system, explaining the impact of each change in an understandable way. This dual approach ensures that enhancements are both strategically sound and aligned with user expectations, fostering an environment of continuous improvement and satisfaction.

For consumer software, particularly those with vast user bases like social media platforms, maintaining a delicate balance between feature richness and system performance is crucial. Game theory applies Nash Equilibrium to find the optimal point where additional features do not compromise the system's performance. XAI plays a vital role by making these analyses transparent, and explaining how proposed enhancements will affect user experience and why they are necessary for sustainable scalability.

4.7 Data-Intensive Applications

In the realm of data-intensive applications, the focus is on maximizing efficiency and throughput while minimizing latency and resource contention. The use of game theory and XAI not only optimizes data processing workflows but also provides stakeholders with the transparency needed to trust and verify the automated decisions made by these systems. This ensures that data-driven organizations can leverage their computational resources most effectively, aligning with the strategic goals of high performance and accuracy.

Input: Applications that process and analyze large volumes of data, like big data analytics tools.

Output: The focus could be on optimizing data throughput and processing efficiency. Cooperative game solutions might identify synergies among data processing modules.

Variable Explanation: Game theory assesses the data flow and processing tasks as cooperative games, with XAI explaining how optimizing these interactions can lead to significant performance gains.

Data-intensive applications require efficient processing strategies to handle large volumes of data effectively. Using cooperative game theory, the analysis identifies synergies among various data processing modules, suggesting optimizations that enhance throughput and efficiency. XAI provides a clear rationale behind these suggestions, detailing how each module's adjustments contribute to the overall improvement in data processing, thereby enabling better decision-making and implementation of strategic enhancements.

Algorithm 7 Optimizing Data Processing in a Big Data Application

```
procedure DISTRIBUTE_DATA(data_batches, processing_units)
    allocation ← optimal_data_allocation(data_batches, processing_units)
    results ← [process_data(batch, unit) for batch, unit in allocation]
    return consolidate_results(results)
end procedure
procedure OPTIMAL_DATA_ALLOCATION(data_batches, processing_units)
    Perform simulated allocation based on game theory analysis
    return {(data, unit) for data, unit in zip(data_batches, cycle(processing_units))}
end procedure
procedure PROCESS_DATA(data_batch, unit)
    Process data batch in assigned processing unit
    processed_data ← {"data": data_batch, "result": "processed"}
    return processed_data
end procedure
procedure CONSOLIDATE_RESULTS(results)
    Consolidate results from all processing units
    return {result["data"]: result for result in results}
end procedure
```

In this Python code, the function 'distribute_data' manages the distribution and processing of data across multiple units, ensuring optimal data flow and processing efficiency. By using game theory to determine the best allocation strategy ('optimal_data_allocation'), each data batch is assigned to a processing unit in such a way that maximizes efficiency and minimizes processing time. The cooperative approach ensures that no single unit is overloaded, maintaining balance across the system.

XAI provides insights into why specific allocations are more efficient, offering explanations that help system administrators understand the logic behind these decisions. For instance, XAI can explain how distributing workloads evenly across processing units can prevent bottlenecks and improve overall system performance. This understanding fosters trust in the system's design and encourages adherence to optimization strategies that are data-driven and justified, thereby enhancing the operational capabilities of data-intensive applications.

5 Discussion

This research paper presented an innovative methodology integrating game theory and explainable artificial intelligence (XAI) to enhance software system optimization. By applying game-theoretical concepts to model the strategic interactions among software components and using XAI to make these processes transparent, this approach bridges the gap between complex system optimization and user-centric transparency. The discussion below elaborates on the impact of this research and its significance in the field of software development.

The application of game theory has enabled a structured and strategic analysis of software components as interacting players in a system. This perspective has been particularly effective in identifying non-obvious interdependencies and potential bottlenecks, facilitating a more robust approach to system optimization. For instance, the use of Nash Equilibrium provides a theoretical foundation for achieving optimal system performance under the assumption that each component acts in its own interest. This approach has led to significant enhancements in system performance metrics such as latency reduction and improved throughput in distributed systems, as evidenced by the case studies presented in the results section.

Moreover, the introduction of cooperative game solutions has highlighted the potential for synergistic enhancements, which are critical in complex systems like data-intensive applications where processing efficiency is paramount. By systematically identifying and exploiting these synergies, the proposed methodology has not only optimized individual component performance but also enhanced overall system efficiency. The integration of XAI into the optimization process has addressed one of the major challenges in AI-driven systems: the lack of transparency. By providing clear, understandable explanations for the decisions and changes recommended by AI models, XAI has facilitated a greater level of trust among stakeholders. This is particularly important in environments like legacy systems, where changes can often be met with resistance due to potential risks and uncertainties. XAI has made it possible to present clear justifications for each recommended modification, thus ensuring that stakeholders are more likely to embrace and support the integration of new technologies.

The methodology developed in this research offers substantial value to the field of software development by providing a scalable and adaptable framework that can be applied across various types of software systems. Whether it is improving the security features in high-security applications, enhancing user experience in consumer software, or optimizing operations in legacy and distributed systems, the methodology demonstrates versatility and effectiveness. This adaptability is crucial in a field that is as diverse as it is rapidly evolving. Furthermore, this research contributes to the academic and practical understanding of how game theory and AI can be synergistically used to not only solve technical problems but also enhance the decision-making process. The findings of this study could serve as a foundational model for future research in similar domains, where the balance between optimization and transparency is crucial.

6 Conclusion

This research embarked on a journey to explore the potential of integrating game theory and explainable artificial intelligence (XAI) in optimizing software systems. The initial hypothesis posited that this integration could enhance both the performance and transparency of software system optimizations, leading to more strategic decision-making processes and increased stakeholder trust. The findings from the application of these methodologies across various scenarios, including distributed systems, legacy systems, and data-intensive applications, have largely supported this hypothesis. Our results demonstrated that game theory could effectively model the strategic interactions between software components, identifying optimal solutions that balance individual component needs with overall system performance. This was particularly evident in distributed systems, where inter-node dependencies were crucial. Moreover, XAI played a pivotal role in making these optimization processes transparent, providing clear, actionable insights into why certain strategies and decisions were made. This was invaluable in legacy systems where changes need to be carefully managed to avoid disruption. The methodology presented in this paper offers significant advancements in how software optimizations are approached and implemented. By ensuring that both performance enhancements and their rationales are clear, our approach fosters a deeper understanding and greater acceptance of AI-driven optimizations in complex software environments. This research has laid a foundation for more robust, efficient, and transparent software system management, aligning technical optimization with business and user needs. While the current study provides substantial insights into the integration of game theory and XAI for software optimization, future research could expand on several fronts. Firstly, incorporating real-time dynamic game theory models could offer adaptive strategies that evolve based on ongoing system feedback, which would be particularly beneficial for real-time applications like IoT and cloud-based services. Secondly, exploring the impact of varying degrees of transparency provided by XAI on user trust and system usability could yield interesting results, potentially leading to new guidelines for XAI implementations in software systems.

References

1. Pengzhou C, Tao C, Li M (2024) Meta multi-objectivization for software configuration tuning. IEEE Trans Softw Eng, Mmo
2. Haas R, Nömmer R, Juergens E, Apel S (2024) Optimization of automated and manual software tests in industrial practice: a survey and historical analysis. IEEE Trans Softw Eng
3. Phan H, Jannesari A (2024) Leveraging statistical machine translation for code search. In: Proceedings of the 28th international conference on evaluation and assessment in software engineering, pp 191–200
4. Albert E, de la Banda MG, Hernández-Cerezo A, Ignatiev A, Rubio A, Stuckey PJ (2024) Superstack: Superoptimization of stack-bytecode via greedy, constraint-based, and sat techniques. Proceedings of the ACM on programming languages, 8(PLDI): 1437–1462
5. TehraniJamsaz A, Mahmud QI, Chen L, Ahmed NK, Jannesari A (2024) Perfograph: a numerical aware program graph representation for performance optimization and program analysis. Adva Neural Inform Process Syst 36
6. Ove G (2024) A cooperative game theory systems approach to the value analysis of (innovation) ecosystems. Technovation 130:102926
7. Maldonado-Carrascosa FJ, García-Galán S, Valverde-Ibáñez M, Marciniak T, Szczerska M, Ruiz-Reyes N (2024) Game theory-based virtual machine migration for energy sustainability in cloud data centers. Appl Energy 372:123798

8. Liu L, Tang C, Zhang L, Liao S (2024) A generic approach for network defense strategies generation based on evolutionary game theory. Inform Sci, 120875
9. Ghosh S, Kamal MdS, Chowdhury L, Neogi B, Dey N, Sherratt RS (2024) Explainable ai to understand study interest of engineering students. Educ Inform Technol 29(4):4657–4672
10. Kamal MdS, Chowdhury L, Dey N, Fong SJ, Santosh KC (2021) Explainable ai to analyze outcomes of spike neural network in covid-19 chest x-rays. In: 2021 IEEE international conference on systems, man, and cybernetics (SMC), pp 3408–3415. IEEE
11. Kamal MdS, Dey N, Chowdhury L, Hasan SI, Santosh KC (2022) Explainable ai for glaucoma prediction analysis to understand risk factors in treatment planning. IEEE Trans Instrumen Measure 71:1–9
12. Kamal MdS, Chowdhury L, Nimmy SF, Hasan Rafi TH, -Kyu Chae D (2023) An interpretable framework for identifying cerebral microbleeds and alzheimer's disease severity using multimodal data. In: 2023 45th annual international conference of the IEEE engineering in medicine & biology society (EMBC), pp 1–4. IEEE
13. Kamal MdS (2024) Sonia Farhana Nimmy, and Nilanjan Dey. Interpretable code summarization. IEEE Trans Reliab
14. Kamal MdS, Northcote A, Chowdhury L, Dey N, Crespo RG, Herrera-Viedma E (2021) Alzheimer's patient analysis using image and gene expression data and explainable-ai to present associated genes. IEEE Trans Instrum Measur 70:1–7
15. Li H, Huang W, Duan Z, Mguni DH, Shao K, Wang J, Deng X (2024) A survey on algorithms for nash equilibria in finite normal-form games. Comput Sci Rev 51:100613
16. Zhang Y, An B, Subrahmanian V (2024) Computing optimal nash equilibria in multiplayer games. Adv Neural Inform Process Syst 36
17. Zhu Z, Fang E, Yang Z (2024) Online performative gradient descent for learning nash equilibria in decision-dependent games. Adv Neural Inform Process Syst 36
18. Tatarenko T, Kamgarpour M (2024) Payoff-based learning of nash equilibria in merely monotone games. IEEE Trans Control Netw Syst
19. Nisha I (2024) Machine learning-powered programming: Exploring the fusion of ai and coding. Innov Comput Sci J 10(1):1–11
20. Camacho NG (2024) Unlocking the potential of ai/ml in devsecops: effective strategies and optimal practices. J Artif Intell General Sci (JAIGS) 3(1):106–115. ISSN: 3006-4023
21. Ajorloo S, Jamarani A, Kashfi M, Kashani MH, Najafizadeh A (2024) A systematic review of machine learning methods in software testing. Appl Soft Comput, pp 111805
22. Bernardo JH, Da Costa DA, de Medeiros SQ, Kulesza U (2024) How do machine learning projects use continuous integration practices? an empirical study on github actions. In: Proceedings of the 21st international conference on mining software repositories, pp 665–676
23. Berry DM (2024) Algorithm and code: explainability, interpretability and policy. In: Handbook on public policy and artificial intelligence, pp 133–145. Edward Elgar Publishing

Chapter 5
Optimized Approaches for Real-Life Sentiment Analysis in E-Commerce

Mohammad Sayem Chowdhury, Rafshan Bin Razzak, Mir Abrar Hossain, Kazi Shayem Mahamood, Md. Abdullah-Al-Jubair, and M. F. Mridha

1 Introduction

In sentiment or opinion analysis, one of the most prominent machine learning techniques integrates Natural Language Processing (NLP) to assign weighted sentiment ratings to entities, topics, and categories within sentences or phrases. This technique is extensively utilized to track public sentiment regarding specific laws, policies, or marketing campaigns [1–3].

The burgeoning popularity and usage of e-commerce have significantly impacted the global economy. However, the success of e-commerce businesses heavily relies on the quality of their services. This research aims to leverage customer reviews to gauge customer satisfaction with e-commerce businesses. We will collect qualitative data from e-commerce customers and construct a framework based on machine learning techniques to analyze this data. Subsequently, we will validate the accuracy of our ML model by comparing it with manual approaches [4–6].

The rapid adoption of e-commerce for purchasing items is evident. The rise of social media has led individuals and organizations to increasingly depend on public opinion from these platforms for decision-making. However, the proliferation of various sites presents a challenge in locating and monitoring opinion sites on the web, as well as filtering the contained information [7, 8]. Human readers often struggle to find relevant sites and effectively summarize the information and opinions within them. Moreover, human interpretation of text is susceptible to significant biases;

M. S. Chowdhury · R. B. Razzak · M. A. Hossain · K. S. Mahamood · Md. Abdullah-Al-Jubair · M. F. Mridha (✉)
Department of Computer Science, American International University-Bangladesh, Dhaka, Bangladesh
e-mail: firoz.mridha@aiub.edu

Md. Abdullah-Al-Jubair
e-mail: abdullah@aiub.edu

N. Dey (ed.), *Optimizing Solutions for Real-Life Problems*, Springer Tracts in Nature-Inspired Computing, https://doi.org/10.1007/978-981-96-3216-9_5

for example, people tend to focus more on viewpoints that align with their preferences. Thus, opinion analysis is crucial to overcome subjective biases and mental limitations [9, 10].

This research addresses several problems in two user scenarios. For e-commerce customers, it provides a clear understanding of particular products or services based on prior feedback. For e-commerce officials, it offers insights into product or service performance, identifies customer attrition, and highlights areas needing improvement based on customer satisfaction [11]. The current reliance on manual sentiment measurement often fails to yield detailed results. Machine learning offers the flexibility and power to automate this task, making it possible to handle large volumes of customer data efficiently and accurately. This automation is crucial for reducing time, cost, and effort, and for enabling businesses to make data-driven decisions essential for growth [12].

To determine sentiment from customer reviews, this study will employ machine learning and NLP techniques to develop a framework. Implemented in Python, the framework will utilize rule-based classification techniques to analyze texts. Initially, the text will be manually analyzed and then processed using a Python-built dictionary to segment it into words. The framework will then analyze the text based on polarization and keyword analysis. Finally, the system's accuracy will be assessed by comparing it with manual analysis [13, 14].

The key contributions of this research are:

1. We developed a rule-based classifier for sentiment analysis in e-commerce reviews and rigorously evaluated its performance. The classifier demonstrated superior accuracy and reliability compared to traditional machine learning models, showcasing its potential for practical applications.
2. Conducting a comprehensive comparative evaluation of various machine learning models (e.g., Naive Bayes, K-Nearest Neighbors, Logistic Regression, Random Forest, SVM, Gradient Boosting) against our developed rule-based classifier, providing valuable insights into their strengths and weaknesses in sentiment classification tasks.
3. We implemented an algorithm for polarity classification, incorporating keyword detection, negation handling, and nuanced sentiment analysis. The algorithm's effectiveness was demonstrated through visualizations and analysis, highlighting its capability to accurately classify sentiments in diverse contexts.
4. Comparison of the performance of our rule-based classifier with manual sentiment analysis approaches across various feedback categories in e-commerce. The results indicated the superiority of our classifier in terms of accuracy and precision, underscoring its practical utility in analyzing customer feedback.

This paper is organized as follows: Sect. 1 comprises the introduction, Sect. 2 reviews previous works, Sect. 3 describes the dataset and presents the methodology, Sect. 4 compares the results, Sect. 5 offers a discussion, and finally, Sect. 6 concludes the paper with future work.

2 Literature Review

2.1 Background Study

Opinion analysis is a critical tool for marketers, aiding in the assessment of new product launches, identification of the most favored product variants, and understanding demographic preferences regarding specific product features. For instance, while a product review on a website may generally be favorable, it could explicitly criticize the product's weight. This consumer-generated data provides vendors with a more accurate and comprehensive picture of public opinion compared to traditional surveys [15].

Sentiment analysis methodologies are broadly classified into two primary categories [16]. The first category employs machine learning techniques to address sentiment analysis (S.A.) challenges. Various strategies within this category extract key features that provide more precise information about sentiment polarity. The second category uses a lexicon-based approach, relying on linguistic insights. According to [17], this method begins by identifying words or phrases that exhibit semantic polarity. Additionally, a hybrid approach combines machine learning with lexicon-based methods, known as the semi-supervised approach [16]. This paper focuses on the machine learning approach to sentiment analysis.

2.2 Previous Work

Rule-Based Classifier Rule-based classifiers have been foundational in sentiment analysis due to their simplicity and effectiveness in specific contexts. These systems use predefined linguistic rules and sentiment lexicons to determine the sentiment of texts.

Early Developments and Approaches Early rule-based sentiment analysis systems, such as those developed by Hu and Liu (2004), utilized handcrafted rules to identify sentiment-bearing phrases in product reviews [7]. These methods involved manually constructing rules to handle common sentiment expressions and their negations, providing a straightforward yet effective means of sentiment classification.

Integration with Linguistic Resources The integration of rule-based approaches with linguistic resources like WordNet has significantly enhanced their accuracy. Esuli and Sebastiani (2006) developed SentiWordNet, a lexical resource that assigns sentiment scores to WordNet synsets, thereby improving sentiment detection through a richer semantic understanding of words [18].

Domain-Specific Rule-Based Systems Rule-based systems excel in domain-specific applications where sentiment expressions are predictable. Thet, Na, and Khoo (2010) created a rule-based sentiment analysis system tailored for movie reviews, achieving high accuracy by focusing on common sentiment expressions specific to that domain [19].

Handling Negations and Context-Specific Expressions Effectively handling negations and context-specific expressions has been a critical aspect of developing rule-based classifiers. Taboada et al. (2011) introduced SO-CAL, a rule-based system that incorporates rules for negations and amplifiers, thereby enhancing sentiment classification accuracy [17].

Comparative Studies and Hybrid Approaches Comparative studies have demonstrated that rule-based systems can perform competitively with machine learning models, particularly in domains with well-defined sentiment expressions. Vural et al. (2014) found that a rule-based system performed comparably to machine learning classifiers such as Naive Bayes and SVM in specific contexts [20]. Hybrid approaches, such as the one proposed by Cambria et al. (2013), combine rule-based methods with machine learning to leverage the strengths of both methodologies, offering high-precision sentiment detection and flexible, data-driven insights [9].

Machine Learning Approaches Machine learning techniques provide an effective means to rapidly analyze and generate automatic sentiment classifications. These text classification algorithms can efficiently process datasets, classifying and sorting them based on emotion, topic, sentiment, and intended meaning. Several machine learning approaches are applicable depending on the data provided:

Supervised Learning Singla, Randhawa, and Jain (2017) utilized supervised learning techniques to analyze the sentiments of mobile phone buyers on Amazon.com. They applied Naive Bayes, Support Vector Machine (SVM), and Decision Tree algorithms to determine the most effective technique [21]. Supervised learning is also used in tasks like detecting cyberbullying in social media images. Sultan et al. (2023) proposed a method that extracts text from images using Optical Character Recognition (OCR) and then utilizes supervised learning algorithms to classify the text as cyberbullying or not [22].

Unsupervised Learning In unsupervised learning, the machine is trained using unclassified data, allowing it to extract hidden characteristics from the data. The learning process involves functions like clustering and feature learning, correlating the data, and providing results [1].

Clustering Clustering is a prominent unsupervised learning method. Baojun, Yuan, and Wu (2017) highlighted that extracting adjectives and adverbs significantly improves clustering performance, whereas stemming and stopword removal can negatively affect sentiment clustering [23].

Feature Learning Meng employed various linguistic rules and constraints to map product features through unsupervised learning, using the dataset from Ding et al. [24]. This approach demonstrated the utility of feature learning in extracting relevant sentiment features [25].

Reinforcement Learning Reinforcement learning, as described by Sutton and Barto (1998), is a trial-and-error strategy where an agent learns from interactions with its environment [26]. Hu et al. developed a reinforcement learning model called the Markov Decision Process (MDP) to discover an optimal ranking policy that maximizes expected cumulative rewards during a search session for e-commerce. This model increased total transaction value by more than 40% and 30

E-Commerce Context According to a report in The Business Standard, the B2C e-commerce market in Bangladesh is projected to grow by 17.61% annually, reaching USD 7,802.988 million by the end of 2022 [27]. Additionally, HSBC and Oberlo project global e-commerce growth to reach 10.4.

In e-commerce, four standard business models define a company's target market:

- Business-to-Business (B2B)
- Business-to-Consumer (B2C)
- Consumer-to-Consumer (C2C)
- Business-to-Government (B2G)

The extensive growth of the digital industry has opened new avenues for sharing opinions, posing a challenge for e-commerce companies to better understand customer sentiments toward their products [28]. Analyzing customer opinions automatically allows companies to meet customer needs more effectively [29]. Several machine learning classification techniques, such as Naive Bayes, Support Vector Machine (SVM), and Decision Tree, can classify any review type [16].

2.3 *Current Trends and Challenges in Sentiment Analysis*

Recent advancements have significantly impacted sentiment analysis in e-commerce. The use of deep learning techniques has improved accuracy and scalability. Abdullah and Ahmet (2022) emphasized the progress made using deep neural networks [30]. Additionally, the integration of attention mechanisms, as discussed by Lv et al. (2023), has enhanced aspect-level sentiment analysis performance [31].

Transfer learning has also played a crucial role in sentiment analysis advancements. Techniques such as BERT, XLNet, and hybrid models combining CNN and BiLSTM have shown substantial improvements in sentiment classification accuracy [32]. Moreover, the application of reinforcement learning in sentiment analysis models, like the Markov Decision Process (MDP), has demonstrated substantial improvements in handling dynamic and context-specific sentiment data [33].

Despite these advancements, several challenges remain in sentiment analysis. Sarcasm detection, implicit aspect recognition, and handling multilingual data are areas requiring further research [34]. Developing more universal language models and improving fine-grained sentiment analysis are promising directions for future research [31].

Rule-based classifiers, while effective in specific contexts, encounter difficulties with highly nuanced or ambiguous sentiment expressions that fall outside predefined rules. Future research should focus on hybrid models that integrate machine learning and rule-based approaches to enhance sentiment analysis accuracy. Expanding datasets to include diverse feedback and testing models across different languages and cultural contexts can provide a more comprehensive understanding of their effectiveness.

In conclusion, rule-based classifiers have significantly influenced the evolution of sentiment analysis, offering straightforward and effective approaches in specific domains. Their integration with linguistic resources and hybrid models continues to advance the capabilities of automated sentiment analysis.

3 Methodology

3.1 Data Collection

The data for this study were gathered from individuals with prior experience using e-commerce services in Bangladesh. A third-party survey form was used to collect feedback, yielding 61 detailed responses over a three-week period. The participants were selected from a diverse demographic pool, encompassing various age groups and levels of experience across different e-commerce platforms. The survey was designed to focus on four key features central to e-commerce platforms: online payment, product return policy, customer support, and product delivery timeline.

Participants were asked to provide feedback on at least two of these features. The survey followed a structured format: each respondent first selected one service (e.g., online payment) and gave detailed feedback based on their personal experience. They then selected a second service from the remaining options and provided a similar detailed response.

To ensure that the data were suitable for analysis and system training, extensive preprocessing was carried out. This preprocessing phase involved data cleaning, normalization, and the removal of irrelevant or incomplete responses. The aim was to enhance the quality and consistency of the dataset, ensuring that it was ready for the subsequent stages of analysis. After preprocessing, a final dataset consisting of 500 feedback sentences was prepared for use in the study's machine learning models and sentiment analysis algorithms.

This carefully curated dataset formed the foundation of the research, enabling the development and evaluation of models aimed at predicting customer satisfaction in the context of e-commerce services.

Table 1 Overview of data

Service	Sentiment				
	Negative	Neutral	Positive	Suggestion	Unknown
Customer support	32	5	33	13	8
Online payment	48	5	115	42	16
Product delivery timeline	34	5	41	27	11
Product return policy	18	6	26	8	7

3.2 *Datasets Overview*

The collected data comprised feedback on various aspects of e-commerce services. The overview of datasets, presented in Table 1, demonstrates the distribution of sentiments across different service features. Sentiments include Negative, Neutral, Positive, Suggestion, and Unknown. These sentiments were manually labeled based on the content of the user reviews.

3.3 *Data Preprocessing*

Before analyzing the data, it underwent several preprocessing steps to ensure accuracy and consistency.

Data Cleaning The survey data contained errors that were rectified through manual intervention. The data cleaning process included:

- **Removal of Mismatched Data**: Data consisting solely of numbers and special characters were eliminated.
- **Missing Data Cleaning**: Incomplete sentences, missing words, and words that were contextually irrelevant were treated as missing values. Some were removed, while others were replaced with appropriate keywords.
- **Noisy Data Reduction**: Reviews containing emojis, symbols, URLs, and extraneous blank spaces were cleaned.
- **Duplication Reduction**: Duplicate reviews and repeated sentences within the same review were eliminated to enhance data quality.

Manual Data Labeling Each review sentence was meticulously labeled into Positive, Negative, Suggestions, or Neutral categories, guided by decision words. For instance:

- Words like "good," "amazing," or "great" suggested positive sentiment.
- Words like "bad" or "worst" indicated negative sentiment.
- The word "improvement" indicated a suggestion.
- Sentences containing both positive and negative words were labeled neutral.

During labeling, we considered negation words.

Our study aims to develop an automated sentiment analysis system. This system will automatically undergo the same preprocessing tasks when provided with reviewer input.

3.4 Algorithm Development

Overview This research methodology aims to address the challenge of predicting customer satisfaction levels in e-commerce platforms by developing a robust machine learning algorithm. The success of e-commerce businesses heavily relies on customer satisfaction, making it imperative to accurately predict and understand customer sentiments. To achieve this goal, a comprehensive approach to algorithm design and development is essential.

At the core of the algorithm development process is the framework of supervised machine learning augmented with a rule-based classifier for sentiment analysis. This hybrid approach leverages the strengths of both machine learning techniques and rule-based systems to achieve more accurate predictions. The algorithm begins by extracting relevant keywords from user reviews, which serve as the input data. These keywords are then categorized into different sentiment labels using if-else conditions, forming the basis of sentiment analysis.

The algorithm's architecture encompasses several interconnected steps, each contributing to the overall process of customer satisfaction prediction. These steps include data preprocessing, feature extraction, model training, and sentiment classification. Figure 1 provides a visual representation of the algorithm's flow, illustrating how data flows through each stage of the process. This structured approach lays the foundation for building a scalable and efficient system for estimating customer satisfaction levels.

The algorithm's theoretical framework outlines a series of sequential steps aimed at analyzing and interpreting customer feedback. These steps include tokenizing feedback sentences, matching keywords with predefined dictionaries, updating variables based on word matches, and ultimately determining the sentiment of the feedback. By following this systematic approach, the algorithm can effectively process and analyze large volumes of customer reviews, providing valuable insights into customer satisfaction trends.

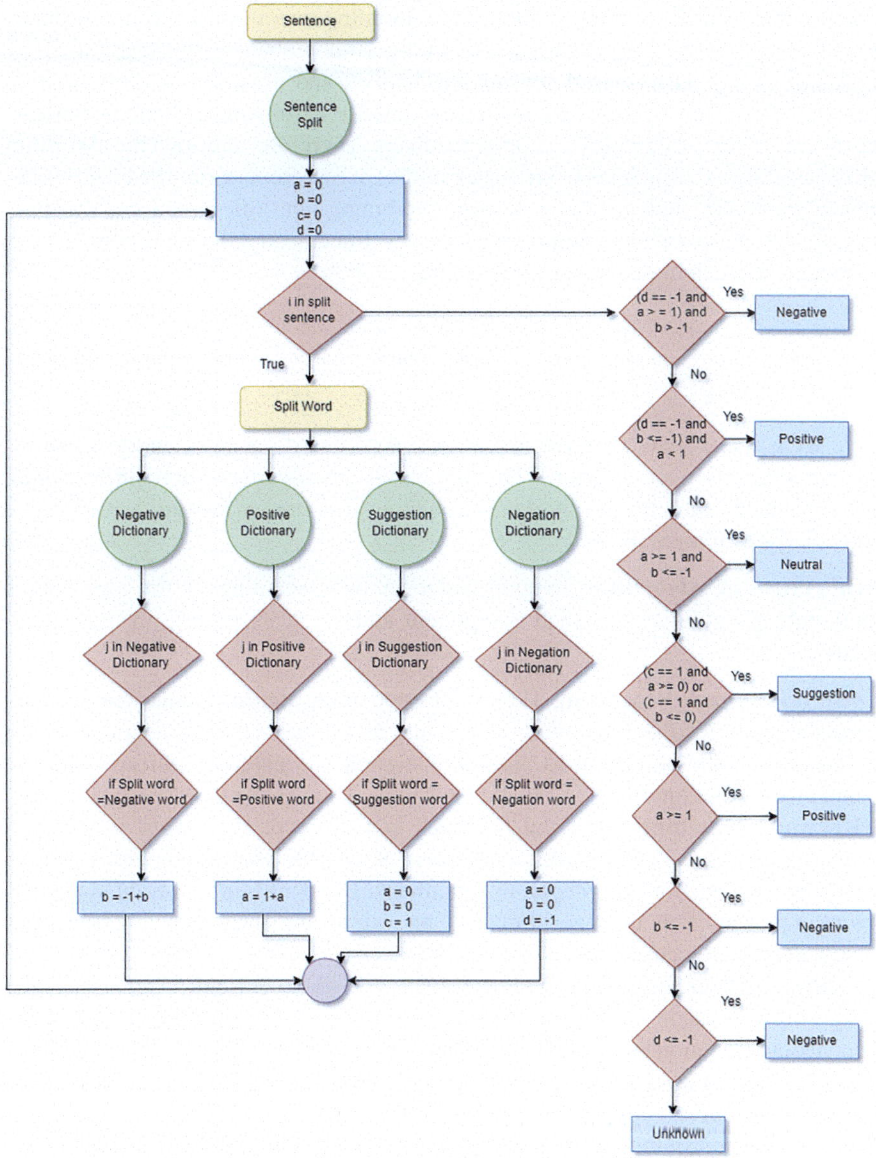

Fig. 1 Flow chart of the proposed algorithm

Rule-Based Classifier Our rule-based classifier employs predefined dictionaries for positive, negative, suggestion, and negation words, drawing inspiration from foundational works in sentiment analysis, such as the study by [35]. However, this study presents a completely new algorithm that integrates these traditional rule-based methods with innovative supervised machine learning techniques.

While the classification logic operates by iterating over each word in a sentence and checking against these dictionaries, our algorithm enhances this framework by incorporating a polarity-based scoring mechanism and novel keyword extraction methods. These modifications allow for more precise sentiment predictions, particularly in the context of complex customer feedback. Therefore, although the algorithm utilizes established principles from earlier studies, it represents a significant advancement in sentiment analysis methodology, combining both rule-based and machine learning elements to achieve improved accuracy. Specifically, the sentiment score S of a sentence is calculated using the formula:

$$S = \sum_{i=1}^{n} w_i \tag{1}$$

where w_i is the weight assigned to the i-th word based on its category (positive, negative, suggestion, or negation). The rule-based classifier does not require training and is evaluated directly on the dataset, making it a straightforward yet effective approach for sentiment analysis.

Architecture of Algorithm To gain a deeper understanding of the algorithm's functionality, we delve into each component and elucidate its role in the overall process:

1. **Given Feedback Processing** The initial stage of the algorithm involves processing the given feedback by tokenizing it into individual words. This tokenization step enables the algorithm to break down complex sentences into manageable units for further analysis.
2. **Keyword Matching** Once the feedback is tokenized, each word is matched against predefined dictionaries containing keywords associated with positive, negative, suggestion, and negation sentiments. By identifying relevant keywords within the feedback, the algorithm can gain insights into the underlying sentiment expressed by the customer.
3. **Sentiment Determination** With the keywords identified and categorized, the algorithm proceeds to determine the overall sentiment of the feedback. This sentiment determination process involves analyzing the accumulated scores of positive and negative keywords and applying predefined rules to classify the feedback as positive, negative, neutral, or suggestive. By incorporating mathematical analysis and logical reasoning, the algorithm can make accurate predictions about customer sentiments, aiding businesses in understanding and addressing customer needs effectively.

The sentiment determination process can be mathematically represented using a set of conditional equations, as shown below:

Table 2 Label-keywords mapping

Label	Keywords
Positive	Excellent, great, wonderful, amazing, fantastic
Negative	Poor, terrible, awful, bad, disappointing
Suggestion	Improve, enhance, add, develop, upgrade
Neutral	Combination of Positive and Negative
Unknown	No specific keywords

$$\text{Sentiment} = \begin{cases} \text{Negative} & \text{if } (d = -1 \wedge a \geq 1) \wedge b > -1 \\ \text{Positive} & \text{if } (d = -1 \wedge b \leq -1) \wedge a < 1 \\ \text{Neutral} & \text{if } a \geq 1 \wedge b \leq -1 \\ \text{Suggestion} & \text{if } (c = 1 \wedge a \geq 0) \vee \\ & \qquad (c = 1 \wedge b \leq 0) \\ \text{Positive} & \text{if } a \geq 1 \\ \text{Negative} & \text{if } b \leq -1 \\ \text{Negative} & \text{if } d \leq -1 \\ \text{Unknown} & \text{otherwise} \end{cases}$$

These equations encapsulate the logic behind sentiment determination and provide a clear framework for interpreting the results obtained by the algorithm.

Moreover, Table 2 provides an insightful depiction of the sentiment labels paired with their respective keywords. This table serves as a crucial reference, facilitating a deeper understanding of the intricate relationship between keywords and sentiment categories.

The machine learning algorithm developed herein adopts a supervised learning approach with a rule-based classifier, employing a polarity-based methodology. Predicated on the polarity assessment, words are systematically labeled into distinct categories. Words eluding the system's detection are pragmatically classified as unknown.

3.5 *Feature Extraction*

Before feature extraction, we partitioned the data into 80% for training and 20% for testing to assess the performance of our machine learning models.

Feature extraction is a crucial step in our methodology, where raw text data is transformed into a format suitable for machine learning analysis. In this study, we employ the Term Frequency-Inverse Document Frequency (TF-IDF) method for feature extraction.

TF-IDF is a numerical measure that evaluates the significance of a word within a document relative to a larger corpus. It serves as a weighting factor in information retrieval and text-mining tasks. The TF-IDF value increases with the frequency of a word within a document but is balanced by the word's frequency across the entire corpus, accounting for the commonality of certain terms.

The TF-IDF for a term t in a document d within a document set D is computed as follows:

$$\text{tfidf}(t, d, D) = \text{tf}(t, d) \times \text{idf}(t, D) \tag{2}$$

where $\text{tf}(t, d)$ represents the term frequency of t in d, calculated by dividing the number of occurrences of t in d by the total number of terms in d.

$\text{idf}(t, D)$ denotes the inverse document frequency of t in D, computed as the natural logarithm of the ratio between the total number of documents in D and the number of documents in D containing term t.

After computing the TF-IDF values, we obtain a matrix where each row corresponds to a document and each column represents a term. This matrix serves as the input feature set for our machine learning models, enabling further analysis and classification.

3.6 Model Training and Evaluation

Following feature extraction, we proceed with model training and evaluation. Several machine learning models are trained on the extracted features, and their performance is evaluated using various metrics such as accuracy, precision, recall, and F1 score. The models evaluated include Naive Bayes, K-Nearest Neighbors, Logistic Regression, Random Forest, Support Vector Machine (SVM), and Gradient Boosting. Additionally, the performance of these models is compared with the rule-based classifier developed in this study.

Machine Learning Models The machine learning models are trained using the TF-IDF features. The models evaluated include:

- Random Forest
- Gradient Boosting
- Support Vector Machine (SVM)
- Naive Bayes
- Logistic Regression
- K-Nearest Neighbors

Evaluation Metrics The models are evaluated using the following metrics:

- **Accuracy**: The proportion of correct predictions over the total number of instances.

$$\text{Accuracy} = \frac{\text{TP} + \text{TN}}{\text{TP} + \text{TN} + \text{FP} + \text{FN}} \tag{3}$$

- **Precision**: The proportion of true positive predictions over the total number of positive predictions.

$$\text{Precision} = \frac{\text{TP}}{\text{TP} + \text{FP}} \tag{4}$$

- **Recall**: The proportion of true positive predictions over the total number of actual positives.

$$\text{Recall} = \frac{\text{TP}}{\text{TP} + \text{FN}} \tag{5}$$

- **F1 Score**: The harmonic mean of precision and recall.

$$\text{F1 Score} = 2 \times \left(\frac{\text{Precision} \times \text{Recall}}{\text{Precision} + \text{Recall}} \right) \tag{6}$$

4 Result and Analysis

4.1 Comparative Evaluation of Machine Learning Models and Rule-Based Classifier

After implementing the rule-based classifier algorithm, we evaluated its performance using 10-fold cross-validation (CV). Cross-validation is a widely recognized technique for assessing the generalization capability of machine learning models by partitioning the dataset into multiple subsets or folds. The model is trained on a subset of the data and evaluated on the remaining folds. This process is repeated multiple times, with each fold serving as the validation set exactly once, providing a robust estimate of model performance. We applied 10-fold cross-validation to rigorously evaluate the performance of our rule-based classifier in sentiment analysis. Common evaluation metrics-accuracy, precision, recall, and F1 score-were employed to comprehensively assess the classifier's effectiveness. Additionally, we visualized the performance metrics across each fold to provide insights into the consistency and reliability of the classifier. The mean values of these metrics offer a summary of the classifier's overall performance. The results indicated that the classifier achieved an average accuracy of 0.90, a precision of 0.91, a recall of 0.90, and an F1 score of 0.90, demonstrating consistent performance across the folds. Some variability was observed, particularly in folds 6 and 9, but the overall trend remained stable. These

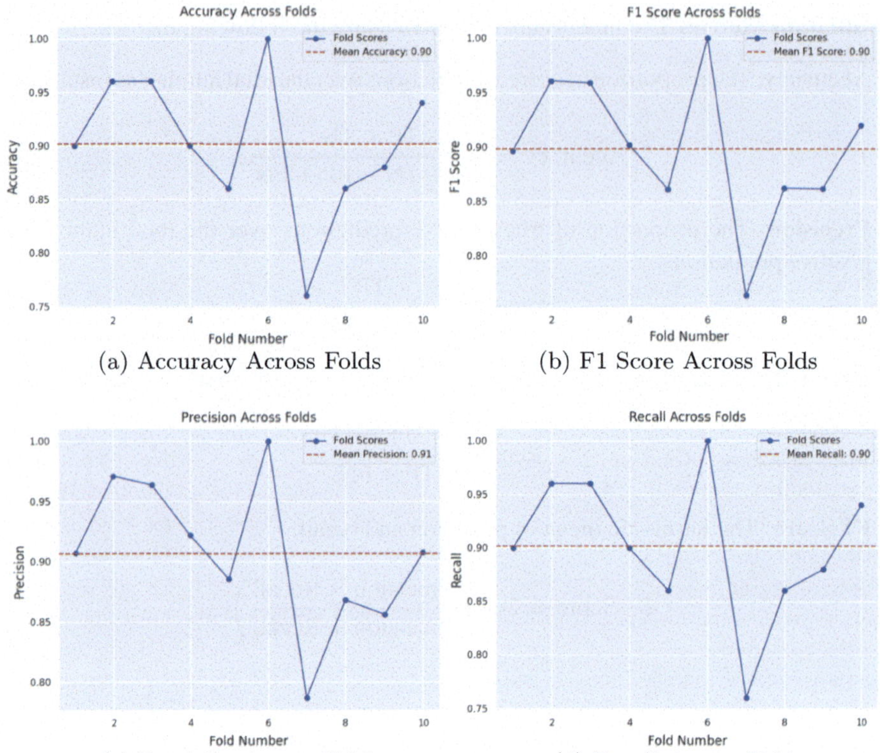

(a) Accuracy Across Folds

(b) F1 Score Across Folds

(c) Precision Across Folds

(d) Recall Across Folds

Fig. 2 10-fold cross-validation to assess the performance of our rule-based classifier

visualizations and metrics provide a comprehensive understanding of the classifier's performance and its consistency across different subsets of the data. The mean values of these metrics underscore the classifier's robustness and reliability in sentiment analysis tasks, validating its potential for practical applications in analyzing customer reviews in e-commerce (Fig. 2).

We evaluated the performance of the rule-based classifier alongside several traditional machine learning models commonly used in sentiment analysis tasks. The machine learning models evaluated include Random Forest, Gradient Boosting, Support Vector Machine (SVM), Naive Bayes, Logistic Regression, and K-Nearest Neighbors. Each model was trained on the TF-IDF features extracted from the text data. Their performance was assessed using metrics such as accuracy, precision, recall, and F1 score, providing insights into different aspects of the model's ability to classify instances correctly.

The rule-based classifier's performance, evaluated using 10-fold cross-validation, was included in the comparison to ensure a robust assessment. The rule-based classifier demonstrated superior performance, achieving an accuracy of 0.90, precision of 0.91, recall of 0.90, and an F1 score of 0.90. In comparison, the Gradient Boosting

Table 3 Evaluation metrics for machine learning models and rule-based classifier

Model	Accuracy	Precision	Recall	F1 Score
Naive Bayes	0.66	0.64	0.66	0.60
K-nearest neighbors	0.74	0.77	0.74	0.72
Logistic regression	0.76	0.68	0.76	0.71
Random forest	0.78	0.81	0.78	0.77
SVM	0.80	0.83	0.80	0.77
Gradient boosting	0.82	0.83	0.82	0.81
Rule-based classifier	0.90	0.91	0.90	0.90

model, which performed the best among the traditional machine learning models, achieved an accuracy of 0.82, precision of 0.83, recall of 0.82, and an F1 score of 0.81. Table 3 and Fig. 3 present the evaluation results for each model, showcasing their performance across different metrics. The rule-based classifier demonstrates superior performance compared to traditional machine learning models, indicating its potential effectiveness for sentiment analysis tasks. The significant improvement in accuracy, precision, recall, and F1 score underscores the robustness of the rule-based

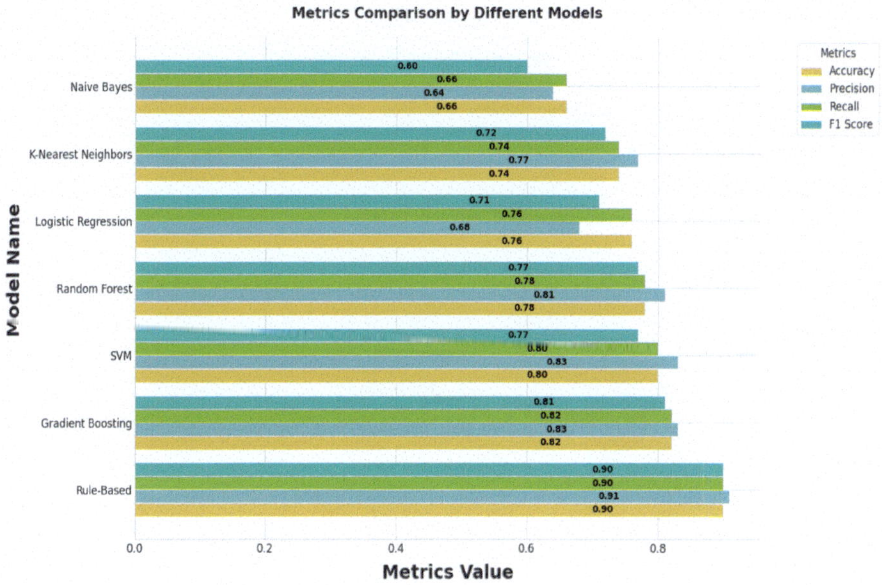

Fig. 3 Metrics comparison across different models

classifier, making it a highly reliable choice for sentiment analysis. To evaluate the performance of our proposed rule-based classifier and compare it with other machine learning models, we analyzed their confusion matrices. A confusion matrix offers a detailed breakdown of a model's performance by displaying the counts of true positives (TP), true negatives (TN), false positives (FP), and false negatives (FN), providing insight into where the model excels and where it may falter.

Figure 4 presents the confusion matrices for each model, illustrating their performance across five sentiment categories: Negative, Neutral, Positive, Suggestion, and Unknown. The Naive Bayes classifier exhibits a notable number of misclassifications, particularly in Positive and Suggestion categories. K-Nearest Neighbors (KNN) performs better but still struggles with classifying Suggestion sentiments correctly.

Logistic Regression shows balanced performance but has difficulties with Neutral sentiments. The Support Vector Machine (SVM) and Random Forest models offer more consistent results, with fewer false positives and false negatives. Gradient Boosting shows improvements, especially in classifying Positive sentiments, but still faces challenges with Neutral and Suggestion categories.

Our proposed Rule-Based classifier outperforms the other models, particularly in the Positive and Suggestion categories, demonstrating fewer false positives and false negatives. This indicates the model's robustness and accuracy in handling complex sentiment classifications, making it a reliable tool for sentiment analysis in diverse and ambiguous contexts. This analysis underscores the proposed model's superior capability in providing precise sentiment classifications compared to other machine learning models.

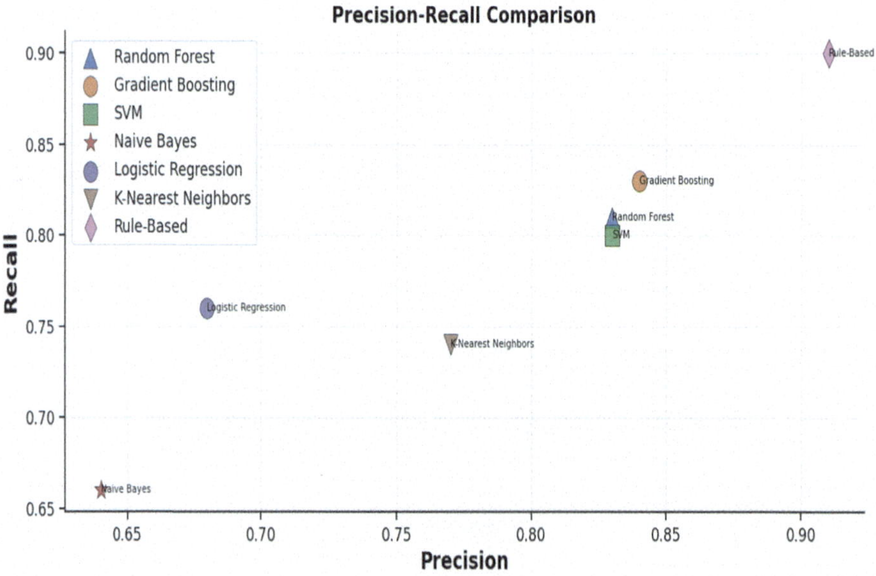

Fig. 4 Precision-recall curve for the proposed rule-based classifier

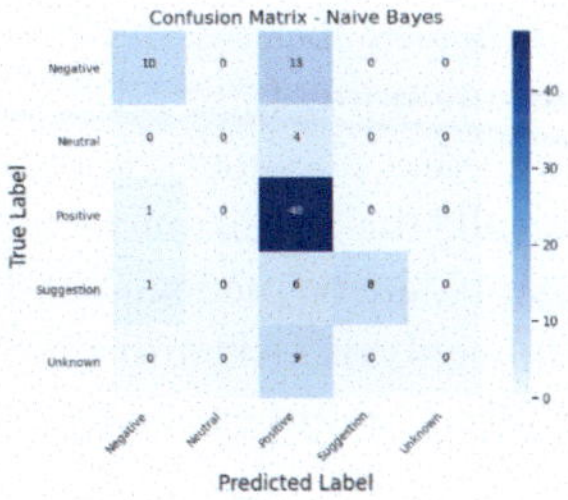

(a) Naive Bayes

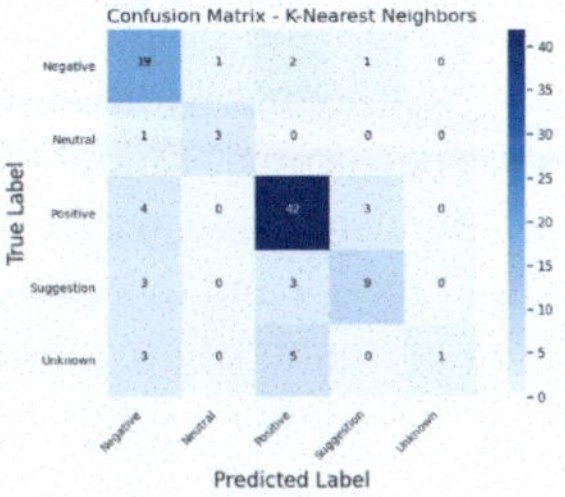

(b) K-Nearest Neighbors (KNN)

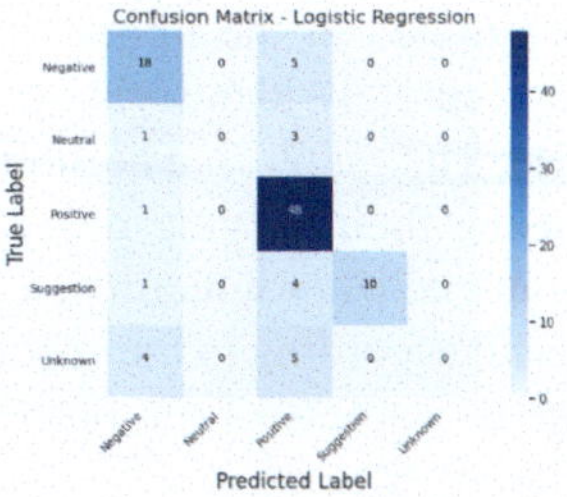

(c) Logistic Regression

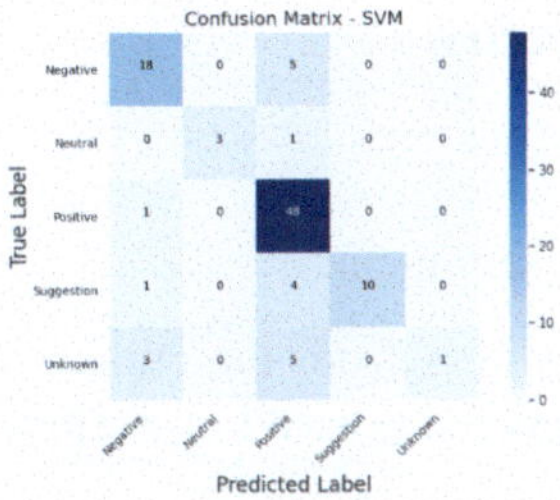

(d) Support Vector Machine (SVM)

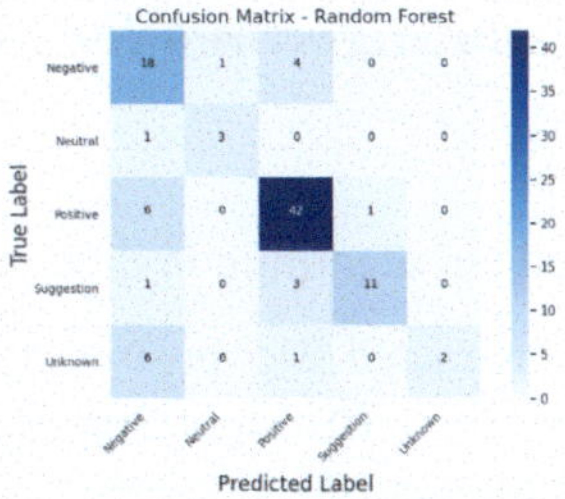

(e) Random Forest

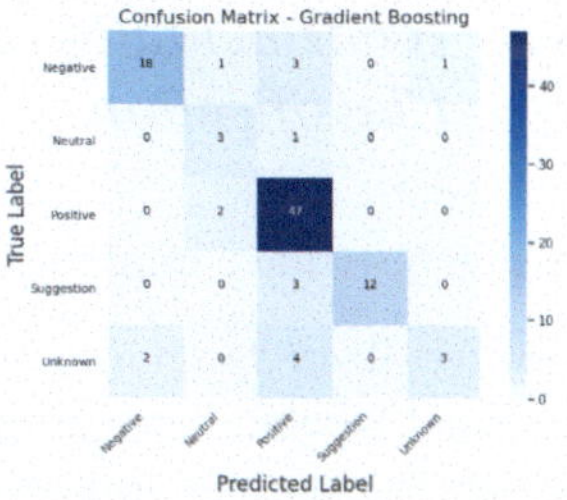

(f) Gradient Boosting

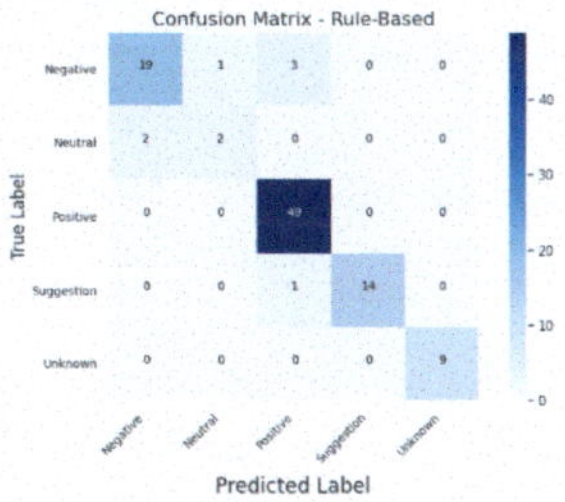

(g) Rule-Based (Pro-posed)

Fig. 5 Confusion matrices for various machine learning models: Naive Bayes, K-nearest neighbors, logistic regression, support vector machine, random forest, gradient boosting, and the proposed rule-based model. These matrices illustrate the performance of each model in predicting the correct labels for the test set

The precision-recall graph in Fig. 5 provides an insightful visualization of the trade-off between precision and recall for various machine learning models, including the proposed rule-based classifier. This graph is especially beneficial when dealing with imbalanced datasets, where correctly identifying the positive class (e.g., positive sentiments in reviews) is often more critical than the negative class.

Figure 5 illustrates the precision-recall curves for several classifiers, such as Random Forest, Gradient Boosting, SVM, Naive Bayes, Logistic Regression, K-Nearest Neighbors, and the proposed rule-based classifier. The proposed rule-based classifier demonstrates superior performance with both high precision and recall values, positioned in the upper right corner of the graph. This indicates its effectiveness in accurately identifying positive sentiments while maintaining a low false positive rate. The graph also highlights the performance trade-offs of other classifiers, showcasing their varying degrees of precision and recall. Overall, the precision-recall graph helps in selecting an optimal model and threshold that balances precision and recall according to the specific requirements of the sentiment analysis task, affirming the robustness of the proposed rule-based classifier in handling sentiment analysis effectively.

4.2 *Implementation of our Algorithm for Polarity Classification*

In Fig. 6 we have implemented our algorithm in different sentiment.

Positive Word Detection We extracted 215 positive keywords from the collected data and labeled them as "positive". The algorithm detects the polarity of positive sentences using the following logic:

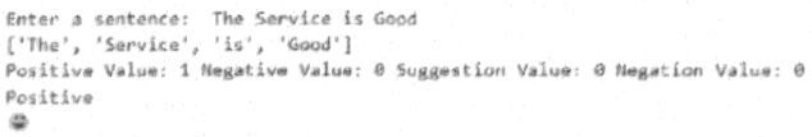

(a) Positive word detection by the algorithm

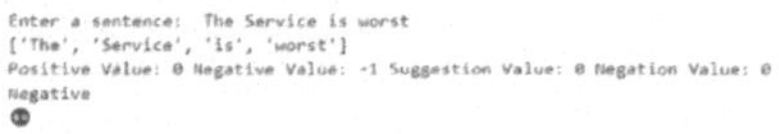

(b) Negative word detection by the algorithm

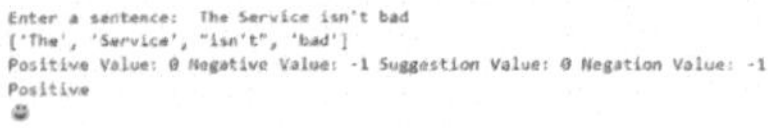

(c) Positive word detection using negation by the algorithm

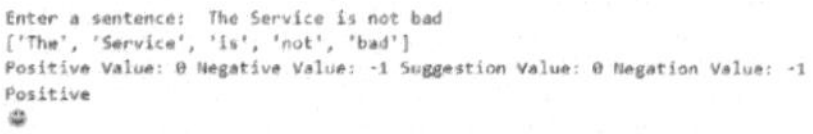

(d) Another positive word detection using negation by the algorithm

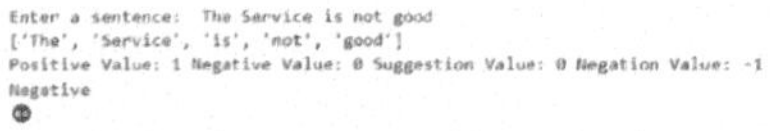

(e) Negative word detection using negation by the algorithm

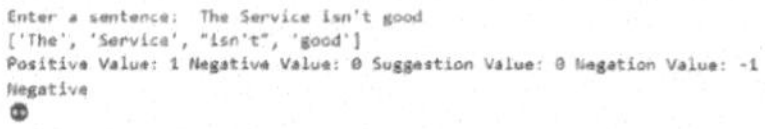

(f) Another negative word detection using negation by the algorithm

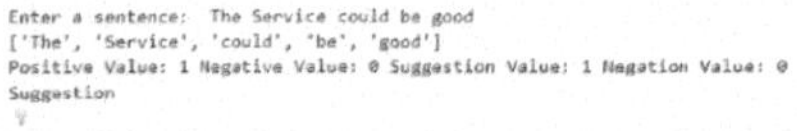

(g) Suggestion word detection by the algorithm

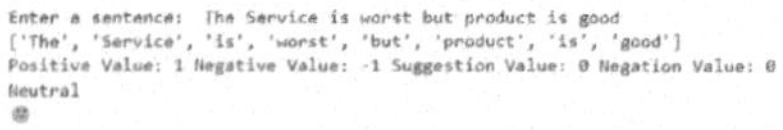

(h) Neutral word detection by the algorithm

Fig. 6 Implementation of our algorithm

$$\text{Word (Positive)} \rightarrow \text{Polarity Positive} \tag{7}$$

Negative Word Detection We extracted 49 negative keywords from the collected data and labeled them as "negative". The algorithm detects the polarity of negative sentences as follows:

$$\text{Word (Negative)} \rightarrow \text{Polarity Negative} \tag{8}$$

Negation Detection Negation words (e.g., no, not, none) can alter the polarity of a sentence. The algorithm detects the polarity of sentences with negation using the following logic:

$$\text{Word (Negation)} + \text{Word (Positive)} \rightarrow \text{Polarity Negative} \tag{9}$$

$$\text{Word (Negation)} + \text{Word (Negative)} \rightarrow \text{Polarity Positive} \tag{10}$$

Suggestion Word Detection Suggestion words include "should," "could," "need," etc. We extracted 89 suggestion keywords from the collected data and labeled them as "suggestions". The algorithm detects these words as follows:

$$\text{Word (Suggestion)} \rightarrow \text{Polarity Suggestion} \tag{11}$$

Neutral Word Detection The algorithm checks words and searches dedicated dictionaries to detect the sentiment of sentences that are neutral. The logic is as follows:

$$\text{Word (Positive)} + \text{Word (Negative)} \rightarrow \text{Polarity Neutral} \tag{12}$$

4.3 Manual Approaches and Our Rule-Based

The results from both the manual and proposed rule-based approaches are presented in Tables 4, 5, 6, and 7. Table 4 shows the sentiment count by service and sentiment label for online payment-related feedback. From the manual approach, it is observed that there are 115 positive sentiments, 48 negative sentiments, 5 neutral sentiments, 42 suggestions, and 16 unknown sentiments. In comparison, the proposed rule-based approach detected 116 positive sentiments, 40 negative sentiments, 7 neutral sentiments, 48 suggestions, and 15 unknown sentiments.

Table 5 displays the sentiment count by service and sentiment label for customer support-related feedback. In the manual approach, there are 33 positive sentiments, 32 negative sentiments, 5 neutral sentiments, 13 suggestions, and 8 unknown sentiments. The proposed rule-based approach identified 34 positive sentiments, 25 negative sentiments, 8 neutral sentiments, 18 suggestions, and 6 unknown sentiments.

Table 4 Sentiment count by service and sentiment label: online payment

Sentiment label	Manual approach	Proposed rule-based
Positive	115	116
Negative	48	40
Neutral	5	7
Suggestion	42	48
Unknown	16	15

Table 5 Sentiment count by service and sentiment label: customer support

Sentiment label	Manual approach	Proposed rule-based
Positive	33	34
Negative	32	25
Neutral	5	8
Suggestion	13	18
Unknown	8	6

Table 6 Sentiment count by service and sentiment label: product delivery timeline

Sentiment label	Manual approach	Proposed rule-based
Positive	41	38
Negative	34	33
Neutral	5	5
Suggestion	27	31
Unknown	11	13

Table 7 Sentiment count by service and sentiment label: product return policy

Sentiment label	Manual approach	Proposed rule-based
Positive	26	27
Negative	18	18
Neutral	6	3
Suggestion	8	10
Unknown	7	7

For feedback related to product delivery timelines, Table 6 provides the sentiment count by service and sentiment label. The manual approach detected 41 positive sentiments, 34 negative sentiments, 5 neutral sentiments, 27 suggestions, and 11 unknown sentiments. On the other hand, the proposed rule-based approach found 38 positive sentiments, 33 negative sentiments, 5 neutral sentiments, 31 suggestions, and 13 unknown sentiments.

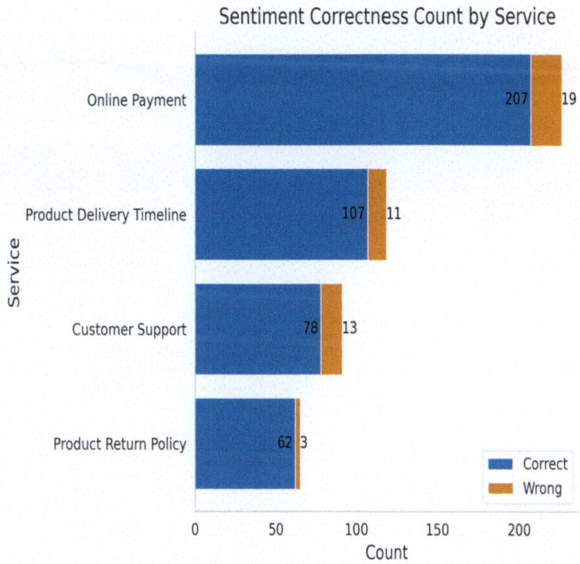

Fig. 7 Sentiment correctness count by service

Finally, Table 7 illustrates the sentiment count by service and sentiment label for product return policy-related feedback. The manual approach identified 26 positive sentiments, 18 negative sentiments, 6 neutral sentiments, 8 suggestions, and 7 unknown sentiments. Similarly, the proposed rule-based approach detected 27 positive sentiments, 18 negative sentiments, 3 neutral sentiments, 10 suggestions, and 7 unknown sentiments.

The analysis of the proposed rule-based system revealed its effectiveness in predicting sentiment correctly. The system achieved a 90.4% correct prediction rate and a 9.6% incorrect prediction rate. This demonstrates the system's capability to provide near-accurate results efficiently.

Figure 7 illustrates the sentiment correctness count by service:

- **Customer Support**: Among 91 feedback instances, 78 were correctly predicted (85.7%), while 13 were incorrect (14.3%).
- **Online Payment**: Out of 226 feedback instances, 207 were correctly predicted (91.6%), while 19 were incorrect (8.4%).
- **Product Delivery Timeline**: Out of 118 feedback instances, 107 were correctly predicted (90.7%), while 11 were incorrect (9.3%).
- **Product Return Policy**: Among 65 feedback instances, 62 were correctly predicted (95.4%), while 3 were incorrect (4.6%).

Figure 8 shows the sentiment correctness count by sentiment label:

- **Positive Sentences**: Out of 216 instances, 209 were correctly predicted (96.8%), and 7 were incorrect (3.2%).

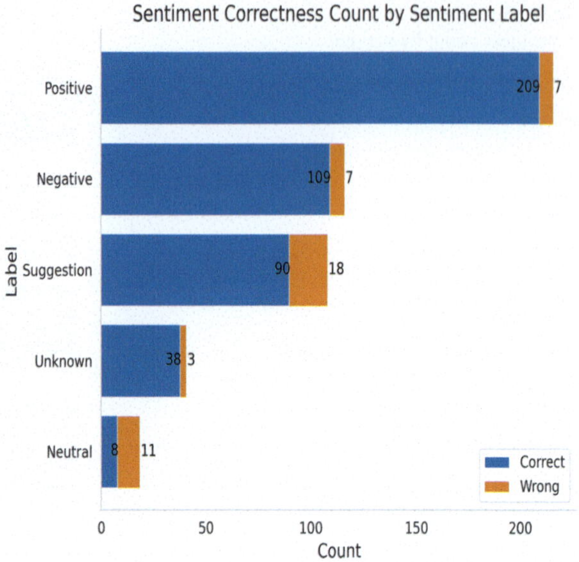

Fig. 8 Sentiment correctness count by sentiment label

- **Negative Sentences**: Out of 116 instances, 109 were correctly predicted (94.0%), and 7 were incorrect (6.0%).
- **Suggestion-Labeled Sentences**: Out of 108 instances, 90 were correctly predicted (83.3%), and 18 were incorrect (16.7%).
- **Unknown Sentences**: Out of 41 instances, 38 were correctly predicted (92.7%), and 3 were incorrect (7.3%).
- **Neutral Sentences**: Out of 19 instances, 8 were correctly predicted (42.1%), and 11 were incorrect (57.9%).

The percentages were derived by dividing the number of correctly or incorrectly predicted instances by the total number of instances for each category and then multiplying by 100. For example, the percentage of correct predictions for customer support was calculated as follows:

$$\text{Correct Percentage} = \left(\frac{\text{Number of Correct Predictions}}{\text{Total Predictions}}\right) \times 100$$

$$\text{Correct Percentage} = \left(\frac{78}{91}\right) \times 100 \approx 85.7\%$$

These results indicate that the proposed rule-based system is effective in predicting sentiment with a high degree of accuracy.

5 Discussion

This study evaluates various machine learning models and a rule-based classifier for sentiment analysis in e-commerce reviews. The machine learning models assessed include Naive Bayes, K-Nearest Neighbors, Logistic Regression, Random Forest, SVM, and Gradient Boosting, with Gradient Boosting achieving the highest accuracy among them at 82%. Despite the effectiveness of these models, our rule-based classifier, designed using linguistic rules and a sentiment-laden keyword dictionary, outperformed the traditional models, achieving a 90% accuracy, recall, F1 score alongside 91% precision. This superior performance is attributed to its ability to incorporate domain-specific knowledge and handle context-specific sentiment expressions effectively. The rule-based approach's success highlights its potential in specific domains where sentiment expressions are predictable and domain knowledge is readily available.

To validate the robustness of our findings, we employed 10-fold cross-validation, which revealed that the rule-based classifier consistently exhibited higher precision, recall, and F1 scores compared to its machine learning counterparts. These results imply that in contexts characterized by well-defined sentiment expressions, rule-based systems can be particularly effective.

The practical implications of our automated sentiment analysis system for e-commerce platforms are profound; by providing real-time insights into customer feedback, businesses can promptly address concerns, enhance product features, and ultimately improve customer satisfaction. While the speed and accuracy of this automated approach offer a significant advantage over traditional manual methods, it is important to acknowledge the limitations of the rule-based classifier, particularly concerning nuanced or ambiguous sentiments that may fall outside predefined rules.

Looking ahead, several promising avenues for future research emerge. Expanding the dataset to include feedback from a broader array of users across different industries and regions would enhance the model's robustness and generalizability. Integrating advanced techniques such as neural networks and deep learning models could also significantly improve the algorithm's accuracy and scalability, allowing for better interpretation of complex sentiments. Furthermore, incorporating multilingual support would enable the analysis of feedback in various languages, broadening the model's applicability, especially in non-English-speaking regions. Finally, tailoring the algorithm for specific applications within e-commerce–such as improving personalized product recommendations or enhancing customer support services–could yield targeted solutions that directly impact customer satisfaction and operational efficiency.

6 Conclusion

In conclusion, this study has demonstrated the effectiveness of a rule-based classifier for sentiment classification in e-commerce reviews, significantly surpassing manual methods and various machine learning models in terms of accuracy and efficiency. The findings highlight the practical utility of automated sentiment analysis systems in processing large datasets and delivering timely insights for businesses. Our rigorous evaluation, which included cross-validation and confusion matrix analysis, reinforces the classifier's reliability and its potential for real-world applications. By enhancing the understanding of customer feedback, this research contributes meaningfully to improving service quality and customer satisfaction in the e-commerce sector. Overall, this work lays the groundwork for further advancements in sentiment analysis techniques tailored to the unique challenges of e-commerce environments.

References

1. Zhang L, Liu B (2012) Sentiment analysis and opinion mining. In: Synthesis lectures on human language technologies
2. Gamon M, Aue A, Corston-Oliver S, Ringger E (2005) Pulse: mining customer opinions from free text. In: International symposium on intelligent data analysis, pp 121–132. Springer
3. Pang B, Lee L (2008) Opinion mining and sentiment analysis. Found Trends R Inform Retr 2(1–2):1–135
4. Kim H, Kim E (2004) Customer satisfaction and dissatisfaction: a new approach to their relationship. Acad Market Sci Rev 1
5. Turney PD (2002) Thumbs up or thumbs down? semantic orientation applied to unsupervised classification of reviews. In: Proceedings of the 40th annual meeting on association for computational linguistics, pp 417–424. Association for Computational Linguistics
6. Wilson T, Wiebe J, Hoffmann P (2005) Recognizing contextual polarity in phrase-level sentiment analysis. In: Proceedings of human language technology conference and conference on empirical methods in natural language processing, pp 347–354
7. Hu M, Liu B (2004) Mining opinion features in customer reviews. In: Proceedings of the tenth ACM SIGKDD international conference on Knowledge discovery and data mining, pp 168–177
8. Ronen F (2013) Techniques and applications for sentiment analysis. Commun ACM 56(4):82–89
9. Cambria E, Schuller B, Xia Y, Havasi C (2013) New avenues in opinion mining and sentiment analysis. IEEE Intell Syst 28(2):15–21
10. Medhat W, Hassan A, Korashy H (2014) Sentiment analysis algorithms and applications: a survey. Ain Shams Eng J 5(4):1093–1113
11. Mohammad SM, Kiritchenko S, Zhu X (2013) Nrc-canada: building the state-of-the-art in sentiment analysis of tweets. In: Proceedings of the seventh international workshop on semantic evaluation exercises (SemEval-2013), pp 321–327
12. Tang L, Tan S, Cheng X (2009) A survey on sentiment detection of reviews. Expert Syst Appl 36(7):10760–10773
13. Ceron A, Curini L, Iacus SM (2014) Every tweet counts? how sentiment analysis of social media can improve our knowledge of citizens' political preferences with an application to Italy and France. New media Soc 16(2):340–358

14. Yang H, Macdonald C, Ounis I (2018) A survey of sentiment analysis in social media. Knowl Inform Syst 50(2):345–397
15. Popescu A-M, Etzioni O (2005) Opine: extracting product features and opinions from reviews. In: Proceedings of the conference on human language technology and empirical methods in natural language processing, pp 339–346
16. Pang B, Lee L, Vaithyanathan S (2002) Thumbs up?: sentiment classification using machine learning techniques. arXiv preprint cs/0205070
17. Taboada M, Brooke J, Tofiloski M, Voll K, Stede M (2011) Lexicon-based methods for sentiment analysis. Computat Linguist 37(2):267–307
18. Esuli A, Sebastiani F (2006) Sentiwordnet: a publicly available lexical resource for opinion mining. In: Proceedings of the fifth international conference on language resources and evaluation (LREC'06), pp 417–422. Citeseer
19. Thet TT, Na J-C, Khoo CSG (2010) Aspect-based sentiment analysis of movie reviews on discussion boards. J Inf Sci 36(6):823–848
20. Vural AE, Cambazoglu BG, Senkul P, Tokgoz ZO (2014) A machine learning approach for sentiment analysis to the occupygezi event on twitter. Proceedings of the international conference on computational linguistics (COLING), pp 553–567
21. Singla P, Randhawa DS, Jain P (2017) Sentiment analysis of customer product reviews using machine learning. Int J Comput Appl 167(10):20–24
22. Sultan T, Jahan N, Basak R, Jony MSA, Nabil RH (2023) Machine learning in cyberbullying detection from social-media image or screenshot with optical character recognition. Int J Intell Syst Appl 15:1–13
23. Liu B, Zhao Y, Wu S (2017) Clustering and classification methods for sentiment analysis. J Inform Sci Eng 33(3):515–528
24. Ding X, Liu B, Yu PS (2008) A holistic lexicon-based approach to opinion mining. In: Proceedings of the 2008 international conference on web search and data mining, WSDM '08, pp 231–240, New York, NY, USA, 2008. Association for Computing Machinery
25. Xuanjing M, Rui W (2012) Unsupervised learning for product feature extraction based on linguistic rules. Inform Process Manag 48(4):640–655
26. Sutton RS, Barto AG (2018) Reinforcement learning: an introduction. MIT Press
27. The Business Standard. B2c e-commerce market in bangladesh. *The Business Standard*, 2022. Retrieved from https://www.tbsnews.net/
28. Mishra A, Reddy M (2018) The evolution of sentiment analysis: a review of research topics, venues, and top cited papers. arXiv preprint arXiv:1807.10939
29. Vinodhini G, Chandrasekaran RM (2012) Sentiment analysis and opinion mining: a survey. Int J Data Mining Knowl Manag Process 2(6):31–48
30. Abdullah Z, Ahmet M (2022) Recent advancements in sentiment analysis using deep learning techniques. J Artif Intell Res 65:231–259
31. Lv X, Zhang W, Wang L (2023) Contextual and aspectual memory network for sentiment analysis. J Big Data 10(1):1–23
32. Yang Y, Li M, Liu H (2023) A hybrid model for sentiment analysis: combining bert, cnn, and bilstm with attention mechanism. Inform Process Manag 60(2):102–115
33. Xiaoyu H, Xia Z, Wang M, Zhang W (2018) A reinforcement learning framework for the optimization of e-commerce recommendation systems. Inform Syst 76:1–15
34. Mustafa A, Wang F (2023) Challenges in sentiment analysis: a review of current trends and future directions. Artif Intel Rev 56(4):267–289
35. Hu M, Liu B (2004) Mining and summarizing customer reviews. In: Proceedings of the Tenth ACM SIGKDD international conference on knowledge discovery and data mining, pp 168–177, New York, NY, USA. ACM

Chapter 6
From Nature to Navigation: A Comparative Study of Biologically Inspired Path Planning

Divya Agarwal

1 Introduction

Mobile robots are intelligent machines programmed with partial human-like capabilities to perform precise movements while navigating paths and avoiding obstacles to reach destinations in familiar or partially unfamiliar environments, ensuring equipment and workforce safety [1]. Path planning requires extensive prior knowledge of work scenarios, which may be dynamic, unstructured, vague, or vast, and aims to find an optimal path. This activity involves planning and directing a safe, shortest, and smooth path for an AMR from the source to the target location [27]. On the other hand, obstacle avoidance operates without prior knowledge of work scenarios and involves approaches to bypass static and dynamic obstacles while shaping the AMR path. It relies on real-time information from advanced sensing techniques to guide mobile robots through the work environment efficiently [61, 70, 111]. Consequently, Path Planning and Obstacle Avoidance (PPOA) are crucial for mobile robots in autonomous operations, as they address the inherent challenges of AMRs [54].

The PPOA problem is more significant in obstacle-cluttered environments, as some robots halt when obstacles are nearby [4, 102]. Thus, algorithms that can monitor a robot's workspace and respond swiftly are needed. The robot should be capable of selecting the shortest path with minimal processing time and computational effort. PPOA is vital in fields such as education, manufacturing, entertainment, aerospace, aeronautics, safety, security, and agriculture [3, 28]. PPOA is a challenging problem that has been extensively studied by researchers, leading to various solutions for navigating obstacles within time constraints without becoming trapped in local minima [9, 14, 76].

D. Agarwal (✉)
Vivekananda Institute of Professional Studies, New Delhi, India
e-mail: er.divya7@gmail.com

N. Dey (ed.), *Optimizing Solutions for Real-Life Problems*, Springer Tracts in Nature-Inspired Computing, https://doi.org/10.1007/978-981-96-3216-9_6

The PPOA problem comprises three elements for effective execution [36] (shown in Fig. 1): (i) Modeling the environment to create a precise virtual simulation platform [50, 54], (ii) path exploration to find collision-free paths on the basis of the objective function and environmental constraints [74, 87], and (iii) path smoothing to optimize the robot's movement by choosing the shortest, smoothest, and safest path; minimizing cost, energy, and computation time; and avoiding obstacles [47, 110].

The PPOA problem is complex and multiconstrained and requires careful evaluation [2]. Robots can choose from multiple paths to reach a target position, but the goal is to find the shortest collision-free path with minimal Euclidean distance in challenging environments [10, 28]. The Euclidean distance measures the robot's path length from the start to the destination [6]. Research in PPOA began in the 1960s, with significant interest sparked by Lozano and Perez's work on robot motion planning in 1979 [6, 33]. The PPOA is typically studied in static and dynamic environments [7, 46]. In static settings, obstacles are fixed and known [22], whereas dynamic environments require real-time updates due to moving obstacles or other robots [89]. Dynamic environment research remains less explored, making it a compelling area of study.

PPOA algorithms must adapt to specific applications, often providing stable solutions for one problem but failing in others [34, 106]. The choice of algorithm depends on factors such as the environment type, performance parameters, robot and obstacle dimensions, and workspace complexity [94]. Algorithms are classified into global and local PPOA algorithms on the basis of initial information requirements [24, 76]. Global algorithms require complete path information initially, which is suitable for known static terrains but has relatively high computation times [4, 45]. Examples include Rapidly Exploring Random Trees (RRTs), the A* algorithm, pure pursuit, and Particle Swarm Optimization (PSO) [12, 18, 82, 86]. Local algorithms, with partial path information, adapt to real-time changes via sensor networks but may become trapped in local minima [17, 45]. Examples include Artificial Neural Networks (ANNs), Artificial Potential Fields (APFs), Genetic Algorithms (GAs), and fuzzy logic algorithms [38, 84, 104, 107].

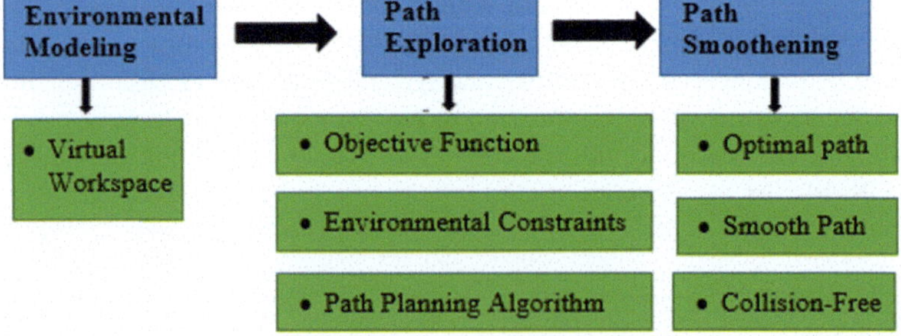

Fig. 1 Steps involved while evaluating the PPOA problem

Researchers further categorize PPOA algorithms into classical (conventional) and heuristic (reactive) types [56, 90, 94]. Classical algorithms, which were prominent until 2007, include A*, APF, bidirectional trees, and cell decomposition [6, 76, 80]. Challenges in conventional algorithms include data noise, variability, slow computation, local minima, limited intelligence, and failure under dynamic conditions [6, 55, 92, 97]. Heuristic algorithms emerged to address these issues, offering better performance in NP-complete problems and dominating PPOA research from 2003 to 2007 [7, 11, 65, 101]. They provide efficient, near-optimal solutions with lower computational complexity and adaptability to uncertain environments [37]. The classification of PPOA problems can be visualized in various categories, as shown in Fig. 2.

Heuristic algorithms are divided into nature-inspired and Artificial Intelligence (AI)-based categories [31, 42, 105]. Nature-inspired algorithms include evolutionary algorithms, Swarm Intelligence (SI) algorithms, bioinspired algorithms without SI, and human activity-inspired algorithms. Evolutionary algorithms derive their models from natural inheritance, evolution, and deviation but struggle in complex environments [4, 76]. Examples include the GA, harmony search, A*, and Differential Evolution (DE). SI-based algorithms are population-based and use shared knowledge for optimal solutions. Examples include Artificial Bee Colony Optimization (ABCO), the Bat Algorithm (BA), the Cuckoo Search Algorithm (CSA), Grey Wolf Optimization (GWO), Harris Hawks Optimization (HHO), PSO, and the Shuffled Frog Leaping Algorithm (SFLA) [76]. Bioinspired algorithms without SI, such as the Firefly Algorithm (FA) and flower pollination algorithm, draw from solitary creatures' behaviors [78]. Human-activity-based algorithms mimic human actions, including Teaching Learning-Based Optimization (TLBO), Simulated Annealing (SA), gravitational search, and tabu search.

AI-based algorithms, such as neural networks, reinforcement learning, machine learning, and deep learning, are robust and adaptable but face issues with generalizability [23, 36]. The effectiveness of these algorithms can degrade with varying

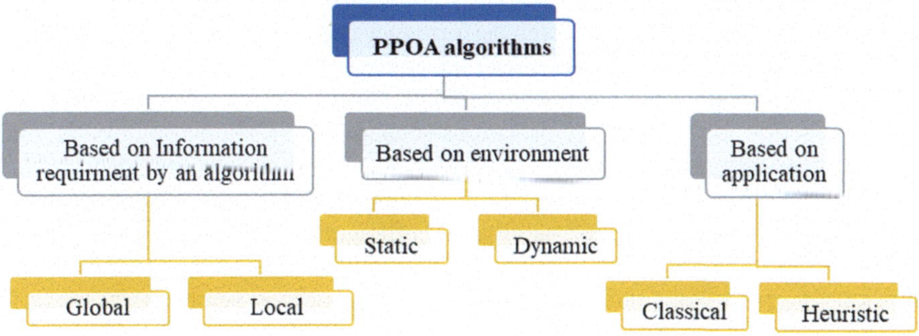

Fig. 2 Classification of PPOA algorithms

control parameters in complex scenarios [2]. Both evolutionary algorithms and AI-based algorithms require complete environmental information, limiting their real-time applicability to dynamic obstacles [6, 68, 92]. Despite significant advancements, some heuristic algorithms still face challenges, such as high computation time, low convergence speed, and accuracy issues in complex environments [19]. Nature-inspired algorithms, however, offer optimal solutions across various complexities, inspired by the survival instincts of organisms such as bacteria, insects, birds, and animals [36, 77, 106]. The adaptability and efficiency of these algorithms make them suitable for diverse environmental situations with reduced computation time and effort [50, 110]. Their dynamic, parallel approach and self-organizing capabilities are particularly beneficial for solving complex problems [73, 94]. This review examines recent nature-inspired algorithms for solving the PPOA problem, highlighting their efficiency in generating optimal paths with minimal computation compared with classical and heuristic methods. It also discusses their strengths, weaknesses, and future research directions, presenting a chronological analysis of work done using these algorithms for PPOA problems.

Motivated by nature's effective exploration and exploitation mechanisms, this chapter introduces the HHO, WOA, and MFO algorithms to this domain for the first time, identifying gaps and suggesting future improvements. Simulation experiments on MATLAB 2020a, which use five different environmental maps, demonstrate the stability and effectiveness of the HHO and MFO algorithms for solving PPOA problems in AMRs, validating their robustness and paving the way for further research and application.

2 Background of the PPOA and Nature-Inspired Algorithms

Nature-inspired algorithms draw from the behavior and strategies of natural elements to solve physical problems, enhancing performance and robotic capabilities in PPOA [1, 32]. These algorithms adapt biological structures and mechanisms for engineering challenges [32]. They enable robots to emulate animal motion, enhancing human–robot interaction for adaptability in various scenarios [45]. Animal motion is vital because of its robustness, dexterity, and adaptability, which aligns with Darwin's survival principle [32]. Autonomous Mobile Robots (AMRs) that use nature-inspired algorithms demonstrate superior stability and efficiency in PPOA tasks [1, 13].

The PPOA remains a key research area in robotics, focusing on continuous, collision-free path computation within constraints such as vehicle size, obstacle nature, and task completion time [46]. Robots must navigate obstacles by understanding their types, locations, and orientations [89]. Various algorithms have been proposed, with the choice depending on the desired outcomes and application constraints. PPOA algorithms are broadly classified into classical and heuristic categories. The following sections discuss different nature-inspired PPOA algorithms, highlighting their strengths, weaknesses, and future potential.

2.1 Classical Algorithms

Classical algorithms, or Geometric Planning Algorithms (GPAs), rely on geometry and consider robot kinematics and dynamics, producing fewer smooth paths [112]. The GPA includes roadmaps, cell decomposition, and APF algorithms. Despite their utility, classical algorithms face several challenges:

- High computation time and complexity for online applications [6, 33, 46]
- APF's assumption of a single repulsion force often fails in narrow paths [71]
- Exponential computational complexity for higher dimensions
- Tendency to become trapped in local minima, missing global information [44, 62, 69]
- Decreased efficiency and increased complexity with more intricate obstacles [112]
- High costs for nodes surrounded by obstacles in Dijkstra's algorithm (Tan and Sloman 2017)
- Suboptimal solutions from sampling-based algorithms [112]
- Infeasibility and inefficiency of roadmaps for complex networks [69]
- Unrealistic computational demands of a Probabilistic Road Map (PRM) [108]
- Suboptimal solutions and potential local minima in RRTs [86]

2.2 Nature-Inspired Algorithms

Recently, interest in nature-inspired PPOA algorithms has increased, particularly because of their effectiveness in dynamic and uncertain environments [72]. These algorithms draw inspiration from a wide range of organisms, ranging from microorganisms such as bacteria and slime molds to large animals such as whales and lions, which are selected on the basis of their complexity and specific application domain (Sierakowski and Coelho 2006) [4]. For example, ABCO mimics the collaborative behavior of bees, PSO simulates fish schooling and bird flocking, the FA replicates firefly mating, HHO emulates the foraging tactics of Harris hawks, and the GWO model's social leadership dynamics among gray wolves [45]. Simulation studies of nature-inspired PPOA algorithms have demonstrated promising results and significant potential in real-time applications, highlighting their ability to adapt and respond to challenges akin to expert systems [45]. These algorithms are further categorized into evolutionary algorithms, SI-based algorithms, bioinspired algorithms without SI, and those inspired by human activities (see Fig. 3). An analysis of some of the examples of PPOA problems evaluated on the basis of biological groups is presented in Table 1.

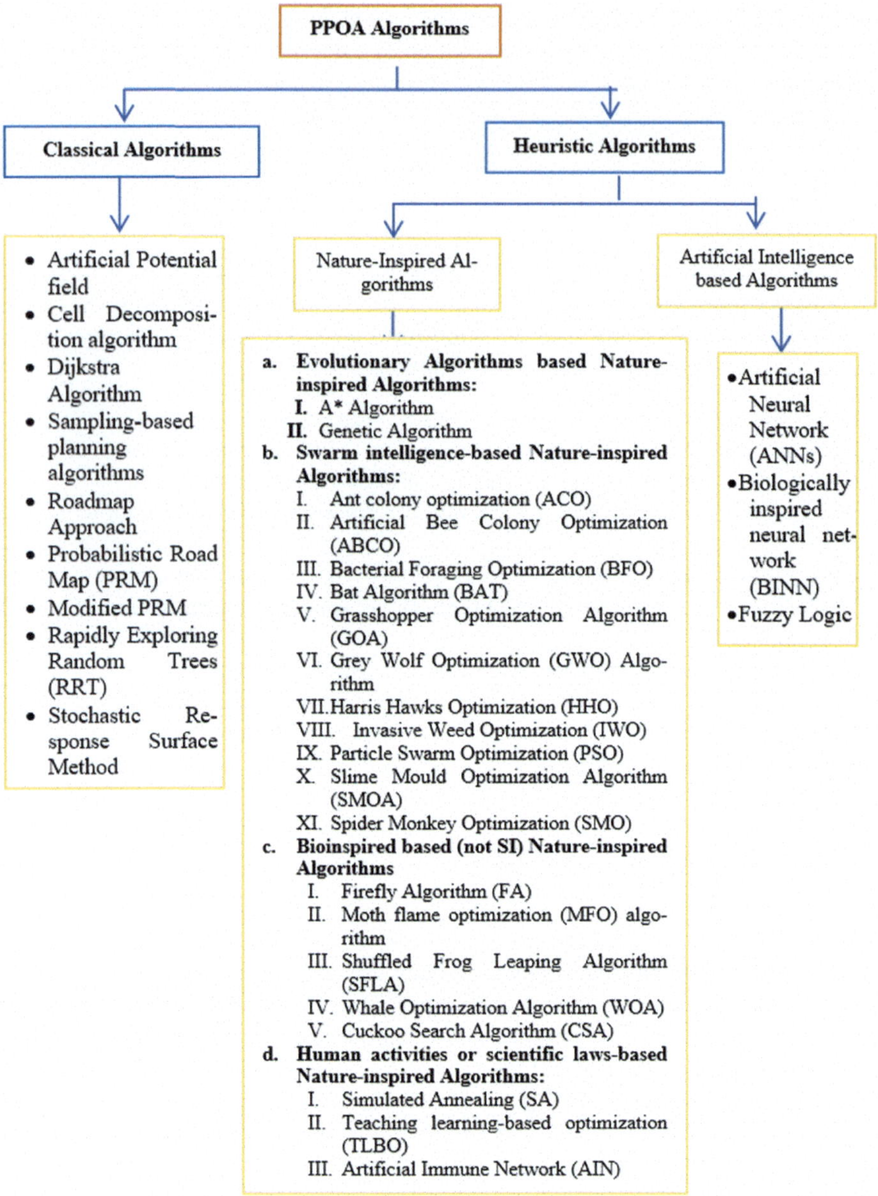

Fig. 3 Types of PPOA algorithms based on application

Table 1 Analysis of some of the examples of PPOA problems evaluated on the basis of biological groups

Biological groups	NIE approaches	Working principle	Advantages	Gaps/future scope
Based on microorganisms	Bacterial Foraging Algorithm (BFO) [1, 16, 48], Hussain and Ferdous (2015)	Chemotactic behavior of bacteria [75]]	Reduced effort, faster convergence, handles multiple objectives and nonlinear optimization [48, 75]	Vulnerable to poor exploitation capabilities [16]; gets trapped in local minima [1]
	Invasive Weed Optimization (IWO) [29, 56, 63, 94]	Weeds colonization and distribution [56]	Robust, efficient, reduces computation time, outperforms BFO [29, 56, 63]	Premature convergence mitigated by integrating with PSO [94]
Based on insects	Bat Optimization (BO) Algorithm [25, 26, 35]	Echolocation behavior of micro-bats [105]	Optimal in harsh conditions; simple, few parameters, fast convergence [60]; high accuracy	Premature convergence; modified to binary BO for sequential problems
	Firefly Optimization (FO) [24, 26, 44, 49, 78]	Bioluminescence and mating behavior of firefly insects [96]	Flexible, adaptive algorithm ensuring safe paths with minimal computation	Struggles with light and attraction variability in multi-objective contexts
	Grasshopper Optimization Algorithm (GOA) [21, 51, 67, 83]	Swarming behavior of grasshoppers [83]	Provides accurate, robust solutions for constrained and non-constrained optimizations; strong local search capability	Slow convergence; poor time complexity. Future: explore multi-objective optimization
Based on aquatic life	Whale Optimization Algorithm (WOA) [15, 58]	Bubble-net foraging behavior of humpback whales [58]	Optimal results for both constraint and non-constraint optimization	

(continued)

Table 1 (continued)

Biological groups	NIE approaches	Working principle	Advantages	Gaps/future scope
Based on birds	Cuckoo Search Optimization (CSO) [53, 64, 98]	Brood reproductive strategy [100, 103]	Efficient in unknown environments; fast convergence; minimal error; handles nonlinearities	Its control parameters are kept constant which decreases its efficiency
Based on animals and humans	Grey Wolf Optimization (GWO) [60, 69, 79, 99]	Hunting behavior and social leadership of gray wolves [60, 69]	High accuracy, simple implementation, efficient energy use, fast chaotic convergence	Early convergence, local optima issue, exploration more than exploitation
	Spider Monkey Optimization (SMO) [3, 5, 8, 30, 43]	Fission fusion social structure [5, 8]	Self-division, faster computation, high convergence, parallelism, avoids local minima	RPP problem improvement potential due to underutilization of this algorithm

I. **Microorganism-based—Invasive Weed Optimization (IWO)**: Inspired by the robust nature of weeds, which thrive in hostile environments, the IWO algorithm effectively navigates through unfamiliar terrains while avoiding obstacles [29, 94]. It leverages the reproductive strategy of weeds to optimize robot path planning, ensuring efficient collision avoidance and target reaching [73].
II. **Aquatic—Whale Optimization Algorithm (WOA)**: Modeled after humpback whales' foraging techniques, the WOA balances exploration and exploitation phases to optimize solutions in single constraint scenarios [58]. It employs a unique bubble-net feeding strategy to enhance the search for optimal solutions and has been applied successfully in robot path planning [15].
III. **Insects—Moth Flame Optimization (MFO) Algorithm**: Inspired by moths' phototactic behavior and Levy flights, the MFO algorithm efficiently explores search spaces and has demonstrated superior performance in heuristic optimization tasks [57, 95]. Its flight path mimics the spiral movements of moths toward light sources, enabling robust solution finding [57].
IV. **Mammal—Grey Wolf Optimization (GWO) Algorithm**: Drawing from the social behaviors of gray wolves during hunting, the GWO algorithm optimizes search processes by iteratively updating wolf positions to locate optimal solutions while avoiding local minima [69, 99]. It has proven effective in various applications, such as robot trajectory planning and path optimization [79].

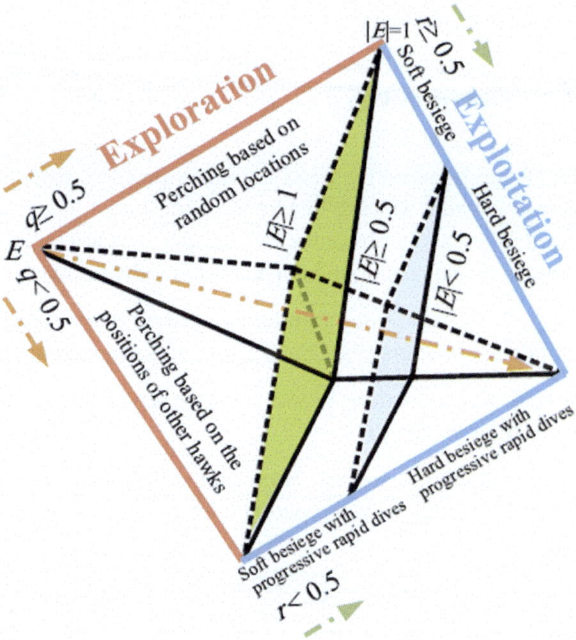

Fig. 4 Different phases of the HHO algorithm [34]

V. **Birds—Harris Hawks Optimization (HHO) Algorithm**: The HHO algorithm, inspired by Harris hawks' cooperative foraging strategies, is versatile for optimization tasks and avoids algorithm-specific parameters [31, 34]]. It employs a group attack strategy to explore for globally optimal solutions, minimizing local optima and computation time [40]. Updating to the global and local best positions per iteration ensures superior performance, maintaining stability between exploration and exploitation [20]. Standard HHO may encounter local minima, prompting the proposal of a quasi-reflected variant for better accuracy and convergence [23]. Different phases of HHO are shown in Fig. 4.

These algorithms showcase how nature-inspired approaches can address complex optimization challenges across various domains, promising efficient solutions with minimal computational overhead and high adaptability to changing environments.

3 Methodology

3.1 Invasive Weed Optimization (IWO)

IWOs are inspired by weed colonization [56]. Robots replicate weed reproduction across terrains, storing locations of the fittest and weakest seeds in their programming. They avoid obstacles and navigate toward targets. Figures 5 and 6 illustrate its flowchart and pseudocode. The IWO mathematical model includes the following:

Step 1: Initialization: Random distribution of weeds/robots, iteration steps, and control parameters [94]. Each weed initializes a path from start to goal [63].

Step 2: Reproduction: Weeds produce seeds on the basis of fitness, with more seeds from fitter weeds [63].

Step 3: Spatial dispersal: Seeds spread near parent weeds, enhancing local search. Fitter seeds stay near the parent, and weaker seeds are eliminated [94] (Eq. 1).

$$\sigma_{iter} = \left(\frac{iter_{max} - iter}{iter_{max}}\right)^n * \left(\sigma_{initial} - \sigma_{final}\right) + \sigma_{final} \tag{1}$$

```
Initialize Source and Goal Position (SP, GP), iter_max
Current Position (curr = SP)
while (curr = GP) do
  Robot proceeds towards the Goal (GP)
  if (obstructed by obstacles)
    Initial Population of the weeds is generated (N solution)
      for iter = 1 to itermax
          Calculate fitness f(k), for each weed
          Evaluate fitness_max and fitness_min in colony
            for each weed (w ε N)
                Find the number of fittest seeds of w,
                Use Random Distribution for dispersal of seeds around parents,
                Include the seeds produced to solution (N).
            end for;
            if (N > Nmax) then;
              Sort (N) in decreasing order of fitness, f(k).
              Eliminate weed, w with lower f(k)
              Update solution (N)
            end if
      next iter;
      end for;
  end if;
  else if curr = GP then;
          Robot reached the Goal
End
```

Fig. 5 Pseudocode for IWO [73]

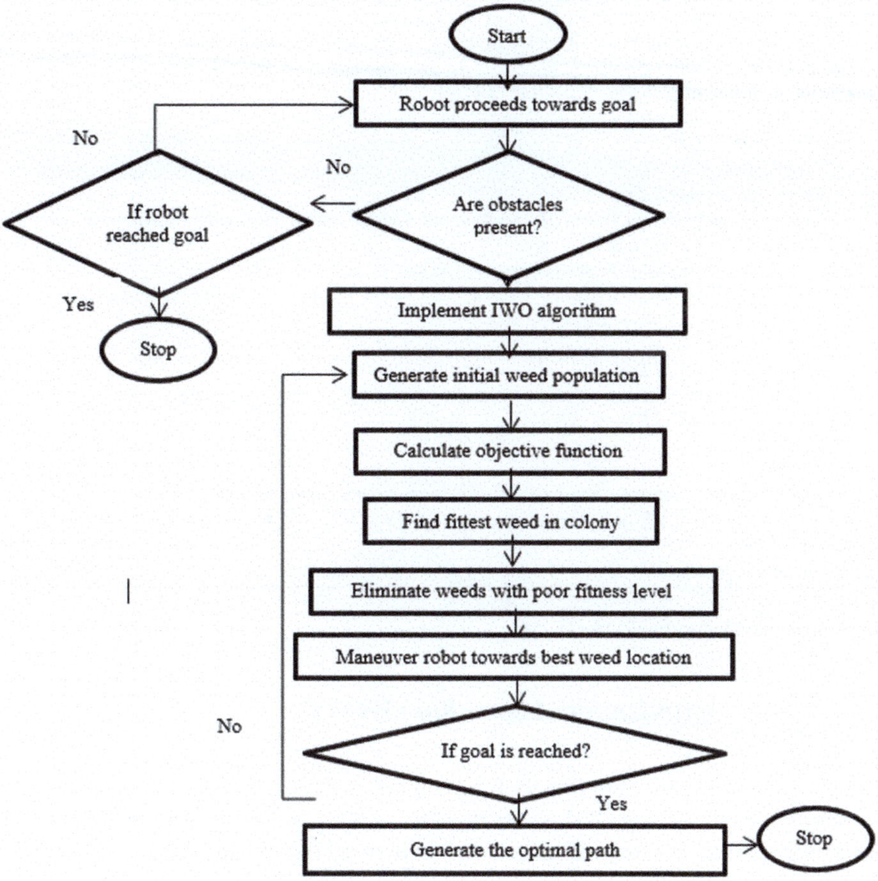

Fig. 6 Flowchart architecture of IWO [85]

where iter and $iter_{max}$ represent the current iteration and maximum iteration count, respectively. σ_{final} and $\sigma_{initial}$ are the final and initial standard deviations, respectively, and n is a nonlinear modulation index [63].

Step 4 *Competitive exclusion*: This means that the reproduction and mutation of fitter seeds stop when the iteration limit is reached. This phase justifies Darwin's principle of survival of the fittest [94].

```
Initialize the whales population Xi = (i = 1,2 ... ... n)
Calculate the fitness of each search agent
X * = the best search agent
while (t < maximum number of iterations)
   for each search agent
   Update a, A, C, I, p
      if 1 (p < 0.5)
         if 2 (|A| < I)
         Update the position of the current search agent by the equation 12.1
         else if 2 (|A| ≥ I)
         Select a random search agent (Xrand)
         Update the position of the current search agent by following equation
          X(t + 1) = Xrand(t) - AD
         end if 2
      else if 1 (p ≥ 0.5)
      Update the position of the current search by following equation
X(t + 1) = Debt cos (2Πt) + X * (t)
      end if
   end for
Check if any search agent goes beyond the search space and amend it
Calculate the fitness of each search agent and Update X * if there is a better solution
T = t + 1
end while and return X*
```

Fig. 7 Pseudocode for the WOA [15, 58]

3.2 *Whale Optimization Algorithm (WOA)*

Humpback whales use bubble nets for foraging, forming bubble circles or spirals to hunt [58]. Figures 7 and 8 show its flowchart and pseudocode. The mathematical model is discussed, starting with prey encirclement equations [4]. (Eqs. 2–6) as **step 1.**

$$\vec{D} = |\vec{C}\vec{X}p(t) - \vec{X}(t)| \tag{2}$$

$$\vec{X}(t+1) = \vec{X}p(t) - \vec{A}.\vec{D} \tag{3}$$

$$A = 2\vec{a}\,\vec{r_1} - \vec{a} \tag{4}$$

$$C = 2\vec{r}_2 \tag{5}$$

$$a = 2 - \frac{2.t}{t_{max}} \tag{6}$$

where A and C = constants; D = the distance between $\vec{X}p(t)$ and $\vec{X}(t)$ at the tth iteration; t = the current iteration count; t_{max} = the maximum iteration count; $\vec{X}p(t)$

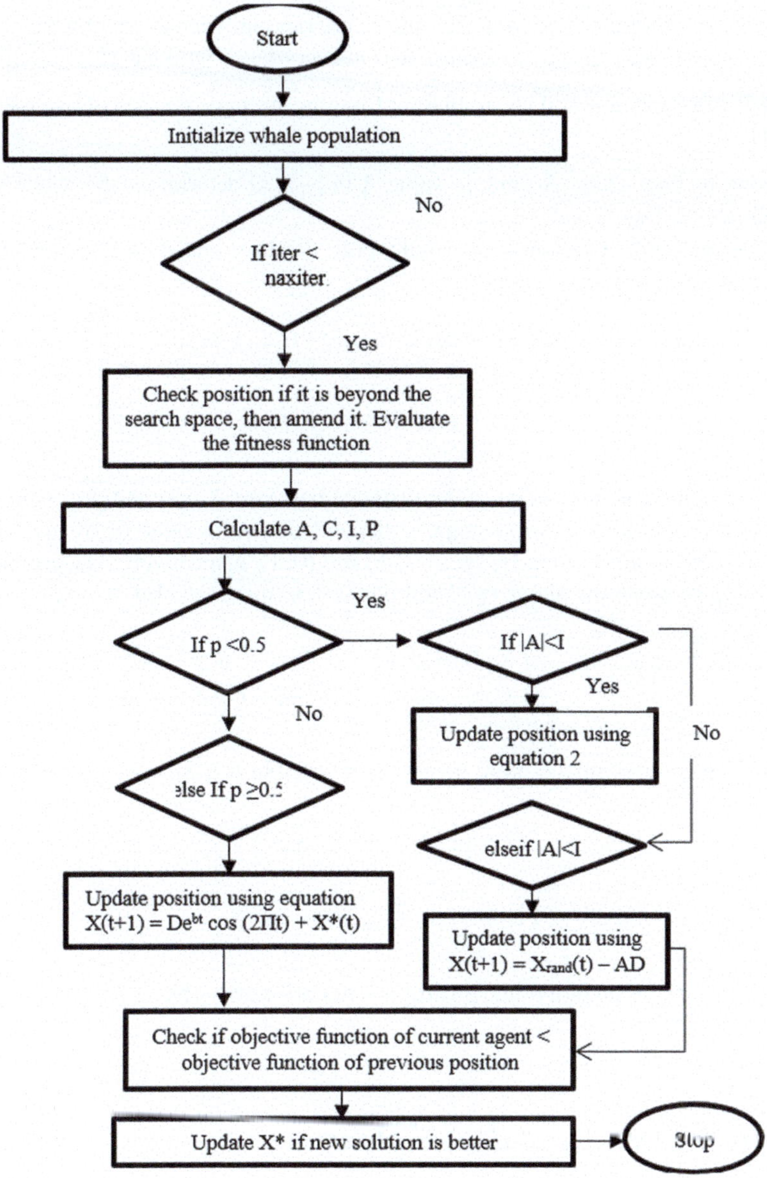

Fig. 8 Flowchart architecture of the WOA [61]

= the position vector of the whale; r_1 and r_2 are random vectors ε (0,1); and a is an element whose value decreases from 2 to 0 in accordance with Eq. (6).

Step 2: Bubble-net attack: The bubble-net foraging strategy consists of two methods [58, 61]:

a. Shrinking encircling: A random value A in [−a, a] determines the whale's new position via Eqs. 3 and 4.
b. Spiral updating: Evaluates the whale–prey distance and then simulates helix-shaped movement (Eq. (7) [15].

$$\overrightarrow{X}(t+1) = \overrightarrow{D}'.e^{bl}.\cos(2\pi l) + \overrightarrow{X}p(t) \tag{7}$$

$$\overrightarrow{D}' = |\overrightarrow{X}p(t) - \overrightarrow{X}(t)| \tag{8}$$

where $\overrightarrow{D}'$ = the absolute value of the distance between $\overrightarrow{X}p(t)$ and $\overrightarrow{X}(t)$ at the tth iteration; the logarithmic spiral shape is dependent on the value of b; and l and p denote random numbers in the ranges (-1, 1) and (0, 1), respectively. The probability of choosing between the above two methods is determined via Eq. 9 [61].

$$\vec{X}(t+1) = \begin{cases} \vec{X}p(t) - \vec{A}.\vec{D} & if\ p < 0.5 \\ \vec{D}'.e^{bl}.\cos(2\pi l) + \vec{X}p(t) & if\ p \geq 0.5 \end{cases} \tag{9}$$

Step 3: Whales randomly explore for prey via variable A [58].

$$\vec{D} = \left|\vec{C}\vec{X}rand\,(t) - \vec{X}(t)\right| \tag{10}$$

$$\vec{X}(t+1) = \vec{X}rand\,(t) - \vec{A}.\vec{D} \tag{11}$$

where $\overrightarrow{X}rand(t)$ is an individual position vector arbitrarily chosen from the existing population.

3.3 *Moth Flame Optimization (MFO)*

The MFO algorithm is inspired by moths navigating by moonlight to find prey [57]. Figures 9 and 10 display the pseudocode and flowchart, respectively. Moths represent candidate solutions, with their positions indicating problem variables [88].

Step 1: Initialize Population—MFO includes moths and flames, with the moth population arranged in a matrix (Eq. (12)) within a d-dimensional search space [91].

```
Initialize the parameters for Moth − flame
Initialize moth position Mi randomly
for i = 1 to n do
    Calculate the fitness function fi
end for
while iteration ≤ Max_iteration do
    Update the position of Mi
    Calculate the number of flames using equation 19
    Evaluate the fitness function fi
    if iteration == 1 then
      F = sort (M) and OF = sort (OM)
    else
      F = sort (Mt − 1, Mt) and OF = sort (Mt − 1, Mt)
    end if
    for i = 1 to n do
      for j = 1 to n do
          Update the values of r and t
          Calculate the value of D respect to its corresponding moth using Di = |Fj − Mi|
          Update M(i, j) respect to its corresponding moth using equation 16
      end for
    end for
end while
```

Fig. 9 Pseudocode for MFO [95]

$$M = \begin{bmatrix} m_{11} & m_{12} & \cdots\cdots & m_{1d} \\ m_{21} & m_{22} & \cdots\cdots & m_{2d} \\ \vdots & \vdots & \vdots \quad \vdots & \vdots \\ m_{n1} & m_{n2} & \cdots\cdots & m_{nd} \end{bmatrix} \tag{12}$$

The fitness function values are in an array (Eq. 13). The flames' population and fitness values initialize in a matrix [88]. In MFO, moths are search agents, and flames are optimal solutions. Moths track flames to retain the best solutions [57].

$$OM = \begin{bmatrix} OM_1 \\ OM_2 \\ \vdots \\ OM_n \end{bmatrix} \tag{13}$$

$$F = \begin{bmatrix} F_{11} & F_{12} & \cdots\cdots & F_{1d} \\ F_{21} & F_{22} & \cdots\cdots & F_{2d} \\ \vdots & \vdots & \vdots \quad \vdots & \vdots \\ F_{n1} & F_{n2} & \cdots\cdots & F_{nd} \end{bmatrix} \tag{14}$$

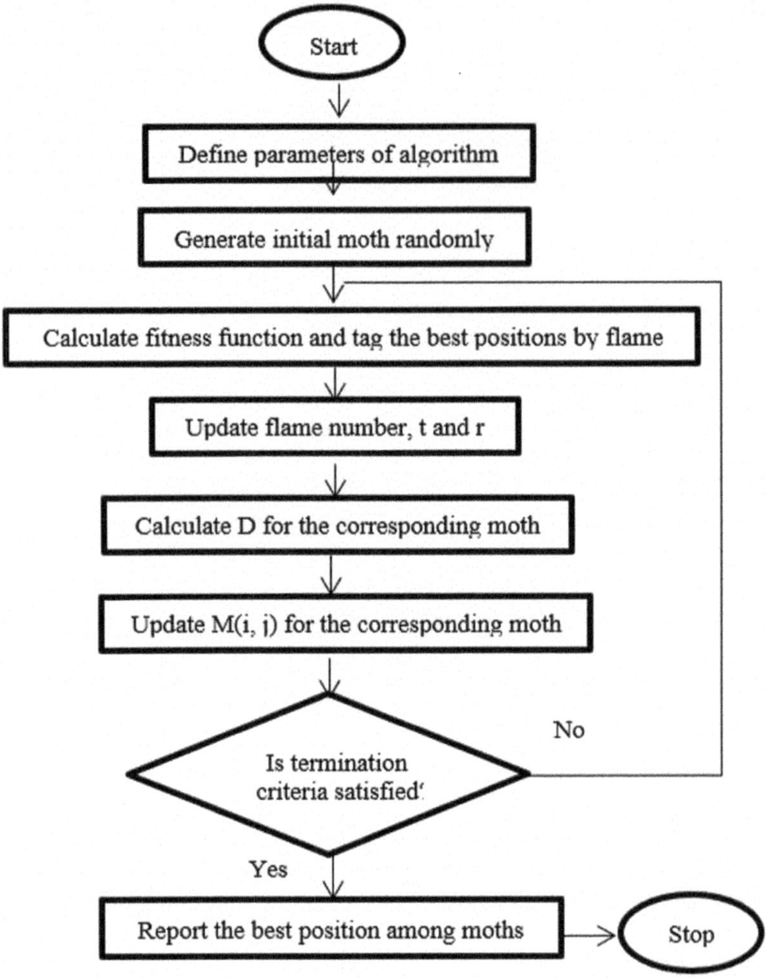

Fig. 10 Flowchart architecture of MFO [88]

$$OF = \begin{bmatrix} OF_1 \\ OF_2 \\ \vdots \\ OF_n \end{bmatrix} \tag{15}$$

Step 2: Update locations—The initial moth population (M) is randomly set in a D-dimensional search space (Eq. 16) [95].

$$M(i, j) = (ub(i) - lb(j)) * rand(0,1) + lb(i) \tag{16}$$

where *lb* and *ub* indicate the lower and upper bounds of the variables, respectively.

Step 3: **Logarithmic spiral—**The spiral starts at the moth's position and ends at the flame's location, varying within the solution limits (Eq. 17). D_i denotes the distance between the ith moth and the jth flame, b shapes the spiral, and t is a random number [−1, 1].

$$S(M_i F_j) = D_i * e^{bt} * \cos(2\pi t) + F_j \tag{17}$$

$$D_i = |F_j - M_i| \tag{18}$$

Step 4: Update flames—Reducing the number of flames enhances solution exploitation as moth positions update [88].

$$\mathit{flame\ no} = round\left(N - l * \frac{N - l}{T}\right) \tag{19}$$

where N is the maximum flame number; l is the current iteration; and T is the maximum iteration count.

3.4 Grey Wolf Optimization (GWO)

The GWO algorithm mimics gray wolves' group foraging behavior [109]. Wolves are categorized as alpha (α) (leader), beta (β) (helper), delta (δ) (patroller), or omega (ω) (followers) [81]. The positions of α, β, and δ are adjusted until ω's position is optimal. Figures 11 and 12 show the pseudocode and flowchart, respectively.

Step 1: Encircling prey involves identifying and encircling prey via specific equations [39, 69] Here, t is the current iteration, A and C are constants, r_1 and r_2 are

```
Initialize the Grey wolf population Xi (I = 1,2, .... n) and algorithmic parameters
Evaluate the fitness of each search agent fiti
Initialize the Xα, Xβ, Xδ as first, second and third best solutions respectively
While (k < maximum number of iteration or stop criteria obtained)
    For i = 1: n
        Update the current search agent position using equation 20 − 25
    End for
    Evaluate the fitness fiti
    Update the coefficient vector α, A and C
    If any better solution is obtained then update the best agent Xα, Xβ, Xδ
    k = k + 1
End while
    Stop the process and visualize the first best agent Xα found so far
```

Fig. 11 Pseudo Code for GWO [39]

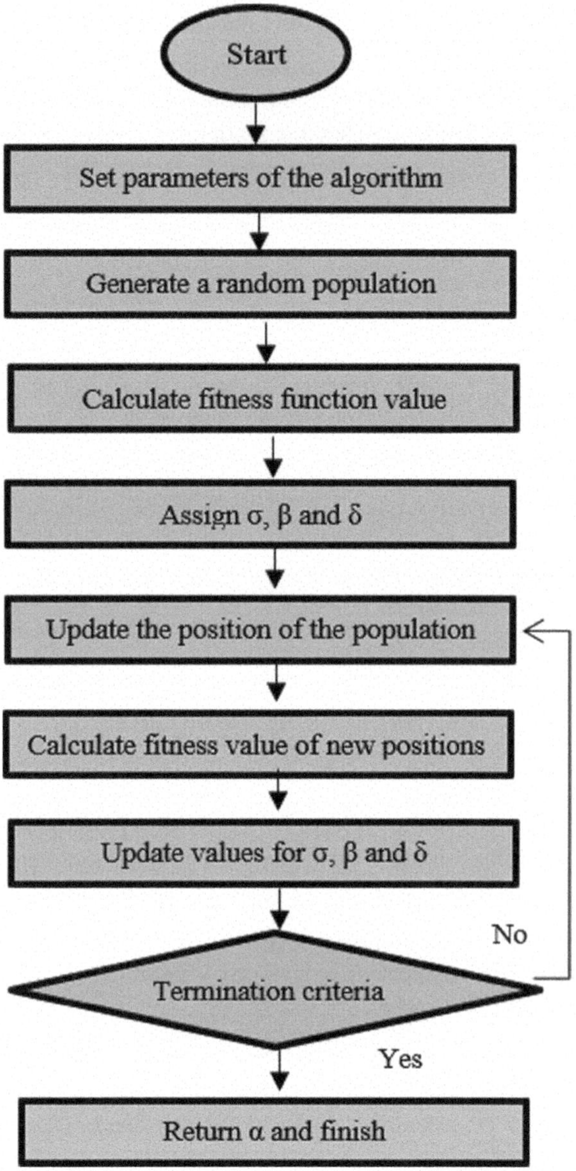

Fig. 12 Flowchart of GWO [41]

random variables, X_p and X are the prey and wolf positions, respectively, and MaxIter is the maximum number of iterations.

$$D = |CXp - AX(t)| \tag{20}$$

$$X(t+1) = Xp(t) - AD \tag{21}$$

$$A = 2 * a * r_1 - a \tag{22}$$

$$C = 2 * r_2 \tag{23}$$

$$a = 2 - \frac{2t}{MaxIter} \tag{24}$$

Step 2: Exploration—Determines the positions of wolves α, β, and δ wolves. In this phase, A ranges from −1 to 1 [93], and C ranges from 0 to 2 [39].

Step 3: Exploitation—Attacking prey. The top three solutions are memorized, and other wolves' positions are updated on the basis of these. Equations 25–27 guide this process, with values ranging from 2 to 0 [52].

$$D_\alpha = |C_1 X_\alpha - X|, D_\beta = \left|C_2 X_\beta - X\right|, D_\delta = |C_1 X_\delta - X| \tag{25}$$

$$X_1 = X_\alpha - A_1 D_\alpha, X_2 = X_\beta - A_2 D_\beta, X_3 = X_\delta - A_3 D_\delta \tag{26}$$

$$X(t+1) = \frac{X_1 + X_2 + X_3}{3} \tag{27}$$

3.5 Harris Hawks Optimization (HHO)—Proposed Algorithm

The HHO algorithm, a gradient-free NISI-based method, can address various optimization problems with proper formulations [2]. Initially, a group of hawks is deployed to find an optimal solution via seven killing strategies. The leader hawk attempts to catch prey, with six additional strategies employed if it fails [106]. This collaborative approach confuses and weakens prey. The HHO algorithm's mathematical model for PPOA problems in AMRs consists of exploration and exploitation phases [34]. The symbols are defined in Table 2, and Figs. 13 and 14 present the pseudocode and flowchart.

Step 1: Exploration phase—Hawks represent candidate solutions, which are iterated to update their best positions until an optimal solution is found. Hawks are randomly initialized, and two tactics are used to spot prey (goal). If $q < 0.5$, hawks group closely to attack; if $q \geq 0.5$, hawks spread randomly within [LB, UB] to capture prey [31, 34]. (Eq. 28)

$$X_i(t) = unifrnd(\mathrm{LB},\ \mathrm{UB},\ 1,\ \mathrm{N}) \tag{28}$$

Table 2 Definition of symbols

S. no	Symbols	Definition
1	$X(t+1)$	Position vector for tth iteration
2	$X_{prey}(t)$	Position of prey animal
3	$X(t)$	Position vector at ith iteration
4	$r_1, r_2, r_3, r_4, r_5, u, v$	Random values within 0 to1 interval
5	q	Equal probability of perching tactics (strategy)
6	UB, LB	Upper bound, Lower Bound
7	$X_{rand}(t)$	Randomly selected hawk from current population at tth iteration
8	$X_m(t)$	Average position of the current population of hawks
9	$X_i(t)$	Each hawks' position at tth iteration
10	N	Total population of hawks
11	E	Escaping strength of the prey
12	Max_Iter	Maximum number of iterations
13	t	Current iteration
14	E_0	Initial state of its energy
15	r	Probability of escaping of prey ($r < 0.5$ = probability of escaping while $r \geq 0.5$ probability of not escaping)
16	$\Delta X(t)$	Difference between position vector and current position of prey in tth iteration
17	J	Represents random leaping behavior of prey
18	Popsize	Problem dimension
19	S	Random vector of size $1 \times D$
20	β	Constant value = 1.5
21	LF	Levy flight function

```
Inputs: Define the population number or size (N) and number of Iteration (T)
Outputs: the location of the rabbit and its fitness value
Initialize: Start within the random point in population Xi (i = 0,1,2, .....N)
while (stopping condition is not met
Calculate the fitness values of hawks
Set or define Xrabbit as location of rabbit (best location)
  for (each hawk (Xi))(do)
Update the starting energy (E0 = 2rand() − 1 and jumping force or strength (J
                    = 2(1 − rand())
                         Escaping Energy, E = 2E0 (1 − t/T)  Update the E
        if (|E ≥ 1|)       // Exploration Phase; Update position using eq. 29
        if (|E < 1|)              // Exploitation Phase
            if (r ≥ 0.5 and |E ≥ 0.5|)       //Update position using eq. 33 − 35
            else if (r ≥ 0.5 and |E < 0.5|)   // Update position using eq. 36
            else if (r < 0.5 and |E ≥ 0.5|)  //Update position using eq. 37 − 41
            else if (r < 0.5 and |E < 0.5|)  //Update position using eq. 42 − 44
End while   and   Return Xrabbit
```

Fig. 13 Pseudocode for HHO [31, 34]

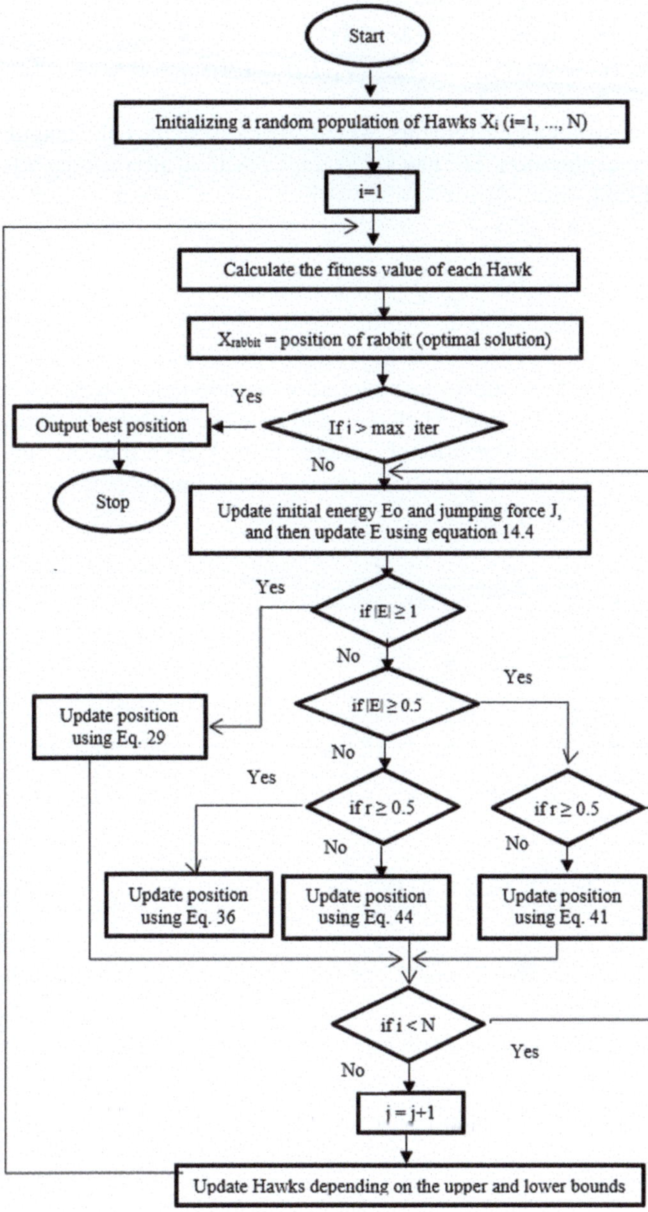

Fig. 14 Flowchart architecture of HHO [66]

$$X(t+1) = \begin{cases} X_{rand}(t) - r_1|X_{rand}(t) - 2r_2X(t)| & q \geq 0.5 \\ (X_{prey}(t) - X_m(t)) - r_3(LB + r_4(UB - LB)) & q \geq 0.5 \end{cases} \quad (29)$$

Equation (29) generates random movements, which are added to the LB's length, enhancing solution space exploration. The hawks' average position, after adding this coefficient, is in Eq. 30 [34].

$$X_m(t) = \frac{1}{N}\sum_{i=1}^{N} X_i(t) \quad (30)$$

Step 2: Switching phase: Hawks switch from exploration to exploitation on the basis of prey escape energy (E). One exploration behavior and four exploitation strategies are used, with E guiding the phase transition. E's value, from 1 to -1, indicates the prey energy level [31]. $|E| \geq 1$ means exploration, whereas $|E| < 1$ means exploitation [34].

$$E = 2E_0(1 - \frac{t}{Max\_Iter}) \quad (31)$$

$$E_0 = 2r_1 - 1 \quad (32)$$

Step 3: Exploitation phase—Hawks intensify attacks on prey via four strategies:

A. *Soft siege* ($r \geq 0.5$, $|E| \geq 0.5$): Prey leaps randomly, misleading hawks until exhausted (Eqs. 33–35) [31].

$$X(t+1) = \Delta X(t) - E\left|JX_{prey}(t) - X(t)\right| \quad (33)$$

$$\Delta X(t) = X_{prey}(t) - X(t) \quad (34)$$

$$J = 2 - (1 - r_5) \quad (35)$$

B. *Hard siege* means $r \geq 0.5$ and $|E| < 0.5$: Prey has little energy; hawks attack directly (Eq. 36) [31].

$$X(t+1) = X_{prey}(t) - E|\Delta X(t)| \quad (36)$$

C. *Soft besiege with the LF concept means* $|E| \geq 0.5$ *and* $r < 0.5$: Prey misleads hawks with zigzag moves, updated by the LF concept (Eqs. 37–41) [31].

$$Y = X_{prey}(t) - E\left|JX_{prey}(t) - X(t)\right| \quad (37)$$

$$Z = Y + S * LF(Popsize) \tag{38}$$

$$LF(x) = 0.1 * \frac{u * \sigma}{|v|^{\frac{1}{\beta}}} \tag{39}$$

$$\sigma = \left(\frac{\Gamma(1+\beta) * \sin\frac{\pi\beta}{2}}{\Gamma\left(\frac{1+\beta}{2}\right) * \beta * 2^{\left(\frac{\beta-1}{2}\right)}} \right)^{\frac{1}{\beta}} \tag{40}$$

$$X(t+1) = \begin{cases} YifF(Y) < F(X(t)) \\ ZifF(Z) < F(X(t)) \end{cases} \tag{41}$$

D. *Hardly besiege with continuous rapid moves* means **that** |E| <0.5 and r < 0.5: Prey is too weak to escape, and hawks capture it (Eqs. 42–44) [31].

$$Y = X_{prey}(t) - E\left|JX_{prey}(t) - X_m(t)\right| \tag{42}$$

$$Z = Y + S * LF(Popsize) \tag{43}$$

$$X(t+1) = \begin{cases} Y\ if\ F(Y) < F(X(t)) \\ Z\ if\ F(Z) < F(X(t)) \end{cases} \tag{44}$$

4 Experimental Simulation

4.1 Simulation Arrangement

This study applies the HHO algorithm to the PPOA problem in AMRs via five diverse environmental maps. The simulations were performed in MATLAB 2020a with the specifications in Table 3.

4.2 Experimental Setup

This section covers constraints and modeling of simulation environment. The PPOA problem assumptions in MATLAB include:

i. Map, with specified obstacles, is given with uniform path conditions
ii. The obstacles are circular and static.

Table 3 Specification of the operating system

S. no	Name		Setting
1	**Windows Edition**		Windows 10 Home Single Language
2	**System**	Processor	Intel ® Core ™ i5-4210U
3		Cache Frequency	2.40 GHz
4		Installed Memory (RAM)	12 GB
5		Hard drive	1 TB
6		System Type	64-bit Operating System, × 64-based processor
7		Pen and Touch	Touch support with 10 touch points
8	**Simulation Platform**		MATLAB 2020a

Table 4 Types of environments

S. no	No. of obstacles	Starting coordinates	Stopping coordinates
1	3	0, 0	4, 6
2	6	0, 0	10, 10
3	13	3, 3	14, 14
4	30	3, 3	14, 14
5	45	0,0	15, 15

iii. The starting and goal points are fixed, and the robot is point-sized.
iv. The map remains unchanged during the PPOA process when an offline method is used.

Experiments were conducted on a 2D grid (10–10 units along x- and y-axes). Various obstacle combinations were introduced for five different environments to test algorithm performance. Tables 4 and 5 detail the obstacle numbers, start/stop coordinates, and obstacle sizes for all the maps. The HHO algorithm's effectiveness for the PPOA problem was compared with those of the IWO, MFO, GWO, and WOA algorithms on the basis of path length and computation time.

4.3 Setting the Control Parameters

Each algorithm has initial parameters, known as control parameters. Fine-tuning these factors is crucial for optimal algorithm performance. Table 6 lists these values for the various algorithms discussed. For unbiased comparison, the specific control parameters were kept the same across all algorithms: 5 handling agents, 500 maximum iterations, and a population size of 150.

Table 5 Coordinates and sizes of obstacles

S. no	X-axis	Y-axis	Obstacle size
Map 1	[1, 1.8, 4.5],	[1, 5.0, 0.9]	[0.8, 1.5, 1]
Map 2	[1.5, 8.5, 3.2, 6.0, 1.2, 7.0]	[4.5, 6.5, 2.5, 3.5, 1.5, 8.0]	[1.5, 0.9, 0.4, 0.6, 0.8, 0.6]
Map 3	[1.5, 4.0, 1.2, 5.2, 9.5, 6.5, 10.8, 5.9, 3.4, 8.6, 11.6, 3.3, 11.8]	[4.5, 3.0, 1.5, 3.7, 10.3, 7.3, 6.3, 9.9, 5.6, 8.2, 8.6, 11.5, 11.5]	[0.5, 0.4, 0.4, 0.8, 0.7, 0.7, 0.7, 0.7, 0.7, 0.7, 0.7, 0.7, 0.7]
Map 4	[10.1, 10.6, 11.1, 11.6, 12.1, 11.2, 11.7, 12.2, 12.7, 13.2, 11.4, 11.9, 12.4, 12.9, 13.4, 8, 8.5, 9, 9.5, 10, 9.3, 9.8, 10.3, 10.8, 11.3, 5.9, 6.4, 6.9, 7.4, 7.9]	[8.8, 8.8, 8.8, 8.8, 8.8, 11.7, 11.7, 11.7, 11.7, 11.7, 9.3, 9.3, 9.3, 9.3, 9.3, 5.3, 5.3, 5.3, 5.3, 5.3, 6.7, 6.7, 6.7, 6.7, 6.7, 8.4, 8.4, 8.4, 8.4, 8.4]	[0.4, 0.4, 0.4, 0.4, 0.4, 0.4, 0.4, 0.4, 0.4, 0.4, 0.4, 0.4, 0.4, 0.4, 0.4, 0.4, 0.4, 0.4, 0.4, 0.4, 0.4, 0.4, 0.4, 0.4, 0.4, 0.4, 0.4, 0.4, 0.4, 0.4]
Map 5	[4, 4, 4, 4, 4, 4, 4, 4, 4, 6, 6, 6, 6, 6, 6, 6, 6, 6, 8, 8, 8, 8, 8, 8, 8, 8, 8, 10, 10, 10, 10, 10, 10, 10, 10, 10, 12, 12, 12, 12, 12, 14, 14, 14, 14]	[3, 3.5, 4, 4.5, 5, 5.5, 6, 6.5, 7, 8, 8.5, 9, 9.5, 10, 10.5, 11, 11.5, 12, 1, 1.5, 2, 2.5, 3, 3.4, 4, 4.5, 5, 6, 6.5, 7, 7.5, 8, 8.5, 9, 9.5, 10, 10, 10.5, 11, 11.5, 12, 10, 10.5, 11, 11.5]	[0.4, 0.4, 0.4, 0.4, 0.4, 0.4, 0.4, 0.4, 0.4, 0.4, 0.4, 0.4, 0.4, 0.4, 0.4, 0.4, 0.4, 0.4, 0.4, 0.4, 0.4, 0.4, 0.4, 0.4, 0.4, 0.4, 0.4, 0.4, 0.4, 0.4, 0.4, 0.4, 0.4, 0.4, 0.4, 0.4, 0.4, 0.4, 0.4, 0.4, 0.4, 0.4, 0.4, 0.4, 0.4]

4.4 *Objective Function*

Metaheuristic algorithms use an objective function to evaluate solutions. This study's objective function aims to minimize path length (PL) and obstacle collisions (D). It is defined as

$$OV(ObjectiveValue) = PL + PL * \gamma * D \tag{45}$$

Here, OV is the objective value, γ is a scaling factor, and D represents the average distance from the interpolant points to the obstacles. Minimizing the OV meets the dual goals of shortest path and obstacle avoidance.

4.5 *Simulation Results of the Proposed Algorithms*

Various PPOA algorithms cater to different applications, research problems, and operational modes. As complexities increase, algorithms incorporate specific parameters and assumptions. This study pioneers the implementation of the IWO, GWO, WOA, MFO, and HHO algorithms for PPOA in AMRs, aiming for globally optimal, collision-free paths with minimal computation time and path length. Implemented in

Table 6 Values of the control parameters

Sr. no	Algorithms	Parameter's values
Common Control Parameters		Number of Handling Points (n) = 5 Maximum no. of Iterations = 500 Number of Population = 150
1	IWO	Initial Population Size = 10 Minimum No. of Seeds = 0 Maximum No. of Seeds = 5 Variance Reduction Exponent = 2 Initial Value of Standard Deviation = 0.5 Final Value of Standard Deviation = 0.001
2	GWO	Reduction Exponent (a) = 2 (decreases from 2 to 0) Coefficient Vectors (A and C), where $\vec{A} = 2 * a * r_1 - a$, and $\vec{C} = 2 * r_2$ (where r_1 and r_2 are random numbers in range (0, 1))
3	MFO	Flame number = 5 Reduction Exponent = decreases from -1 to -2 Constant to define the shape of the logarithmic spiral = 1;
4	WOA	Reduction Exponent (a) = 2 (decreases from 2 to 0) Coefficient Vectors (A and C), where $\vec{A} = 2 * a * r_1 - a$, and $\vec{C} = 2 * r_2$ (where r_1 and r_2 are random numbers in range (0, 1)) Constant to define the shape of the logarithmic spiral = 1 Random number (l) = [-1, 1]
5	HHO	* no specific control parameters required

MATLAB 2020a, the algorithms were evaluated under standardized control parameters and conditions—start and finish points, obstacle numbers, and dimensions (Results in Tables 7 and 8). The HHO algorithm stands out for its parameterless nature, which reduces the computation time and uncertainty. The performance metrics included path length and processing time across static environments of varying complexity. The results, presented in the tables, highlight algorithmic differences in achieving optimal paths. HHO demonstrated superior performance over other algorithms, notably the WOA, in generating efficient paths swiftly (Figs. 15, 16, 17, 18, 19, 20, 21, 22, 23, 24, 25, 26, 27 and Tables 9 and 10).

Table 7 Summary of path lengths obtained after simulation via the PPOA algorithm in MATLAB 2020a

	IWO	MFO	GWO	WOA	HHO
MAP 1	**9.6557**	7.7352	9.6418	7.7325	7.4618
	11.3748	9.6534	7.7404	**7.4179**	**7.413**
	11.4005	**7.7325**	7.4647	7.7365	7.5107
	10.6178	9.6557	**7.7306**	7.7713	7.4179
	9.6606	9.6303	7.7325	7.732	7.5311
MAP 2	**14.3093**	14.3083	14.3078	14.3025	**14.2988**
	16.281	14.3137	14.3034	14.3008	14.3047
	15.8927	**14.3073**	**14.3**	14.302	14.3015
	15.2831	14.3293	14.3097	**14.2998**	14.303
	14.5695	14.3483	14.3024	14.3065	14.3021
MAP 3	16.9925*	18.7747*	17.5146*	17.6939	15.8663
	17.6116*	19.1936	18.7759*	16.1424	16.7866
	19.0375	**18.0796**	**18.0332**	**15.85**	**15.8522**
	22.7471	18.8139	19.1582	16.8695	15.9175
	17.2873*	16.7787*	18.8204	18.0116	16.9206
MAP 4	16.4568	16.2235	16.1977	16.1887	**16.1883**
	15.5633*	16.1985	16.1938	16.2648	16.1902
	16.2022	16.2305	**16.1922**	16.2015	16.1888
	16.1993	**16.1932**	16.2338	16.1889	16.1901
	16.3373	16.1958	16.2018	**16.1902**	16.1884
MAP 5	22.0251*	21.8838	**21.849**	21.8678	21.5559
	21.9556	21.8761	21.8561	21.8806	**21.526**
	21.5604*	21.9094	21.901	21.8358	21.6064
	22.0055*	**21.8296**	21.8769	21.6621	21.6677
	22.1971	21.5166*	21.8607	**21.6145**	21.5424

Table 8 Summary of computation times obtained after the simulation of the PPOA algorithms on MATLAB 2020a

	IWO	MFO	GWO	WOA	HHO
MAP 1	**115.44**	79.21	91.45	60.47	42.25
	120.08	90.77	79.77	**59.14**	**36.69**
	104.61	**81.42**	81.75	63.58	48.03
	119.11	89.36	**87.42**	68.33	39.14
	101.17	88.88	88.64	71.34	53.22
MAP 2	**118.94**	128.96	127.06	99.31	**81.16**
	116.44	137.55	114.36	95.89	86.02
	127.29	**129.82**	**121.12**	93.54	85.50
	132.71	130.85	122.53	**90.47**	82.13
	138.07	135.71	117.98	105.60	82.99
MAP 3	188.14	148.23	132.88	128.05	93.17
	175.76	141.87	140.07	136.56	93.44
	164.27	**128.15**	**127.70**	**122.24**	**92.67**
	191.55	197.15	132.91	140.40	94.88
	180.78	157.12	128.47	197.98	93.83
MAP 4	198.14	213.76	178.12	186.19	**120.05**
	176.27	175.67	177.47	177.11	123.79
	185.66	292.71	**181.16**	174.12	122.47
	169.21	**183.81**	184.95	191.32	124.85
	188.05	231.81	186.80	**176.28**	124.82
MAP 5	221.95	258.12	**218.86**	201.60	149.73
	234.61	265.44	221.70	214.32	**142.26**
	229.70	280.08	274.32	203.91	156.47
	287.11	**277.39**	251.33	214.64	161.21
	273.59	299.18	264.47	**199.86**	148.01

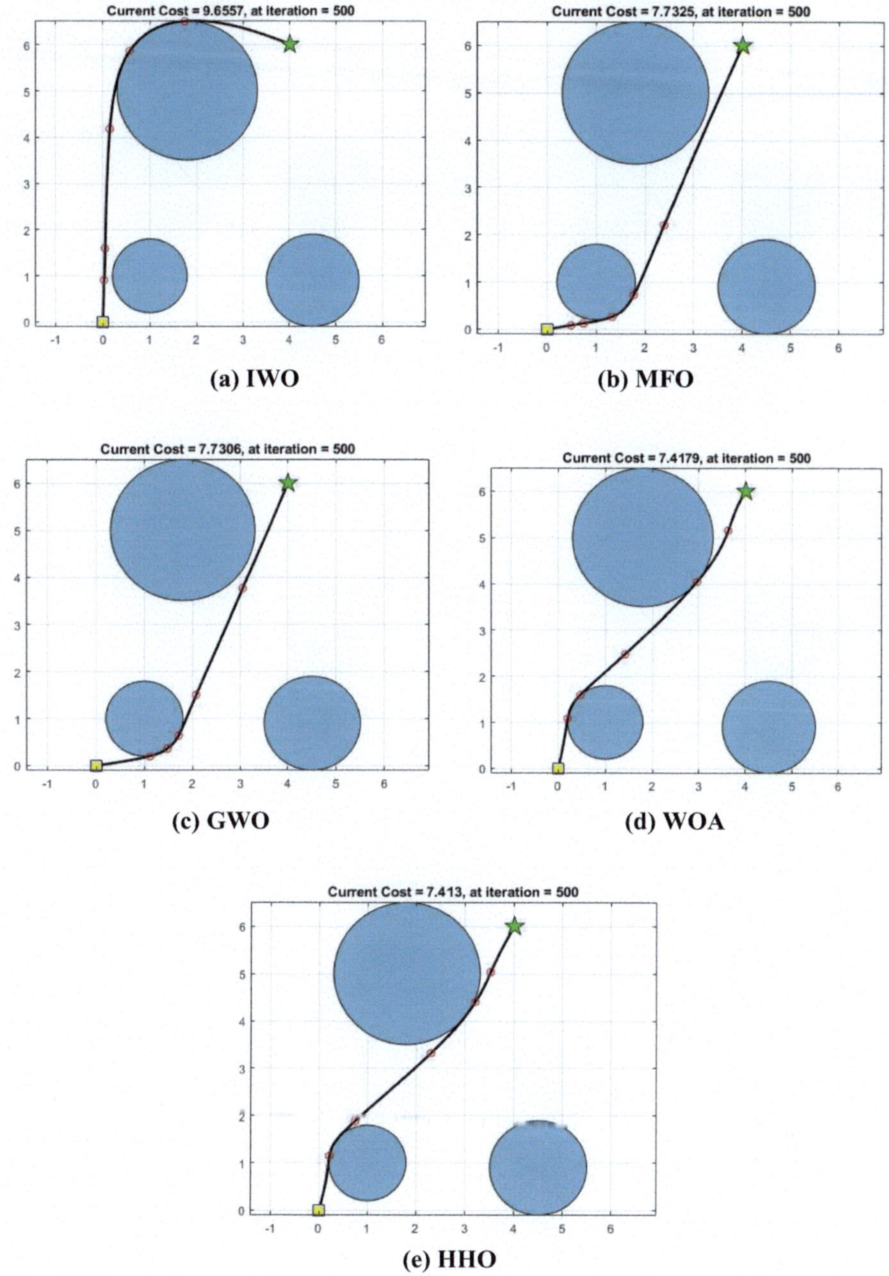

Fig. 15 Obstacle Map 1 for all algorithms in comparison **a** IWO **b** MFO **c** GWO **d** WOA **e** HHO

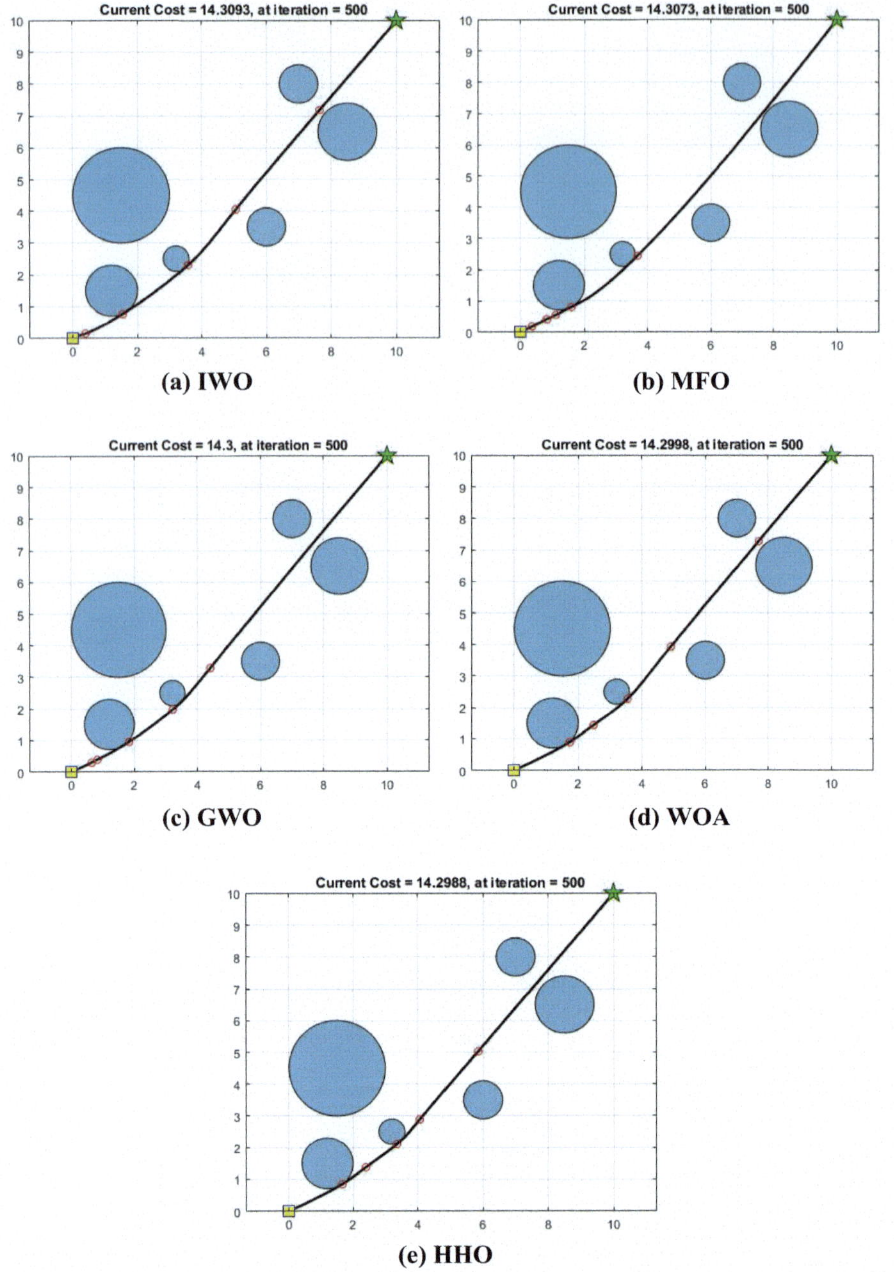

Fig. 16 Obstacle Map 2 for all algorithms in comparison **a** IWO **b** MFO **c** GWO **d** WOA **e** HHO

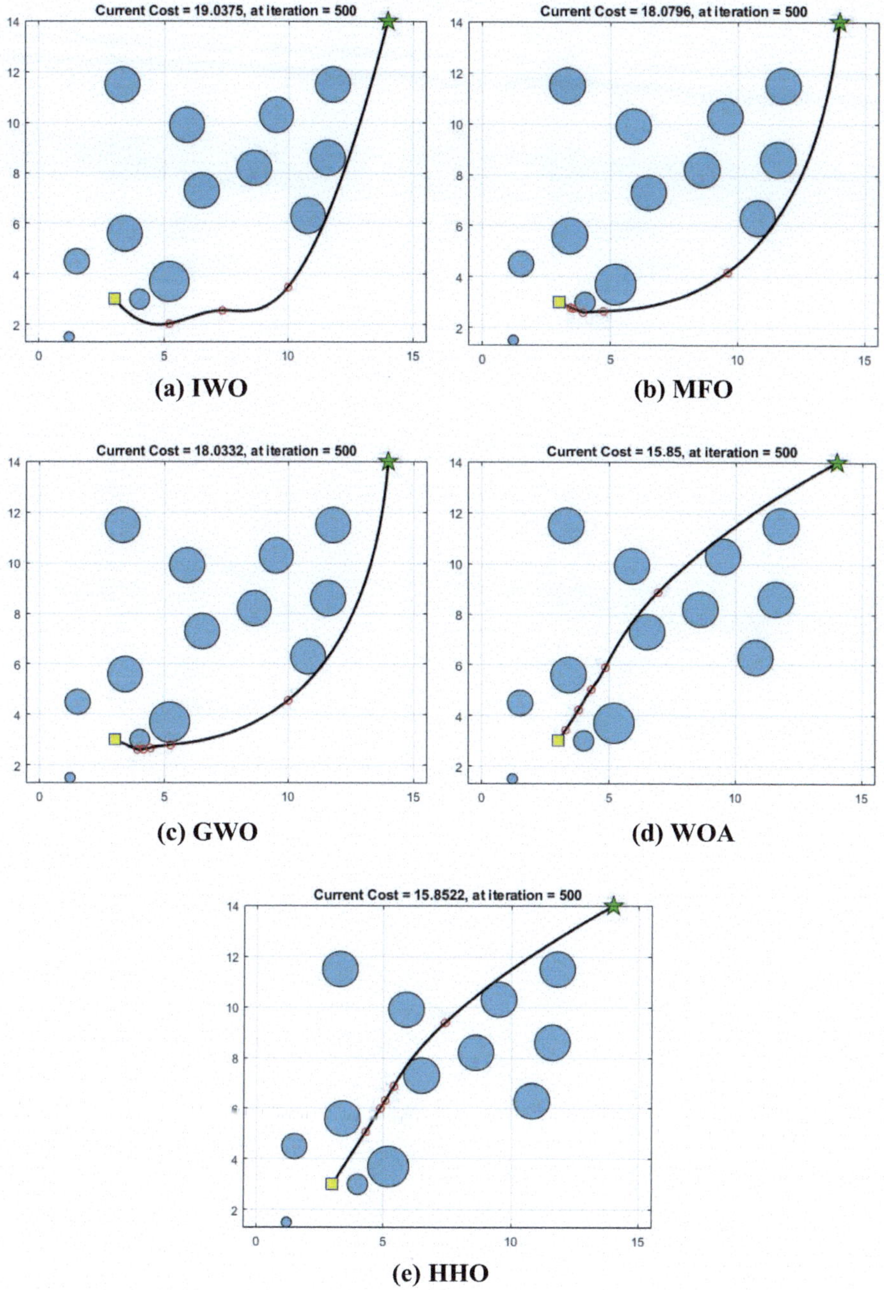

Fig. 17 Obstacle Map 3 for all algorithms in comparison **a** IWO **b** MFO **c** GWO **d** WOA **e** HHO

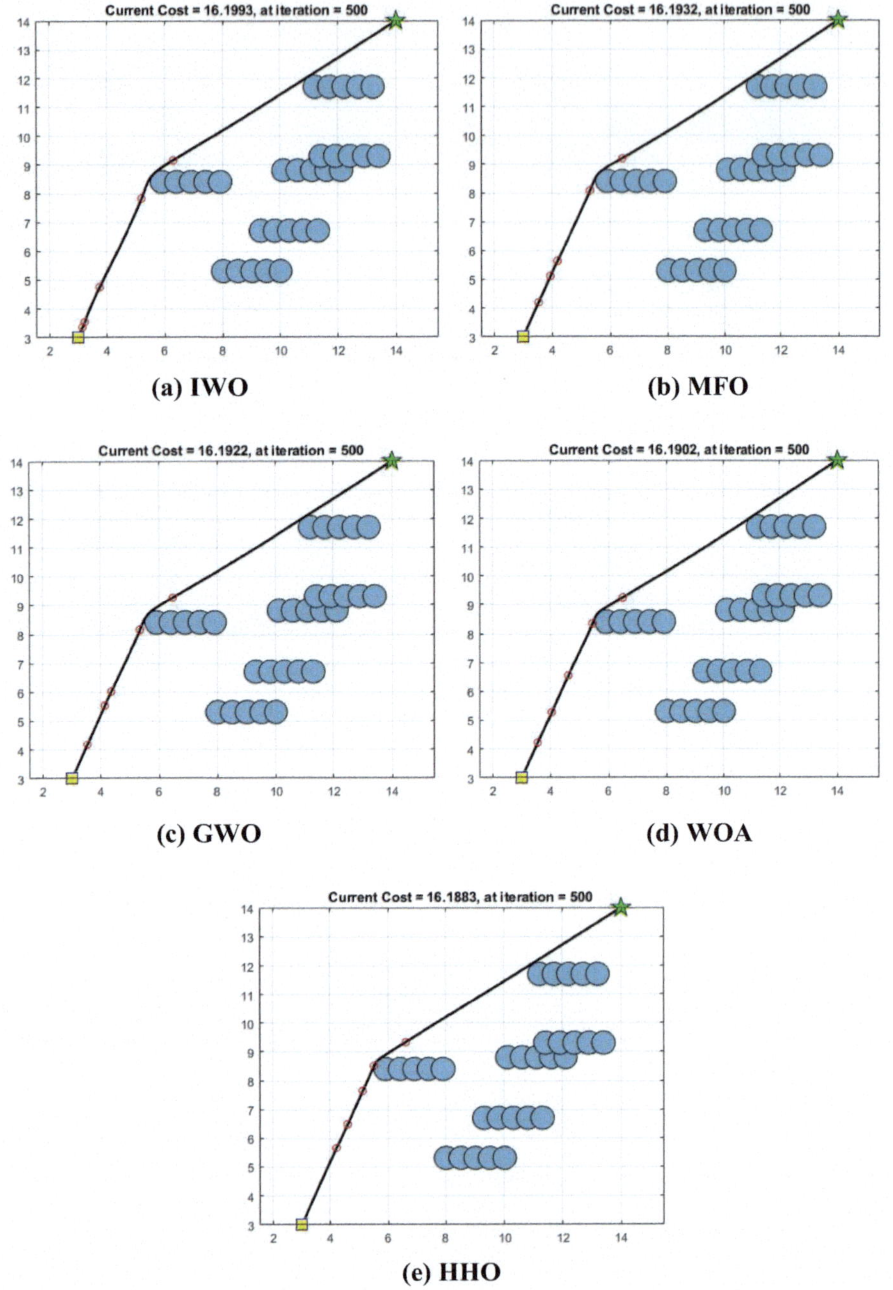

Fig. 18 Obstacle Map 4 for all algorithms in comparison **a** IWO **b** MFO **c** GWO **d** WOA **e** HHO

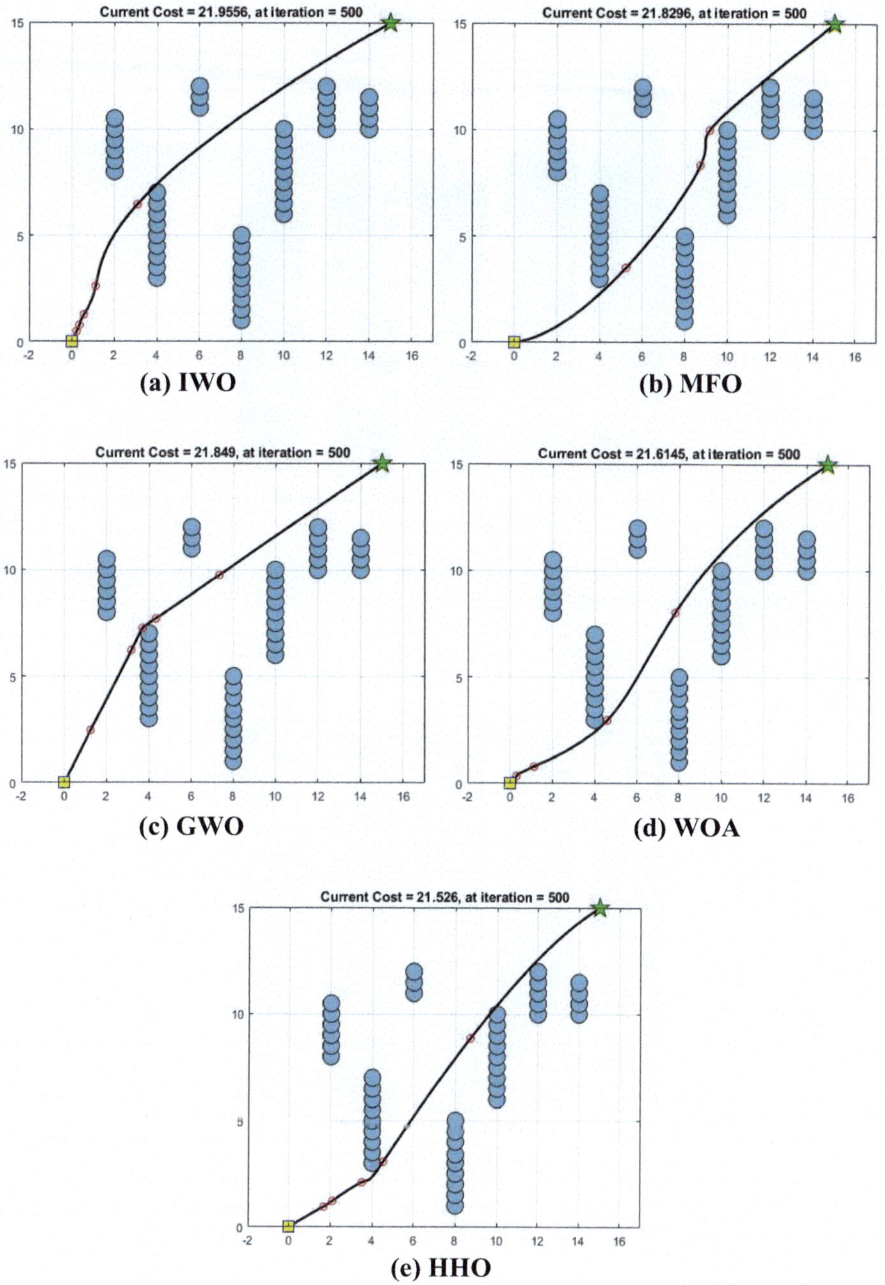

Fig. 19 Obstacle Map 5 for all algorithms in comparison **a** IWO **b** MFO **c** GWO **d** WOA **e** HHO

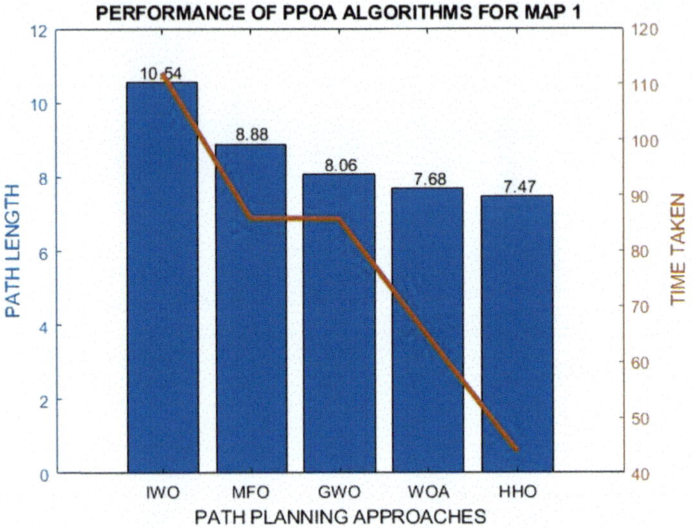

Fig. 20 Bar graph illustrating the performance comparison of all algorithms for Map 1 with respect to path length and computation time

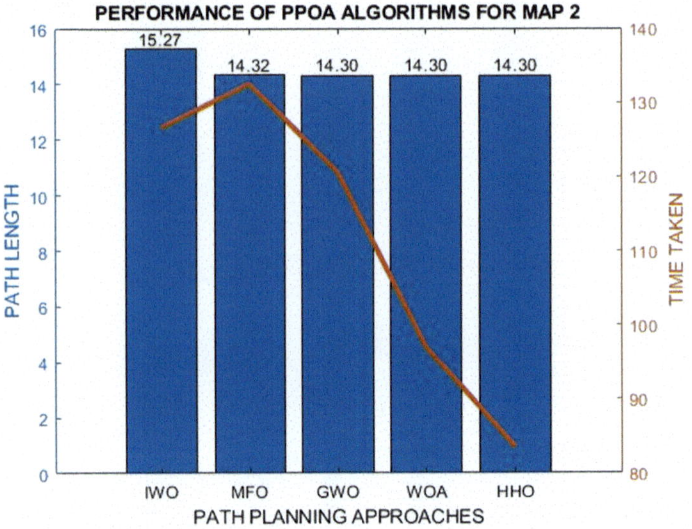

Fig. 21 Bar graph illustrating the performance comparison of all algorithms for Map 2 with respect to path length and computation time

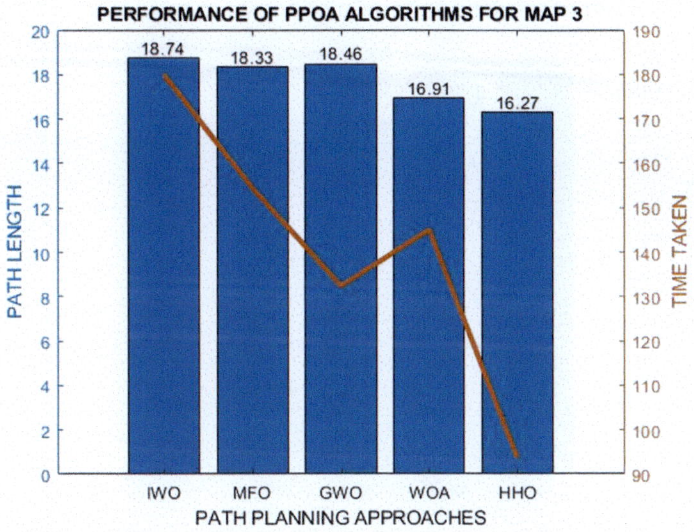

Fig. 22 Bar graph illustrating the performance comparison of all algorithms for Map 3 with respect to path length and computation time

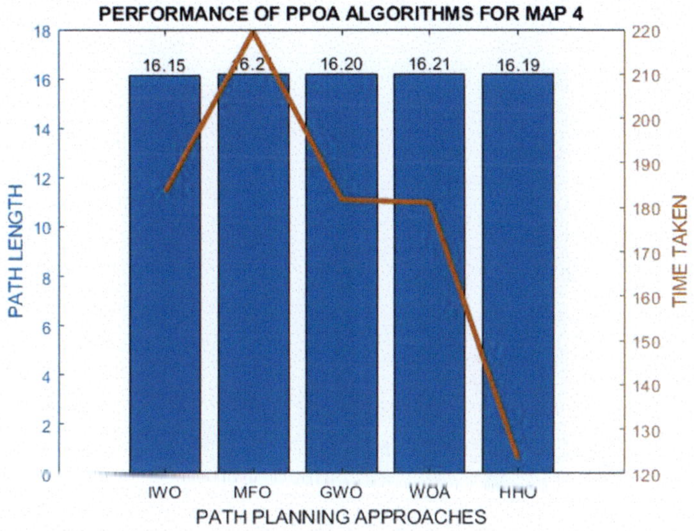

Fig. 23 Bar graph illustrating the performance comparison of all algorithms for Map 4 with respect to path length and computation time

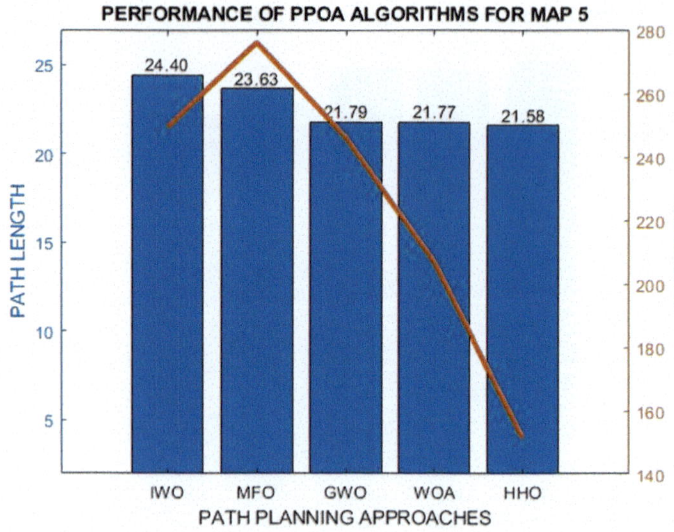

Fig. 24 Bar graph illustrating the performance comparison of all algorithms for Map 5 with respect to path length and computation time

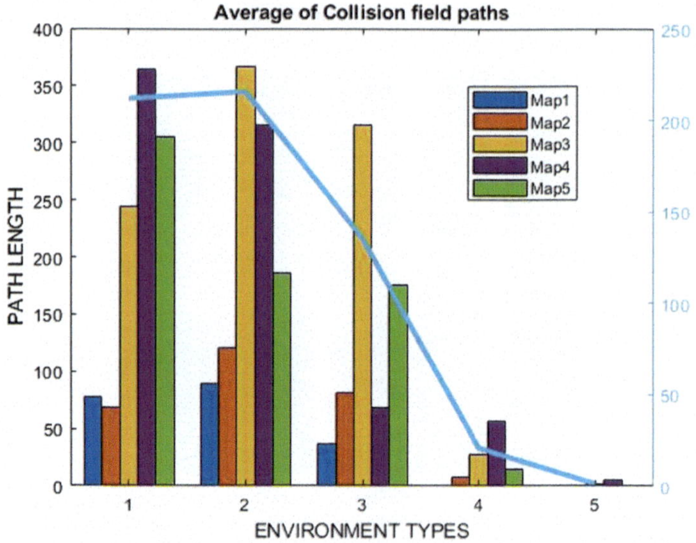

Fig. 25 Bar graph illustrating the average number of collision-filled paths obtained after successful runs of all the PPOA algorithms for all the environments

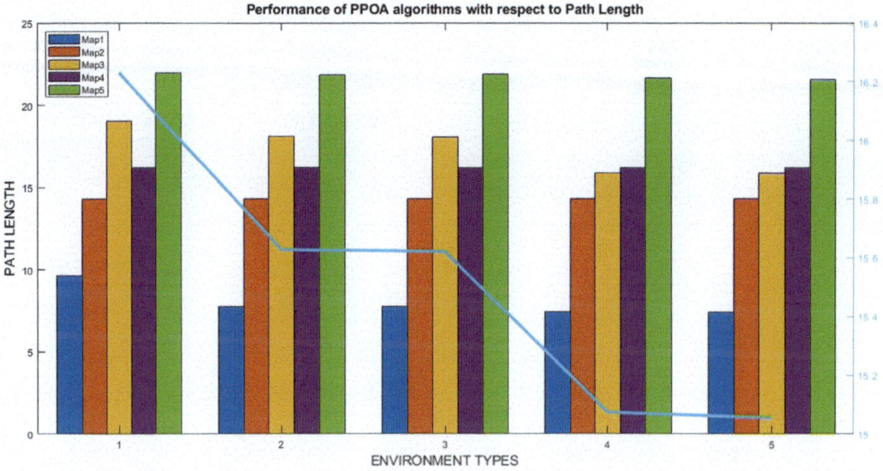

Fig. 26 Bar graph illustrating the performance comparison of all the PPOA algorithms for all the environments with respect to path length

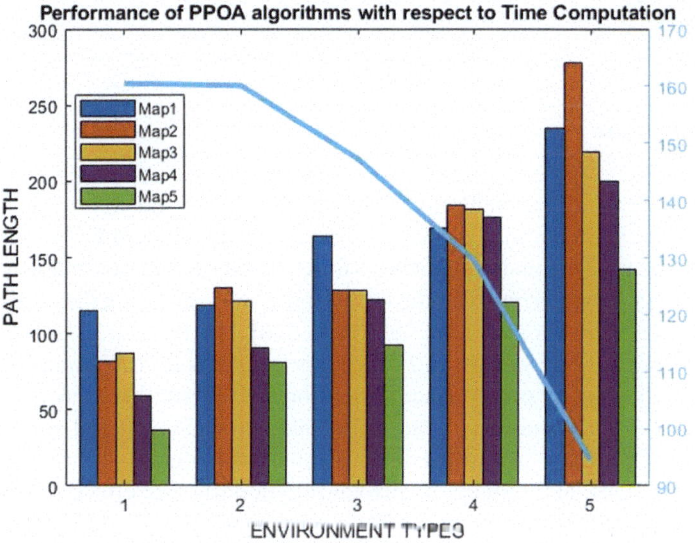

Fig. 27 Bar graph illustrating the performance comparison of all the PPOA algorithms for all the environments with respect to computation time

Table 9 Comparison of percentage differences in path length (% PL) and computation time (% time) with respect to HHO

Maps	% PL	% Time	% PL	% Time	% PL	% Time	% PL	% Time
	IWO–HHO		MFO–HHO		GWO–HHO		WOA–HHO	
1	23.23	68.22	4.13	54.94	4.11	58.03	0.07	37.96
2	0.07	31.76	0.06	37.48	0.01	32.99	0.01	10.29
3	16.73	43.59	12.32	27.69	0.03	34.69	1.39	48.71
4	0.07	29.05	12.09	27.43	0.02	33.73	1.48	35.0
5	1.96	39.36	−0.01	24.19	0.01	31.90	0.41	28.82

Table 10 Comparison of the results of all the algorithms based on the collision-filled paths on Map 1–Map 5 (M1–M5)

	IWO	MFO	GWO	WOA	HHO
M1	78	89	37	0	0
M2	60	121	81	7	0
M3	244	397	315	27	2
M4	364	315	68	56	5
M5	305	186	176	15	0

5 Conclusion

The effectiveness of the HHO algorithm in addressing the PPOA problem was assessed in comparison with the IWO, MFO, GWO, and WOA algorithms, with a focus on path length and computation time across diverse environmental maps. HHO consistently outperforms all the other algorithms, achieving the shortest paths: 7.413 units in 36.69 s for map 1, 14.2988 units in 81.16 s for map 2, 15.8522 units in 92.67 s for map 3, 16.1883 units in 120.05 s for map 4, and 21.526 units in 142.26 s for map 5. This efficient performance underscores HHO's ability to swiftly generate optimal paths while minimizing computational overhead compared with the WOA and SMOA. The study confirmed HHO's reliability, precision, and repeatability across all the maps, highlighting its pivotal role in designing effective PPOA algorithms. The parameter-less nature of HHO significantly reduces the computational effort and mitigates the uncertainties associated with parameter assumptions.

Graphical analyses depicted HHO's superiority in path smoothness and its ability to avoid local minima, distinguishing it from other algorithms. These findings underscore HHO's robustness in path planning for autonomous mobile robots (AMRs), emphasizing its ability to consistently generate optimal solutions in various environmental settings. The comparison clearly demonstrated HHO's substantial advantages in reducing path length and enhancing computation time efficiency compared with competing algorithms, thus justifying its implementation for tackling PPOA challenges effectively.

Overall, the study's comprehensive evaluation across multiple maps and rigorous comparative analysis solidified HHO's position as a leading algorithmic approach for PPOA in AMRs. Its ability to deliver optimal paths swiftly and efficiently underscores its practical applicability and effectiveness in real-world scenarios requiring complex path planning and navigation tasks.

References

1. Abbas NH, Ali FM (2016) Path planning of an autonomous mobile robot using enhanced bacterial foraging optimization algorithm. Al-Khwarizmi Eng J 12(4):26–35
2. Abbasi A, Firouzi B, Sendur P (2019) On the application of Harris hawks optimization (HHO) algorithm to the design of microchannel heat sinks. Eng Comput, 1–20
3. Agarwal D, Bharti PS (2018) A case study on AGV's alternatives selection problem. Int J Inform Technol, 1–13
4. Agarwal D, Bharti PS (2020) Nature-inspired evolutionary approaches for robot navigation: Survey. J Inf Optim Sci 41(2):421–436
5. Agarwal V, Rastogi R, Tiwari DC (2018) Spider monkey optimization: a survey. Int J Syst Assur Eng Manag 9(4):929–941
6. Atyabi A, Powers DMW (2013) Review of classical and heuristic-based navigation and path planning approaches. Int J Adv Comput Technol (IJACT) 5(14):1–14
7. Ayari B, Bouamama S (2017) (2017) A new multiple robot path planning algorithm: dynamic distributed particle swarm optimization. Robot Biomimetics 4(8):1–15
8. Bansal JC, Sharma H, Jadon SS, Clerc M (2014) Spider monkey optimization algorithm for numerical optimization. Memetic Comput 6(1):31–47
9. Barbera HM, Perez DH (2010) Autonomous navigation of an autonomous guided vehicle in industrial environment. Robot Comput Integr Manuf 26:296–311
10. Buniyamin N, Wan Ngah WAJ, Sariff N, Mohamad Z (2011) A simple local path planning algorithm for autonomous mobile robots. Int J Syst Appl Eng Dev 5(2):151–159
11. Castillo O, Trujillo L, Melin P (2007) Multiple objective genetic algorithms for path-planning optimization in autonomous mobile robots. Soft Comput 11:269–279
12. Chung H, Hou C, Liu S (2013) Automatic navigation of a wheeled mobile robot using particle swarm optimization and fuzzy control. IEEE international symposium on industrial electronics, Taiwan, pp 1–6
13. Clemente E, Meza-Sánchez M, Bugarin E, Aguilar-Bustos AY (2018) Adaptive behaviors in autonomous navigation with collision avoidance and bounded velocity of an omnidirectional mobile robot. a control theory with genetic programming approach. J Intell Robot Syst 92(2):359–380
14. Cong YZ, Ponnambalam SG (2009) Mobile robot path planning using ant colony optimization. IEEE/ASME international conference on advanced intelligent mechatronic, pp 851–856
15. Dao T, Pan T, Pan J (2016) A multi-objective optimal mobile robot path planning based on whale optimization algorithm. IEEE 13th international conference on signal processing, pp 337–342
16. Das S, Biswas A, Dasgupta S, Abraham A (2009) Bacterial foraging optimization algorithm: theoretical foundations, analysis and applications, foundations of. Comput Intell 3(3):23–55
17. Diéguez AR, Sanz R, López J (2003) Deliberative on-line local path planning for autonomous mobile robots. J Intell Rob Syst 37:1–19
18. Du Z, Qu D, Xu F, Xu D (2012) A hybrid approach for mobile robot path planning in dynamic environments. In: Proceedings of IEEE international conference on robotics and biomimetics, pp 1058–1063

19. Duan X, Niu T, Huang Q (2018) An improved shuffled frog leaping algorithm and its application in dynamic emergency vehicle dispatching. Mathem Problems Eng, Article ID 7896926, pp 1–34
20. Elaziz MA, Heidari AA, Fujita H, Moayedi H (2020) A competitive chain-based Harris Hawks Optimizer for global optimization and multi-level image thresholding problems. Appl Soft Comput J 95, Article ID 106347, pp 1–32
21. Elmi A, Efe MO (2018) Multi-objective grasshopper optimization algorithm for robot path planning in static environment. IEEE international conference on industrial technology, pp 244–249
22. Faisal M, Algabri M, Abdelkader BM, Dhahri H, Al-Rahhal MM (2017) Human expertise in mobile robot navigation. In IEEE Access 6:1694–1705
23. Fan Q, Chen Z, Xia Z (2020) A novel quasi-reflected Harris hawks optimization algorithm for global optimization problems. Soft Comput 24:14825–14843
24. Fan T, Wang J, Feng M, Zhang X, Wang J, Wu R (2019) Application of multi-objective firefly algorithm based on archive learning in robot path planning. Int J Intell Inf Database Syst 12(3):199–211
25. Fister I Jr, Fister D, Yang XS (2013) A hybrid bat algorithm. Elektrotehniˇski vestnik 80:1–7
26. Fister I, Fister I Jr, Yang XS, Brest J (2013) A comprehensive review of firefly algorithms. Swarm Evol Comput 13:34–46
27. Galceran E, Carreras M (2013) A survey on coverage path planning for robotics. Robot Auton Syst 61(12):1258–1276
28. Garcia MAP, Montiel O, Castillo O, Sepulveda R, Melin P (2009) Path planning for autonomous mobile robot navigation with ant colony optimization and fuzzy cost function evaluation. Appl Soft Comput 9(3):1102–1110
29. Ghalenoei MR, Hajimirsadeghi H, Lucas C (2009) Discrete invasive weed optimization algorithm: application to cooperative multiple task assignment of UAVs. Proceedings of the 48th IEEE conference on decision and control (CDC) held jointly with 28th IEEE Chinese control conference, pp 1665–1670
30. Gupta K, Deep K, Bansal JC (2017) Spider Monkey optimization algorithm for constrained optimization problems. Soft Comput 21(23):6933–6962
31. Gupta S, Deep K, Heidari AA, Moayedi H, Wang M (2020) Opposition-based learning Harris hawks optimization with advanced transition rules: principles and analysis, Expert systems with application, 158. Article ID 113510:1–23
32. Habib MK (2011) Biomimetics: innovations and robotics. Int J Mechatron Manuf Syst 4(2):113–134
33. Hachour O (2008) Path planning of autonomous mobile robot. Int J Syst Appl Eng Dev 2(4):178–190
34. Heidari AA, Mirjalili S, Faris H, Aljarah I, Mafarja M, Chen H (2019) Harris hawks optimization: algorithm and applications. Futur Gener Comput Syst 97:849–872
35. Ibraheem IK, Ajeil FH, Maher RA (2018) Path planning of an autonomous mobile robot in a dynamic environment using modified bat swarm optimization. Cornell University, Robotics, pp 1–8
36. Islam MZ, Wahab NIA, Veerasamy V, Hizam H, Mailah NF, Guerrero JM, Nasir MNM (2020) A harris hawks optimization based single and multi-objective optimal power flow considering environmental emission. Sustainability 12(13):1–26
37. Islam R, Tajmiruzzaman, Haque M, Muftee, Hossain S (2014) Autonomous robot path planning using particle swarm optimization in dynamic environment with mobile obstacles & multiple target. International conference on mechanical, industrial and energy engineering, pp 1402821–1402826
38. Kala R, Shukla A, Tiwari R, Roongta S, Janghel RR (2009) Mobile robot navigation control in moving obstacle environment using genetic algorithm. Artificial Neural Networks and A* Algorithm, Proceedings of the IEEE world congress on computer science and information engineering, pp 705–713

39. Kamalova A, Navruzov S, Qian D, Lee SG (2019) Multi-Robot exploration based on multi-objective grey wolf optimizer. Appl Sci 9(2931):1–16
40. Kamboj VK, Nandi A, Bhadoria A, Sehgal S (2020) An intensify Harris Hawks optimizer for numerical and engineering optimization problems, Applied Soft Computing, 89. Article ID 106018:1–35
41. Karakoyun M, Ozkis A, Kodaz H (2020) A new algorithm based on gray wolf optimizer and shuffled frog leaping algorithm to solve the multi-objective optimization problems. Appl Soft Comput 96(106560):1–48
42. Kashyap AK, Pandey A (2020) Optimized path planning for three-wheeled autonomous robot using teaching–learning-based optimization technique. Lecture Notes in Mechanical Engineering, Springer, pp 49–57
43. Kaur A, Sharma H, Sharma N (2017) Disruption operator-based spider monkey optimization algorithm. 2017 International conference on computing, communication and automation (ICCCA), Greater Noida, pp 216–221
44. Kim HC, Kim J, Ji YK, Park JH (2013) (2013) Path planning of swarm mobile robots using firefly algorithm. J Inst Control Robot Syst 19(5):435–441
45. Kovacs B, Szayer G, Tajti F, Burdelis M, Korondi P (2016) A novel potential field method for path planning of mobile robots by adapting animal motion attributes. Robot Auton Syst 82:24–34
46. Kunchev V, Jain L, Ivancevic V, Finn A (2006) Path planning and obstacle avoidance for autonomous mobile robots: a review, knowledge-based intelligent information and engineering systems. Lect Notes Comput Sci 4252:537–544
47. Liang JH, Lee CH (2015) Efficient collision-free path planning of multiple mobile robots system using efficient artificial bee colony algorithm. Adv Eng Softw 79:47–56
48. Liang X, Li L, Wu J, Chen H (2013) Mobile robot path planning based on adaptive bacterial foraging algorithm. J Central South Univ 20(12):3391–3400
49. Liu C, Zhao Y, Gao F, Liu L (2015) Three-dimensional path planning method for autonomous underwater vehicle based on modified firefly algorithm, Mathematical Problems in Engineering, 2015. Article ID 561394:1–10
50. Liu L, Song Y, Ma H, Zhang X (2012) Physarum optimization: a biology inspired algorithm for minimal exposure path problem in wireless sensor networks. Proceedings of IEEE INFOCOM, pp 1296–1304
51. Luo J, Chen H, Zhang Q, Xu Y, Huang H, Zhao X (2018) An improved grasshopper optimization algorithm with application to financial stress prediction. Appl Math Model 64:654–668
52. Malayjerdi E, Yaghoobi M, Kardan M (2017) Mobile robot navigation based on fuzzy cognitive map optimized with grey wolf optimization algorithm used in augmented reality. 5th IEEE/RSI international conference on robotics and mechatronics (ICRoM), Tehran, pp 211–218
53. Mareli M, Twala B (2018) An adaptive Cuckoo search algorithm for optimization. Appl Comput Inform 14:107–115
54. Masehian E, Katebi Y (2014) Sensor-based motion planning of wheeled mobile robots in unknown dynamic environments. J Intell Robot Syst 74(3–4):893–914
55. Masehian E, Sedighizadeh D (2007) Classic and heuristic approaches in robot motion planning—a chronological review, international journal of mechanical and mechatronics engineering. World Acad Sci Eng Technol 1(5):228–233
56. Mehrabian AR, Lucas C (2006) A novel numerical optimization algorithm inspired from weed colonization. Eco Inform 1(4):355–366
57. Mirjalili S (2015) Moth-flame optimization algorithm: a novel nature-inspired heuristic paradigm. Knowl-Based Syst 89:228–249
58. Mirjalili S, Lewis A (2016) The whale optimization algorithm. Adv Eng Softw 95:51–67
59. Mirjalili S, Mirjalili SM, Lewis A (2014) Grey wolf optimizer. Adv Eng Softw 69:46–61
60. Mirjalili S, Mirjalili SM, Yang XS (2014) Binary bat algorithm. J Neural Comput Appl 25(3):663–681

61. Mohammed HM, Umar SU, Rashid TA (2019) A systematic and meta-analysis survey of whale optimization algorithm, Computational Intelligence and Neuroscience, Hindawi, 2019. Article ID 8718571:1–25
62. Mohanty PK, Parhi DR (2014) Navigation of autonomous mobile robot using adaptive network based fuzzy inference system. J Mech Sci Technol 28(7):2861–2868
63. Mohanty PK, Parhi DR (2014) A new efficient optimal path planner for mobile robot based on Invasive Weed Optimization algorithm. Front Mech Eng 9(4):317–330
64. Mohanty PK, Parhi DR (2014) Optimal path planning for a mobile robot using cuckoo search algorithm. J Exp Theor Artif Intell 28(1–2):35–52
65. Moreno L, Armingol JM, Garrido S, Escalera ADL, Salichs MA (2002) A genetic algorithm for mobile robot localization using ultrasonic sensors. J Intell Rob Syst 34:135–154
66. Mossa MA, Kamel OM, Sultan HM, Diab AAZ (2020) Parameter estimation of PEMFC model based on Harris Hawks' optimization and atom search optimization algorithms. Neural Comput Appl, 1–16
67. Neve AG, Kakandikar GM, Kulkarni O (2017) Application of grasshopper optimization algorithm for constrained and unconstrained test functions. Int J Swarm Intell Evolut Comput 6(3):1–7
68. Nguyen VQ, Kim H, Jun SC, Boo K (2018) A study on real-time detection method of lane and vehicle for lane change assistant system using vision system on highway. Eng Sci Technol Int J 21(5):822–833
69. Ni J, Wu W, Shen J, Fan X (2014) An Improved VFF approach for robot path planning in unknown and dynamic environments, Hindawi Publishing corporation, mathematical problems in engineering, 2014. Article ID 461237:1–10
70. Nishitani I, Matsumura T, Ozawa M, Yorozu A, Takahashi M (2015) Human-centered X-Y–T space path planning for mobile robot in dynamic environments. Robot Auton Syst 66:18–26
71. Ortiz JAH, Vázquez KR, Castañeda MAP, Cosío FA (2013) Autonomous robot navigation based on the evolutionary multi-objective optimization of potential fields. Eng Optim 45(1):19–43
72. Otte MW, Richardson SG, Mulligan J, Grudic G (2009) Path planning in image space for autonomous robot navigation in unstructured environments. J Field Robot Wiley Publishers 26(2):212–240
73. Panda MR, Das PK, Dutta S, Pradhan SK (2018) Optimal path planning for mobile robots using oppositional invasive weed optimization. Comput Intell 34(4):1072–1100
74. Parhi DR, Singh MK (2009) Navigational strategies of mobile robots: a review. Int J Autom Control 3(2/3):114–134
75. Passino KM (2002) Biomimicry of bacterial foraging for distributed optimization and control. IEEE Control Syst 22(3):52–67
76. Patle BK, Babu LG, Pandey A, Parhi DRK, Jagadeesh A (2019) A review: on path planning strategies for navigation of mobile robot. Defence Technol 15(4):582–606
77. Patle BK, Pandey A, Jagadeesh A, Parhi DRK (2018) Path planning in uncertain environment by using firefly algorithm. Defence Technol 14(6):691–701
78. Patle BK, Parhi DR, Jagadeesh A, Kashyap SK (2017) On firefly algorithm: optimization and application in mobile robot navigation. World J Eng 14(1):65–76
79. Petrović M, Miljković Z (2017) Grey wolf optimization algorithm for single mobile robot scheduling. Proceedings of 4th international conference on electrical, electronics and computing engineering, pp 1–6
80. Qiu Q, Han JD (2011) 2.5-dimensional angle potential field algorithm for the real-time autonomous navigation of outdoor mobile robots. Sci China Inform Sci 54(10):2100–2112
81. Rao AM, Ramji K, Kumar TN (2018) Intelligent navigation of mobile robot using grey wolf colony optimization. Mater Today Proc 5(9):19116–19125
82. Samuel M, Hussein M, Mohamad MB (2016) A review of some pure-pursuit based path tracking techniques for control of autonomous vehicle. Int J Comput Appl 135(1):35–38
83. Saremi S, Mirjalili S, Lewis A (2017) Grasshopper optimization algorithm: theory and application. Adv Eng Softw 105:30–47

84. Sedighi K, Ashenayi K, Wainwright R, Tai H (2004) Autonomous local path planning for a mobile robot using a genetic algorithm. IEEE Congress Evolut Comput 2:1338–1345
85. Sengupta A, Chakraborti T, Konar A, Nagar A (2011) Energy efficient trajectory planning by a robot arm using invasive weed optimization technique. IEEE Third world congress on Nature and biologically inspired computing (NaBIC), pp 311–316
86. Shan EH, Dai B, Song J, Sun JP (2009) A dynamic RRT path planning algorithm based on B-spline. International symposium on computational intelligence and design, pp 25–29
87. Shareef Z, Trächtler A (2016) Simultaneous path planning and trajectory optimization for robotic manipulators using discrete mechanics and optimal control. Robotica 34(6):1322–1334
88. Shehab M, Abualigah L, Al Hamad H, Alabool H, Alshinwan M, Khasawneh AM (2020) Moth–flame optimization algorithm: variants and applications. Neural Comput Appl 32:9859–9884
89. Shukla A, Tiwari R, Kala R (2009) Mobile robot navigation control in moving obstacle environment using genetic algorithms and artificial neural networks. Int J Artif Intell Comput Res 1(1):1–12
90. Sood M, Panchal VK (2020) Meta-heuristic techniques for path planning: recent trends and advancements. Int J Intell Syst Technol Appl 19(1):36–77
91. Srivastava S, Sahana SK (2019) Application of bat algorithm for transport network design problem, applied computational intelligence and soft computing, 2019. Article ID 9864090:1–12
92. Tang SH, Khaksar W, Ismail NB, Ariffin MKA (2012) A review on robot motion planning approaches. Pertanika J Sci Technol 20(1):15–29
93. Tsai PW, Nguyen TT, Dao TK (2017) Robot path planning optimization based on multiobjective grey wolf optimizer. Genetic Evolut Comput Adv Intell Syst Comput 536:166–173
94. Verma D, Saxena P, Tiwari R (2019) Robot navigation and target capturing using nature-inspired approaches in a dynamic environment. 10th international conference on cloud computing, data science & engineering (Confluence), India, pp 629–636
95. Wang G (2018) Moth search algorithm: a bioinspired metaheuristic algorithm for global optimization problems. Memetic Comput 10:151–164
96. Wang G, Guo L, Duan H, Liu L, Wang H (2012) A modified firefly algorithm for UCAV path planning. Int J Hybrid Inform Technol 5(3):123–144
97. Wu L, Li Y, Liu J (2018) Based on improved bioinspired model for path planning by multi-AUV. Proceedings of the 2018 international conference on electronics and electrical engineering technology, pp 128–134
98. Xie C, Zheng H (2016) Application of improved Cuckoo search algorithm to path planning unmanned aerial vehicle. In: International conference on intelligent computing, pp 722–729
99. Xu J, Yan F, Ala OG, Su L, Li F (2018) Chaotic dynamic weight grey wolf optimizer for numerical function optimization. J Intell Fuzzy Syst 37(2):2367–2384
100. Yang B, Miao J, Fan Z, Long J, Liu X (2018) Modified cuckoo search algorithm for the optimal placement of actuators problem. Appl Soft Comput J 67:48–60
101. Yang G, Hou Z, Liang Z (2006) Distributed visual navigation based on neural Q-learning for a mobile robot. Int J Veh Auton Syst 4(2–4):225–238
102. Yang L, Qi J, Song D, Xiao J, Han J, Xia Y (2016) Survey of robot 3D path planning algorithms, Hindawi publishing corporation. J Control Sci Eng, Article ID 7426913, pp 1–22
103. Yang XS, Deb S (2009) Cuckoo search via Lévy flights. IEEE world congress on nature and biologically inspired computing, pp 210–214
104. Yang XS, Luo C (2004) A neural network approach to complete coverage path planning. IEEE Trans Syst Man Cybern 34(1):718–724
105. Yang XS (2008) Nature-inspired metaheuristic algorithm. Luniver Press
106. Yu J, Kim CH, Rhee SB (2020) The comparison of lately proposed Harris hawks optimization and Jaya optimization in solving directional overcurrent relays coordination problem. Hindawi Complexity, Article Id 3807653, pp 1–22

107. Zarate LE, Becker M, Garrido BDM, Rocha HSC (2002) An artificial neural network structure able to obstacle avoidance behaviour used in mobile robots. IEEE Ann Conf Indus Electron Soc 28:2457–3246
108. Zhang P, Lv T (2004) Soccer robot path planning based on artificial potential field approach with simulated annealing. Robotica 22:563–566
109. Zhang X, Ming Z (2019) Trajectory planning and optimization for a Par4 parallel Robot based on energy consumption. Appl Sci 9(13):1–19
110. Zhang X, Wang Q, Adamatzky A, Chan FTS, Mahadevan S, Deng Y (2014) An improved Physarum polycephalum algorithm for the shortest path problem, Hindawi Publishing corporation. Scientific World J, Article ID 487069, pp 1–9
111. Zhu Y, Zhang T, Song J (2009) An improved wall following method for escaping from local minimum in artificial potential field based path planning. Joint 48th IEEE conference on decision and control and 28th Chinese control conference, pp 6017–6022
112. Zuo L, Guo Q, Xu X, Fu H (2015) A hierarchical path planning approach based on A* and least-squares policy iteration for mobile robots. Neurocomputing 170(2015):257–266

Chapter 7
Optimum Solution for System Stability of Interconnected Power Grid Using Hybrid Genetic Firefly Algorithm Optimized PID Controller

D. Boopathi, K. Jagatheesan, Sourav Samanta, Kanendra Naidu, and B. Anand

1 Introduction

Load Frequency Control (LFC) is a crucial control approach in electric power system design for dependable operation. LFC in multiple area power systems aims to maintain a set tolerance for tie-line power flows and regulate each area's frequency within allowed limits. Oscillations and variability in load demand or generation are two of the most serious issues to the power system. This causes fluctuations in system frequency and tie-line power flow. LFC is dangerous for controlling fluctuations and

D. Boopathi (✉)
Department of Electrical and Electronics Engineering, Paavai Engineering College, Namakkal, India
e-mail: boopathime@gmail.com

K. Jagatheesan
Department of Electronics and Communication Engineering, M. Kumarasamy College of Engineering, Karur, India
e-mail: Jaga.ksr@gmail.com

S. Samanta
Department of Computer Science and Engineering, University Institute of Technology, The University of Burdwan, Burdwan, West Bengal, India
e-mail: Sourav.uit@gmail.com

K. Naidu
School of Electrical Engineering, Universiti Teknologi MARA, Shah Alam, Malaysia
e-mail: kanendra@uitm.edu.my

B. Anand
Department of Electronics and Instrumentation Engineering, Hindusthan College of Engineering and Technology, Coimbatore, India
e-mail: anand.eie@hicet.ac.in

N. Dey (ed.), *Optimizing Solutions for Real-Life Problems*, Springer Tracts in Nature-Inspired Computing, https://doi.org/10.1007/978-981-96-3216-9_7

ensuring stable functioning of the power system in order to deliver quality electricity by monitoring power generation and demand. Depending on the power demand, LFC calculates the net change in necessary generation. Maintaining the system frequency and inter-area flows at the designated nominal values is the main responsibility of the LFC. The LFC's job in a grid-connected power system is to ensure that tie-line power flows within specified tolerances and that each region's frequency stays within predetermined bounds. Variations in tie-line power flows and control frequency are measured by LFC. Several control approaches were developed to maintain LFC frequency and tie-line power flow at desirable levels in both normal and disturbed power systems. Recently, researchers have focused on designing load frequency controllers and created several control algorithms.

A fresh cascade fuzzy-based PID—fractional order PIDN (FPIDN-FOPIDN) controller has been appointed in two area thermal, hydro, and gas power plants for LFC, and the results were compared with several optimization techniques. In conclusion, the Imperialist competitive algorithm is better than others [1]. A three-area thermal power network was investigated using a grey wolf optimization (GWO) optimized PID controller, the results of Particle Swarm Optimization (PSO), Ant Colony Optimization (ACO) and Genetic Algorithm (GA) [2]. A multi-area system analyzed for LFC using the Grasshopper optimization algorithm, Salp swarm algorithm, and Collective decision optimization algorithm. In the comparison, the Salp swarm algorithm produces a better solution for the frequency oscillation [3]. A coal-based power system was examined by an Artificial Bee Colony-based PID regulator for LFC, and the result of Ziegler Nichols, GA and Bacterial Foraging Optimization Algorithm was compared [4]. The PSO technique optimized fractional PID (FOPID) controller is employed as a secondary controller in an interconnected thermal, gas, and hydropower system. By contrasting the outcomes of the tilted integral derivative (TID) controller with the same technique, the superiority of the suggested controller was demonstrated [5].

A hybrid power system with thermal and photovoltaic (PV) sources for power production. FA-based PI controller suggested for frequency regulation problem, results were compared with GA [6]. The real-time Egyptian electrical grid has been investigated for frequency regulation using a tuned PID controller, consisting of thermal, nuclear, hydro, wind and gas power plants. The result was compared with the classical PID controller [7]. ACO–PID controller involved in the process of LFC in a microgrid power system, which consists of wind as the major energy source with energy storage systems. The sovereignty of the proposed controller is confirmed by comparing the result with the PSO technique [8]. Koopman model predictive control (MPC) method is introduced for LFC in thermal power network, the Koopman method-based MPC produced better results over classical MPC [9]. Teaching Learning-Based Optimization (TLBO) based Sliding Mode Control (SMC) appointed as a secondary controller in a microgrid. The response was compared with GWO, PSO, and SSA-tuned SMC, from the result TLBO–SMC is performed around 42% over other methods [10]. PID and FOPID controllers are utilized as secondary controllers in two area thermal power networks with nonlinear [11]. Brown bear optimization algorithm tuned PID and PI–PD controllers are utilized for LFC of the interconnected thermal power plant in [12], the suggested controller's performance in comparison to PSO and GWO method-tuned controllers.

Thermal power plant incorporated with PV and energy storage systems was examined by the Mayfly Algorithm (MA)-tuned PID controller, and the proposed controller supremacy was proved by comparing the result with the PI controller [13]. Gradient-based optimizer-tuned PID controller employed as a secondary controller for four-area power network comprised of thermal, gas, hydro, and solar power plants. The PID controller result compared with the Tilt integral controller [14]. When implementing LFC in a two-area power network using a fuzzy controller, the efficacy of the suggested regulator is determined by contrasting the outcome with that of a traditional PID controller. [15]. The Bees algorithm tuned SMC was appointed as a subordinate controller for a thermal power network LFC, the response was compared with previous published results [16]. GA tuned PID controller performance in contrast to frequency oscillation tested in a microgrid power system [17]. A hybrid gravitational search-based FA-tuned PI controller was employed for the LFC thermal power networks, the response has been compared with PSO, GA, and FA. The hybrid algorithm tuned controller performed better than other tuned controllers [18]. An isolated air power plant with an energy storage system has been investigated for LFC using an FOPID controller, which is tuned by the Mayfly algorithm. The proposed controller was compared with the PID controller response for supremacy [19].

The literature research on the LFC problem suggests that several controllers are created and various optimization strategies are applied during the design process. Each optimization approach has unique advantages and downsides. This study proposes using a PID controller in thermal, gas, wind, and hydropower plants with energy storage devices, as discussed above. The controller gain values are tuned using the GA–FA approach, and their efficacy is evaluated by comparing performance to the GA and FA adjusted PID controller response. Summary of the review is in Table 1.

The Sects. 1.1 and 1.2, are contribution of the chapter and construction of the chapter respectively.

1.1 Contribution of the Chapter

- To design a multiple-source interlinked power network consisting of thermal, wind, hydro, and gas (THGW) power plants.
- Develop a secondry control loop for proposed power network to prform LFC.
- Sugeested controller is tuned by hybrid GA-FA, to increase the performance of the controller against frequency oscillation.
- The response is compared with a GA-tuned PID controller in order to assess how well the hybrid GA-FA regulator has improved.

Table 1 Summary of review

Ref	System	Controller	Technique	Result comparison
[1]	Two are -Thermal, hydro, gas	FPIDN-FOPIDN	Imperialist competitive algorithm	PS and PSO
[2]	Three area—Thermal	PID	ACO	GA and PSO
[3]	Two area—Thermal, hydro, gas	Salp swarm algorithm	I, ID, and PID	Grasshopper optimization algorithm
[4]	Two area—Thermal	PID	Artificial Bee Colony	Ziegler Nichols, GA and Bacterial Foraging Optimization Algorithm
[5]	Two area—Thermal, hydro, gas	FOPID	PSO	PSO–TID, PID
[6]	Two area—thermal, PV	PID	FA	GA–PID
[7]	Five area—thermal, hydro, gas, nuclear, wind	PID	GA	Classical PID controller
[8]	Single area—wind, energy storage systems	PID	ACO	PSO–PID
[10]	Micro grid	SMC	TLBO	GWO, PSO, and SSA–SMC
[12]	Two area—thermal	PID & PI–PD	Brown bear optimization	PSO & GWO
[13]	Two area—thermal, PV	PID	Mayfly optimization	Mayfly optimization PI controller
[14]	Four area—thermal, hydro, gas, solar	PID	Gradient based optimizer	Tilt integral controller
[16]	Two area—thermal	SMC	Bees algorithm	TLBO–FPID
[19]	Single area—wind	FOPID	Mayfly	Mayfly–PID

1.2 Construction of the Chapter

This chapter is proposed as Sect. 1, about the introduction and detailed literature review. Section 2 is deals about the invesigatied mathametical model and its parameters discussion. The secondary control loop, controller design, and cost fucntion is discussed in Sect. 3. Section 4, is about the proposed optimizer (GA–FA). The performance of the proposed controller for LFC in the proposed power sysem explined in Sect. 5. Result and improvement of the proposed controller exaplined in Sect. 6. conclusion of the chapter is in Sect. 7.

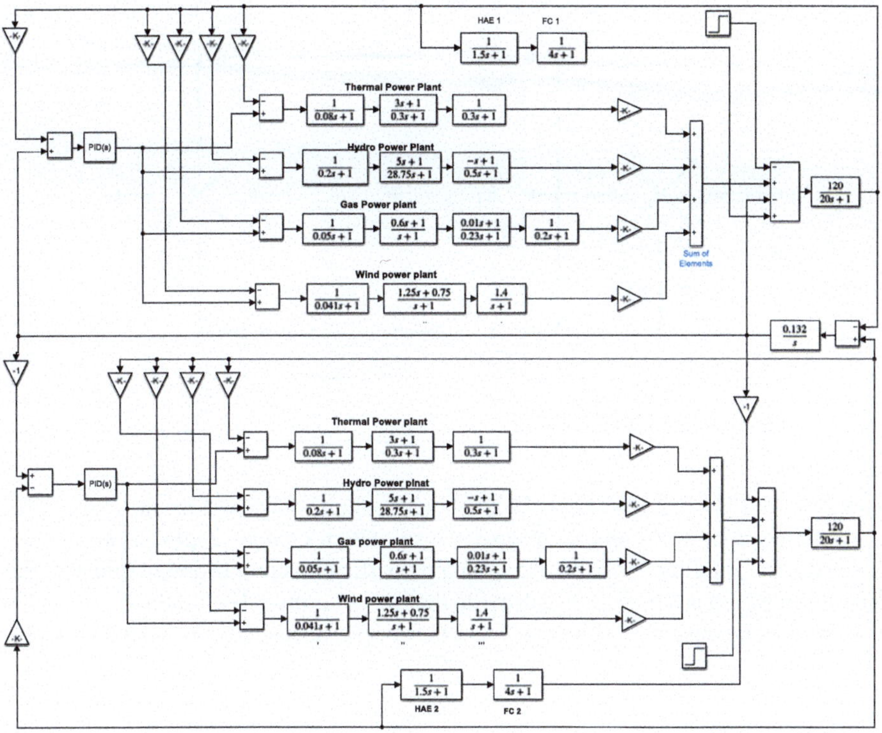

Fig. 1 Simulink mathematical model

2 Investigated Mathematical Model

An identical two-area grid connected power network was developed and investigated for LFC. Each area comprises of thermal, wind, gas, and hydropower plants. To manage the uncertainties, the renewable source energy storage units HAE and FC are utilized in this power network. The mathematical model of the proposed power network is given in Fig. 1 [20].

3 Secondary Controller and Cost Fucntion

Power quality of an interconnected network is manitained by the primary control loop which consists primary regulator (mechanical value). During the unexpected loading time primary controller is not able to maintain the power quality, so a secondary controller is must needed one. In this chapter the LFC is performed by utilizing PID controller as a subordinate controller in the proposed power network. It is widely known that PID control is the most often utilized control approach in industry. The

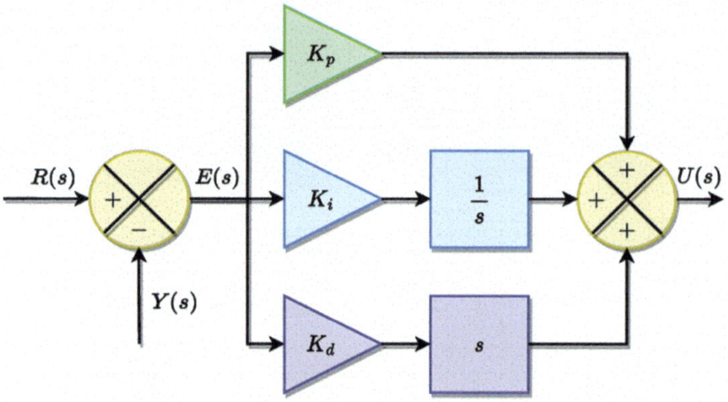

Fig. 2 Simple structure of PID controller

popularity of PID controllers may be attributed to their robust performance in a range of operating conditions and their ease of use, which enables engineers to operate them simply. The PID consists of three main coefficients: proportional, integral, and derivative, which are modified to provide the best response. The Laplace expression of the PID controller is given in Eq. 1 and the simple structure is shown in Fig. 2 [21, 22].

$$G(S)_{PID} = K_p + \frac{K_i}{S} + K_d \cdot S \tag{1}$$

The cost function for ITAE is stated in the equation below (2).

$$J_{ITAE} = \int_0^t t.|\Delta f_1 + \Delta f_2 + \Delta P_{tie}|dt \tag{2}$$

where the above equation, Δf_1, Δf_2, and ΔP_{tie} represents deviations of system frequency in the area—1, area—2, and tie-line power flow between interconnected systems, J indicates the performance index and t—time given to simulate the model.

4 Genetic Algorithm Based Firefly Algorithm

Using conventional approaches to maximize the objective function requires many liberations. In this study, hybrid optimization approaches such as hybrid GA–FA were used in place of more conventional optimizers to maximize the controller gain parameters in the power network's LFC/AGC.

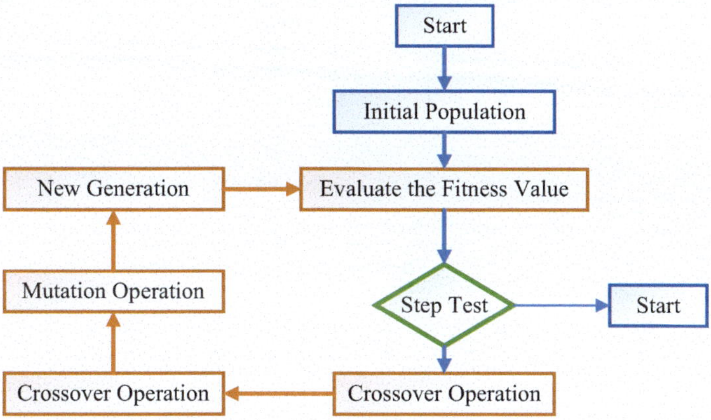

Fig. 3 Flow chart of GA

4.1 GA Method

Holland first presented GA for identifying the worldwide or nearly global optimal solution. A set of chromosomes serves as the starting point (solutions). Next, a fresh set of solutions is produced by using the genetic selection, crossover, and mutation operators one at a time. It is expected that the newly created chromosomes would be of higher quality than the first generation. These phases are maintained until the termination criterion is satisfied. Find the GA flow chart in Fig. 3 [17]. The following are the main steps of the GA algorithm:

Step I: Group initialization randomly.

Step II: Evaluation of suitability of every result.

Step III: Replace old group and generate solutions with new group.

Step IV: If conditions are satisfied, stop; otherwise, go to step II.

4.2 FA Method

Yang (2008) proposed the FA, a metaheuristic optimization technique. FA uses a model of social behavior similar to that of fireflies to address optimization problems. FA's primary benefit is that all local and global modes will be investigated. This is because, depending on the absorption coefficient, the light intensity decreases with distance, causing either a global or local attraction among fireflies. It has thus fascinated a lot of interest and been effectively used to a range of optimization problems. Figure 4 contains the flow chart. The following are the main FA steps:

Step 1: Initialization.

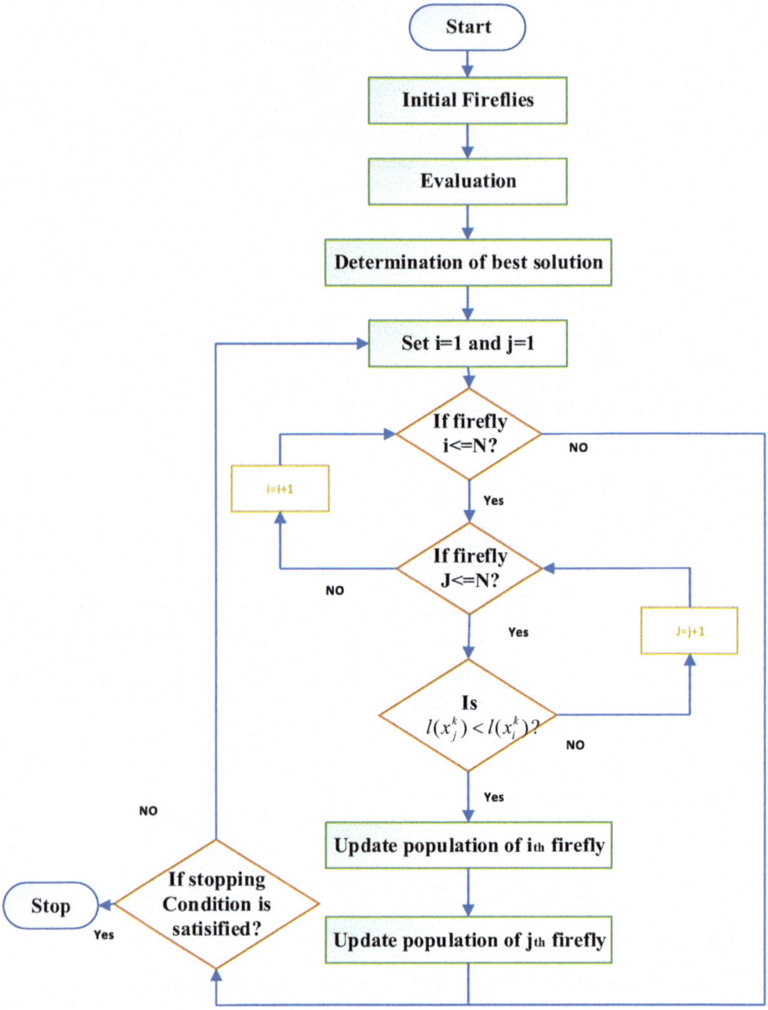

Fig. 4 Flow chart of FA

Step 2: Evaluation.

Step 3: Resolve of best solution.

Step 4: Informing positions of fireflies.

Step 5: Informing the best solution.

Step 6: Stopping condition.

The best features of both the GA and FA algorithms are combined in hybrid GA-FA. While GA has excellent exploitation capabilities because its operators completely change the structure of the solutions, enabling the suggested algorithm to abandon

Table 2 Optimized controller parameters

Optimizer/optimized parameter	Area 1 controller			Area 2 controller		
	K_P	K_I	K_D	K_P	K_I	K_D
GA	0.983035	0.254564	0.52694	0.514869	0.923587	0.085022
FA	0.947381	0.995606	0.46856	0.868069	0.999842	0.469134
GA-FA	0.999999	0.999999	0.01304	0.999999	0.999999	0.433548

local optima when necessary, FA has strong exploration capabilities since it explores all local and global modes and finds good solutions. The suggested controller gain values are optimized with the aforementioned advantage in mind [23]. In Table 2, optimized controller settings are presented.

The above optimized parameters are applied and performance is investigated. The upcoming session will discuss the work efficiency of the proposed controller and optimizer against the frequency oscillations.

5 Performance of the Proposed Controller for LFC in the Proposed Power System

The response of the suggested controller in contradiction of frequency difference from the normal limits at the time of unexpected loading in the developed power system. The performance is studied using MATLAB 2021a working environment with 1% load difference. In this session the presentation of the proposed controller studied with area 1 & 2 frequency difference, grid line power flow, and area controller error (ACE). The area 1 frequency response comparison is shown in Fig. 3 and Ts bar chart comparison is in Fig. 4. Numerical analysis is reported in Table 3.

Figures 5, 6, and Table 3 demonstrated that the proposed technique-based PID controller provided a better response over other compared responses. The setting time is faster than others i.e. GA:120 s > FA:100 s > GA-FA: 88 s.

Figures 5, 6, and Table 4 demonstrated that the proposed technique-based PID controller provided a better response over other compared responses. The setting time is faster than others i.e. GA:90 s > FA:65 s > GA-FA: 40 s (Fig. 7).

Table 3 Optimized parameters in area 1 frequency

Optimizer/optimized parameter	T_S (Sec)	PoS (Hz)	PuS (Hz)
GA	120	75.64×10^{-3}	1.589×10^{-3}
FA	100	2.65×10^{-3}	9.33×10^{-3}
GA-FA	88	1.243×10^{-3}	1.73×10^{-3}

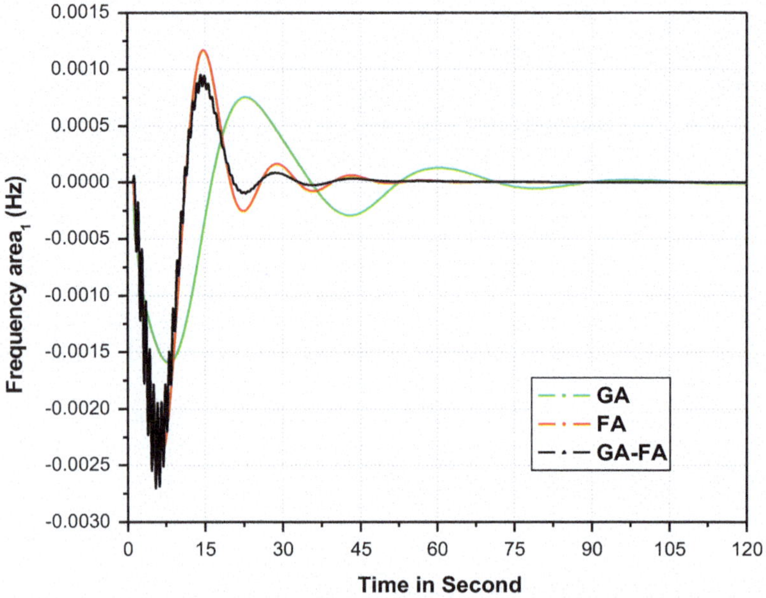

Fig. 5 Area 1 frequency response comparison

Fig. 6 Ts comparison of area 1

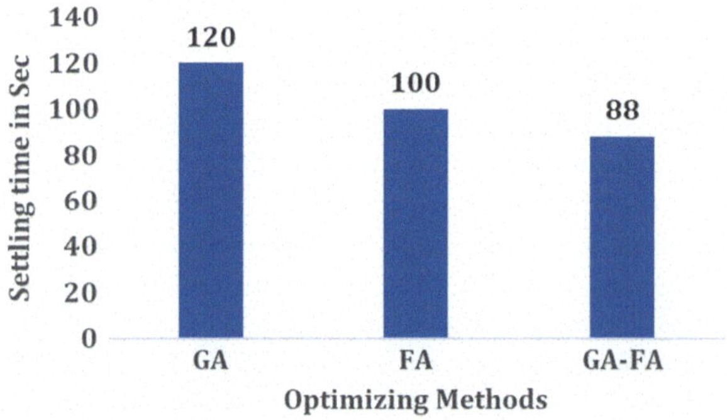

Figures 9, 10, and Table 5 demonstrated that the proposed technique-based PID controller provided a better response over other compared responses. The setting time is faster than others i.e. GA:94 s > FA:65 s > GA-FA: 45 s (Fig. 11).

Figures 10, 12, and Table 6 demonstrated that the proposed technique-based PID controller provided a better response over other compared responses. The setting time is faster than others i.e. GA:120 s > FA:62 s > GA-FA: 50 s.

Figures 13, 14, and Table 7 demonstrated that the proposed technique-based PID controller provided a better response over other compared responses. The setting time is faster than others i.e. GA:80 s > FA:60 s > GA-FA: 42 s.

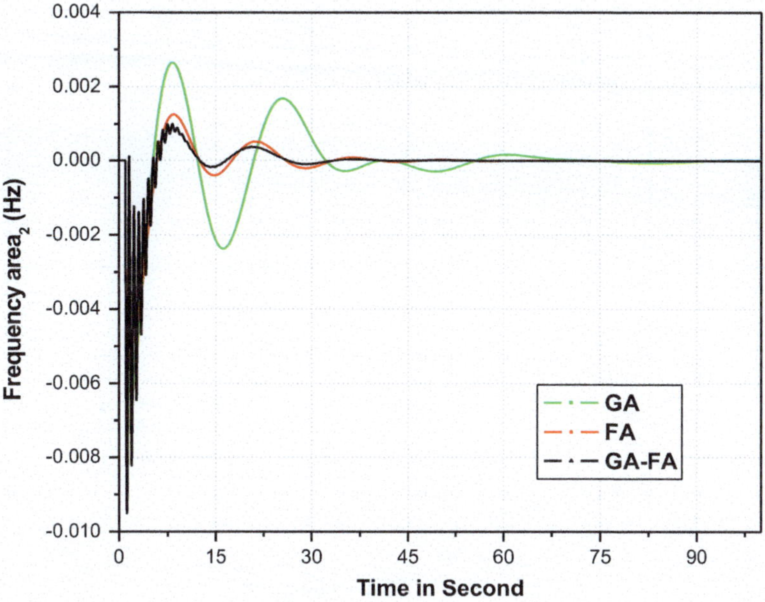

Fig. 7 Area 2 frequency response comparison

Fig. 8 Ts comparison of area 2

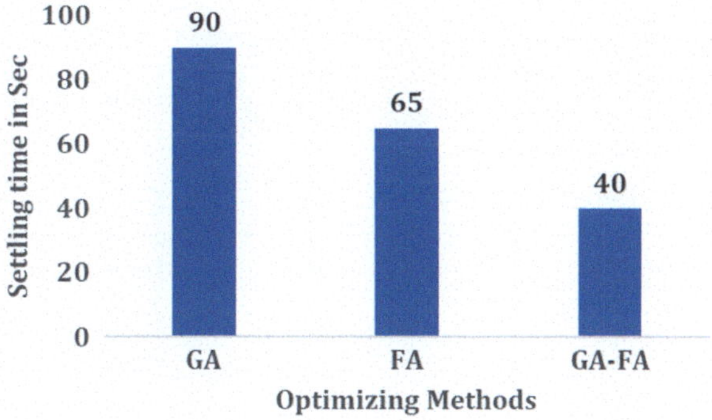

Table 4 Optimized parameters in area 2 frequency

Optimizer/optimized parameter	Ts (Sec)	PoS (Hz)	PuS (Hz)
GA	90	2.65×10^{-3}	9.33×10^{-3}
FA	65	1.26×10^{-3}	4.88×10^{-3}
GA-FA	40	1×10^{-3}	6.46×10^{-3}

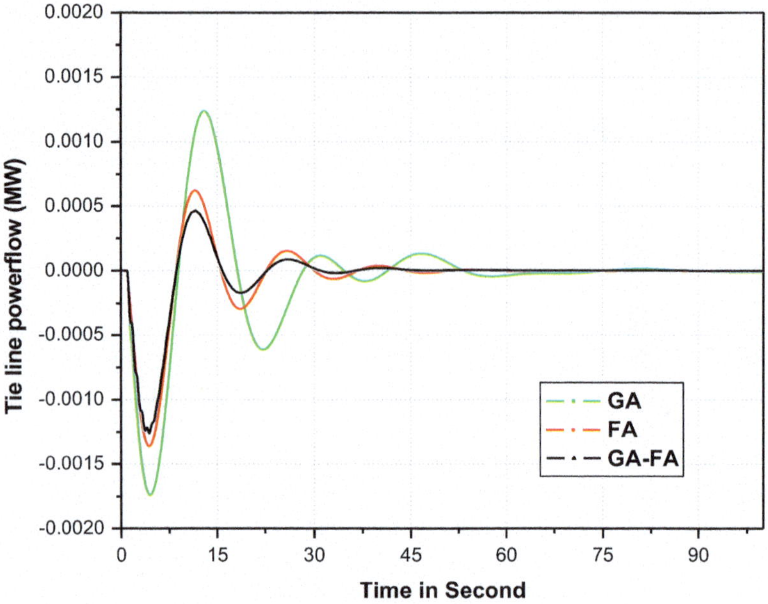

Fig. 9 Tieline power flow response comparison

Fig. 10 Ts comparison of tie line power flow

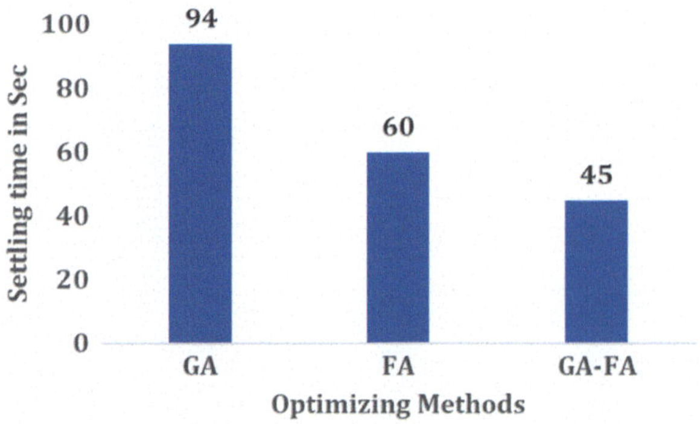

6 Result Discussion and Improvement of the Proposed Controller

The developed power network is investigated with a hybrid GA–FA tuned PID controller to regulate the system stability. In the previous session 7.5, the performance of the proposed controller is explained in detail. The results of GA and FA-tuned controller responses were compared. GA–FA tuned PID controller provides better results than other optimization method-based controller results. Improvement of the hybrid GA–FA technique in terms of settling time is reported in Table 7 and Fig. 15.

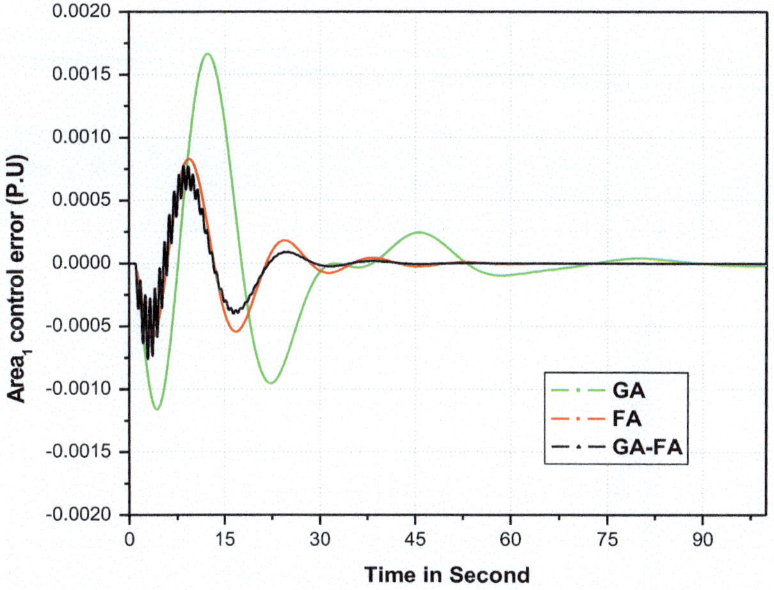

Fig. 11 ACE 2 response comparison

Table 5 Optimized parameters in tieline power flow

Optimizer/optimized parameter	T_S (Sec)	PoS (MW)	PuS (MW)
GA	94	1.24×10^{-3}	1.74×10^{-3}
FA	60	6.22×10^{-4}	1.35×10^{-3}
GA-FA	45	4.63×10^{-4}	1.25×10^{-3}

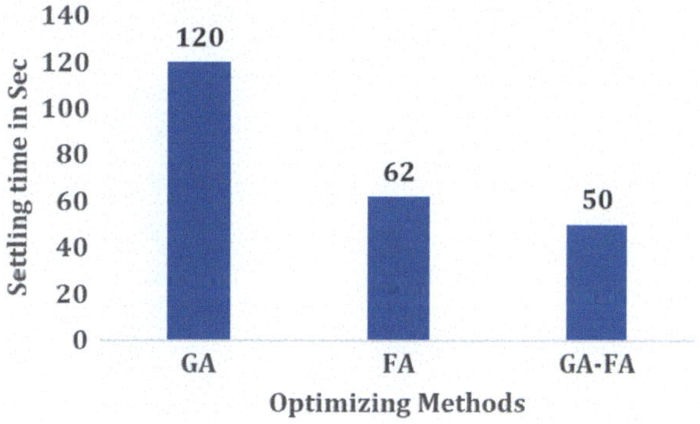

Fig. 12 Ts comparison of ACE 1

Table 6 Optimized parameters in ACE1

Optimizer/pptimized parameter	T_S (Sec)	PoS (P.U)	PuS (P.U)
GA	120	1.67×10^{-3}	1.15×10^{-3}
FA	62	8.32×10^{-4}	5.94×10^{-4}
GA-FA	50	7.72×10^{-4}	7.60×10^{-4}

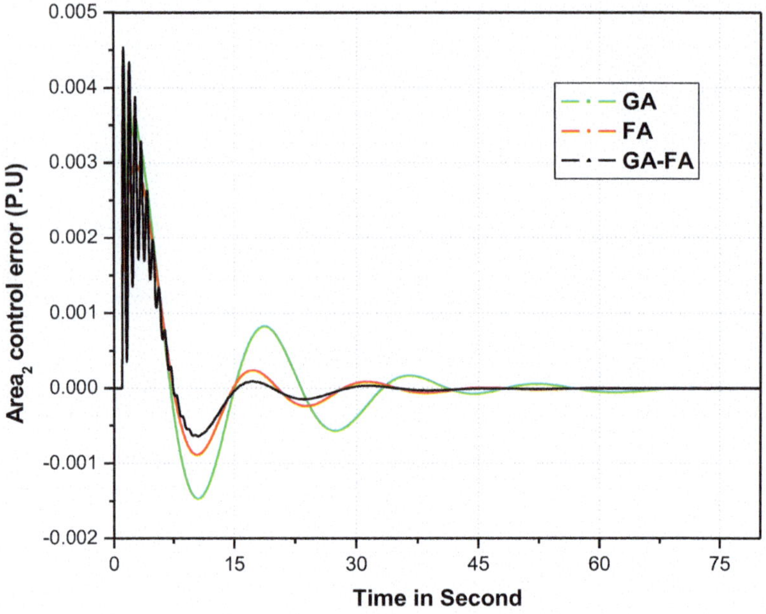

Fig. 13 ACE 1 response comparison

Fig. 14 Ts comparison of ACE 2

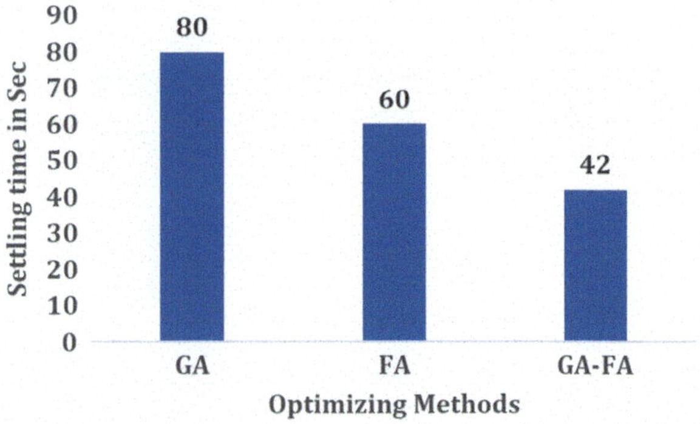

Table 7 Optimized parameters in ACE2

Optimizer/optimized parameter	T_S (Sec)	PoS (P.U)	PuS (P.U)
GA	80	4.53×10^{-3}	1.46×10^{-3}
FA	60	2.98×10^{-3}	8.82×10^{-4}
GA-FA	42	4.53×10^{-3}	6.39×10^{-4}

Table 8 % of Improvement of hybrid GA–FA tuned PID controller

Optimizer/ optimized parameter	% of Improvement				
	First Area frequency	Second Area frequency	Tie line power flow	ACE 1	ACE 2
Hybrid GA–FA over GA	38.4	55.5	52.1	48.3	47.5
Hybrid GA–FA over FA	24.1	27.7	36.1	58.3	42.8

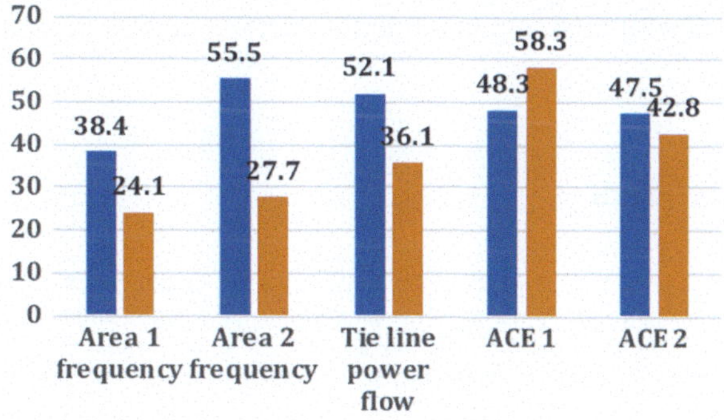

Fig. 15 % of improvement of Ts

Table 8 and Fig. 15, clearly demonstrate that the proposed hybrid GA–FA method tuned PID controller produced noticed an improvement over the other two methods GA and FA tuned PID controller response.

7 Conclusion

This chapter is about the performance analysis of hybrid GA–FA-based PID controller performance analysis over GA and FA tune PID controller. The investigation is conducted in an equal two-area power network, comprised of thermal, gas, hydro, and wind power plants. The performance of the analysis is as follows:

GA–PID controller consumes more time to settle the frequency oscillation than FA and hybrid GA–FA technique in both area and tieline power flow. In area 1 frequency 120 s, area 2 frequency 90 s takes to settle the oscillation.

FA–PID controller consumes more time to settle the frequency oscillation than the hybrid GA–FA technique and lesser time GA, in both area and tieline power flow. In area 1 frequency 100 s, area 2 frequency 65 s takes to settle the oscillation.

GA–FA–PID controller is consuming less time to settle the frequency oscillation than FA and GA technique in both area and tieline power flow. In area 1 frequency 88 s, area 2 frequency 40 s takes to settle the oscillation.

7.1 Challenge and Future Scope of the Study

- When we use the complex power network the selection process of the secondary controller is too difficult task.
- Combined optimization algorithms to optimize the secondary controller need too much time.
- In the future different controllers and many upcoming algorithms may be used to control the oscillation in the proposed power network.

References

1. Arya Y, Dahiya P, Çelik E, Sharma G, Gözde H, Nasiruddin I (2021) AGC performance amelioration in multi-area interconnected thermal and thermal-hydro-gas power systems using a novel controller. Eng Sci Technol Int J 24(2):384–396
2. Jagatheesan K, Boopathi D, Samanta S, Anand B, Dey N (2023) Grey wolf optimization algorithm-based PID controller for frequency stabilization of interconnected power generating system. Soft Comput, pp 1–14
3. Dey P, Saha A, Srimannarayana P, Bhattacharya A, Marungsri B (2022) A realistic approach towards solution of load frequency control problem in interconnected power systems. J Electr Eng Technol 17(2):759–788
4. Elsisi M, Soliman M, Aboelela MAS, Mansour W (2015) ABC based design of PID controller for two area load frequency control with nonlinearities. TELKOMNIKA Indonesian J Electr Eng 16(1):58–64
5. Morsali J, Zare K, Hagh MT (2018) Comparative performance evaluation of fractional order controllers in LFC of two-area diverse-unit power system with considering GDB and GRC effects. J Electr Syst Inform Technol 5(3):708–722
6. Abd-Elazim SM, Ali ES (2018) Load frequency controller design of a two-area system composing of PV grid and thermal generator via firefly algorithm. Neural Comput Appl 30:607–616
7. Metwally MA, Ali MA, Kutb SA, Bendary FM (2024) A genetic algorithm for optimum design of PID controller in multi area load frequency control for Egyptian electrical grid. Int J Eng Res Technol (IJERT) http://www.ijert.org. ISSN, pp 2278–0181
8. Boopathi D, Jagatheesan K, Anand B, Jaya J, Satheeshkumar R (2023) Investigation of a microgrid power system for frequency regulation by implementing ant colony optimization

technique optimized secondary controller. In: Renewable energy optimization, planning and control: proceedings of ICRTE 2022 (pp 229–236). Singapore: Springer Nature Singapore
9. Zhou J, Jia Y, Yong P, Liu Z, Sun C (2024) Robust deep Koopman model predictive load frequency control of interconnected power systems. Electric Power Syst Res 226:109948
10. Dev A, Mondal B, Verma VK, Kumar V (2024) Teaching learning optimization-based sliding mode control for frequency regulation in microgrid. Electr Eng, 1–13
11. Sati MM, Kumar D, Singh A, Raparthi M, Alghayadh FY, Soni M (2024) Two-area power system with automatic generation control utilizing PID control, FOPID, particle swarm optimization, and genetic algorithms. In: 2024 fourth international conference on advances in electrical, computing, communication and sustainable technologies (ICAECT), pp 1–6. IEEE
12. Ojha SK, Maddela CO (2023) Load frequency control of a two-area power system with renewable energy sources using brown bear optimization technique. Electr Eng, 1–25
13. Boopathi D, Jagtheesan K, Samanta S, Naidu K (2024) Performance analysis of a multi-objective function-based PID controller for system frequency regulation. In: Applied multi-objective optimization (pp 115–132). Singapore: Springer Nature Singapore
14. Ali T, Malik SA, Daraz A, Adeel M, Aslam S, Herodotou H (2023) Load frequency control and automatic voltage regulation in four-area interconnected power systems using a gradient-based optimizer. Energies 16(5):2086
15. Tur MR, Wadi M, Shobole A, Ay S (2018) Load frequency control of two areas interconnected power system using fuzzy logic control and PID controller. In: 2018 7th international conference on renewable energy research and applications (ICRERA) (pp 1253–1258). IEEE
16. Shouran M, Anayi F, Packianather M (2021) The bees algorithm tuned sliding mode control for load frequency control in a two-area power system. Energies 14(18):5701
17. Boopathi D, Jagatheesan K, Samanta S, Anand B, Jaya J (2023) Application of genetic algorithm-based controllers in wind energy systems for smart energy management. In: Applied genetic algorithm and its variants: case studies and new developments (pp 139–160). Singapore: Springer Nature Singapore
18. Gupta DK, Soni AK, Jha AV, Mishra SK, Appasani B, Srinivasulu A, Bizon N, Thounthong P (2021) Hybrid gravitational–firefly algorithm-based load frequency control for hydrothermal two-area system. Mathematics 9(7):712
19. Boopathi D, Jagatheesan K, Samanta S, Dey N, Jaya J, Anand B (2023) Mayfly algorithm optimized FOPID regulator for frequency improvement in wind power systems. In: 2023 IEEE Pune section international conference (PuneCon) (pp 1–5). IEEE
20. Hakimuddin N, Nasiruddin I, Bhatti TS, Arya Y (2020) Optimal automatic generation control with hydro, thermal, gas, and wind power plants in 2-area interconnected power system. Electric Power Compon Syst 48(6–7):558–571
21. Boopathi D, Jagatheesan K, Samanta S, Anand B, Satheeshkumar R (2023) Energy storage units for frequency management in nuclear generators-based power system. Energy storage technologies in grid modernization, pp 105–134
22. Boopathi D, Jagatheesan K, Samanta S (2024) Application of ant colony optimizer in thermal power system for frequency management. In: Applications of ant colony optimization and its variants: case studies and new developments (pp 189–210). Singapore: Springer Nature Singapore
23. El-Shorbagy MA, El-Refaey AM (2022) A hybrid genetic–firefly algorithm for engineering design problems. J Comput Design Eng 9(2):706–730

Chapter 8
Optimization and Machine Learning Methods in Multivariate Geochemical Analysis for Anomaly Detection

F. Moradpouri and M. B. Dolatshahi

1 Introduction

Optimization is one of the favorite topics in many fields, especially engineering fields [1]. In many cases, researchers and engineers need to find the optimal parameter to answer their problem, that is, an optimal answer, which minimizes the behavior of a given plan or the value of a given procedure. Optimization is the tool that differentiates the engineering perspective on a problem [2].

Actually, optimization is the action of getting the most suitable result under specific occasions. In the process of designing, modeling, building, and maintaining any engineering system, many decisions must be made at various stages, with the ultimate goal of minimizing the effort required or maximizing a desired benefit [3]. Both effort required and benefit desired can be expressed by functions of decision variables. Therefore, optimization is known as a way to find specific conditions that can optimize the behavior of a system or the response of a function.

Some optimization methods are appropriate for various types of optimization problems such as linear or non-linear, simple or multi-objective, and discrete or continuous. These methods are divided into two major categories: deterministic optimization and stochastic optimization. Deterministic optimization is a classical technique of algorithm optimization that uses calculations and operators in algebraic methods, such as gradients and Hessians. Stochastic optimization employs randomization in its search procedures for the optimal solution [1, 4–7].

F. Moradpouri (✉)
Faculty of Engineering, Department of Mining Engineering, Lorestan University, Khorramabad, Iran
e-mail: moradpouri.fa@lu.ac.ir

M. B. Dolatshahi
Faculty of Engineering, Department of Computer Engineering, Lorestan University, Khorramabad, Iran
e-mail: dowlatshahi.mb@lu.ac.ir

N. Dey (ed.), *Optimizing Solutions for Real-Life Problems*, Springer Tracts in Nature-Inspired Computing, https://doi.org/10.1007/978-981-96-3216-9_8

On the other side, geochemical exploration is a branch of earth sciences that wants to explore the existence of mineral deposits in the earth's subsurface. Prospecting techniques are based on the premise that the combination of materials from the earth's crust, in the vicinity of a mineral deposit, is different from the composition of similar materials located in regions without any type of mineralization. By the action of chemical and physical processes on or near the surface, these anomalous concentrations are incorporated into the weathering cycle with the dispersion of its components, thus producing a plume-shaped dispersion halo that shapes the mineralized body or an exploratory target [8–10].

Geochemical exploration techniques have always had a bias towards mineral research, mainly metallic ores, but have been undergoing adaptations to the new needs of the society. Exploration methods have been used in the study of places with abundance and lack of natural or artificial substances, relating them to the different areas of the knowledge, such as agronomy, veterinary, public health and environmental monitoring. Geochemical surveys are called multi-purpose. The application of techniques exploration is being facilitated by the advancement of methodological field research and laboratory, with the development of new sampling techniques and the isolation of specific fractions. In addition, the advance of multielement analytical techniques and the reduction of detection limits have contributed to reveal important subtle geochemical anomalies and significant. Allied to this, modern computer techniques have made possible the treatment and interpretation of a huge mass of geochemical data producing important results in the study of the physical environment [8–10]. This modern approach to geochemical exploration has made possible the execution of projects of regional or national scope in several regions from the earth. Countries like Canada, United States of America, Finland, Norway, England, Germany, and China, already in the 1970s, and more recently several other countries are preparing low-density geochemical surveys. This approach of geochemical exploration is one of the main methods of identifying the natural environment. The amplitude application of these operations reaches since the projects on environmental changes continental and global to regional and local works in mining exploration research. Geochemical surveys on a national and continental scale contribute to the characterization of the metallogenetic vocation of the diverse geological environments identified in geological cartography [11, 12]. Darnley et al. (1995) emphasized: "Some say that a good geological map can predict the geochemistry of the area" [13]. But what is geochemistry information on the same scale? The response is that, this type of survey contributes a decisive way for geological mapping of large regions with scarcity of outcrops or with dense vegetation cover [14–16]. Therefore, the current chapter aims to discuss some of the popular machine learning methods for the optimization of multivariate geochemical data analysis, which will be discussed by describing two practical examples of the two exploratory geochemical projects.

2 Concepts of Statistics Applied to Geochemical Prospecting

The statistical tools aim to simplify the mass of information obtained in a geochemical survey. There are rare cases that the simple visual examination of the list of results provides the questions raised in a prospective work. The techniques statistics allow to clarify features that were hidden, to characterize the behavior of metallic and non-metallic elements and their relationships in order to make predictions and extrapolations to explain geological and geochemical phenomena [17].

The measurements made on each sample represent the observations made in the field or in the laboratory and can be expressed in different scales, from a simple notation presence or absence to some numerical value. The results of the instrumental geochemical analyses are usually reported quantitatively. Numerical information acquired at the sampling site is called variables and their respective values are generically called data. When conducting a sampling operation; geochemistry is defined as the target population of interest, or simply population, which is the object of interest to be investigated (for example, geological environment or geographical area). Collected samples will be the sampled population for statistical studies. The statistical sample is finite and constitutes part of the population. The statistical procedures performed on the data set will allow interpretations and inferences about the sample population or even about the target population. The higher the number of samples, the greater the degree of reliability in the results of statistical operations and the greater the proximity of the statistical sample to the sampled population.

2.1 *Univariate Statistics*

In the univariate statistical studies, statistical parameters (minimum, maximum, mean, standard deviation), and frequencies, histograms and boxplot for the main variables are calculated and drawn. The applications of univariate methods define levels of background (background content or predominant content), thresholds, anomalies, and outliers (content aberrant). The main purpose of univariate statistics is to identify normal distributions which are represented in the set of analytical results [17].

The analytical results of the samples for a given variable form originally a chaotic mass of values ordered according to the delivery note of the laboratory. To start your interpretation, it is necessary to organize them logically. Listing the results in ascending order a list is obtained with the maximum and minimum values in the extremities, the difference of which is called data amplitude. Counting repetitive analytical results of these data lists allows organizing a table of frequency containing, in each line, one of the determined values and the number of times in that is found in the data set. This count is normally presented in the form of histograms. The main components of the frequency table and the histogram are the class range, constant added to each class to find the next and the point medium, central value that characterizes each of the classes, being the fashion or modal class the one with the highest number of observations; in the same data set can also be other classes with higher frequencies than contiguous classes. The critical phase for the construction of the frequency table and histogram is to find the class range and its origin that best fit the data to represent faithfully the natural phenomena pertaining to the sampled population.

A widely used representation of frequencies is the boxplot or box-and-whisker (Fig. 1). This graph is obtained through the growing list of values regardless of data classification. The accumulated frequency table identifies the values corresponding to quartiles 25%, 50%, and 75%. Quartiles 25% and 75% (Q_1 and Q_3, respectively) form the sides of a rectangle (the box), sectioned by the 50% quartile, and called the median for separating the data set into two identical portions (50% of lower values and 50% higher values). The lines (whiskers) extend from both sides of the box to the extreme values or the values 1.5 times lower and higher (LIF and UIF) than the values of the 25–50% and 50–75% interquartile ranges, respectively. In this case, the results below in addition to whiskers can be considered anomalous values (outliers). The boxplot has the advantage of allowing quick and effective visual comparisons and the disadvantage, of the histogram, of not showing the possible existence of two or more fashions.

A serious problem in data interpretation is to determine the limit values in such a way that the background and anomaly can be separated based on them [18]. Reimann et al. (2005) make a comparison of methods and recommend the establishment of the threshold only after inspection using statistical and geographical data displays [19]. The definition of geochemical background is also subject to variation and depends on the location and scale of work [20].

2.2 *Significance Tests and Hypothesis Tests*

Geochemical prospecting works often have to respond to questions about methodological aspects, such as whether the distribution of a variable in one population is similar or different to the distribution of the same variable in another population (adjustment of the acquired data compared to the Gaussian distribution, equivalence of soil depths or horizons); and if there is similarity or divergence between one individual and another, despite belonging to different data sets (confrontation between field or laboratory replicas). Decisions can be made after performing some statistical tests, between groups in the first case and between pairs in the second, to a certain degree of reliability or significance. Such tests of significance are still divided into two categories: parametric (Chi-square, Student's t, and Snedecor's F among others), that is, performed using parameters such as the mean and variance and requiring association with some kind of theoretical distribution and non-parametric ones that do not use them and are generally independent of the distribution of the data to be compared.

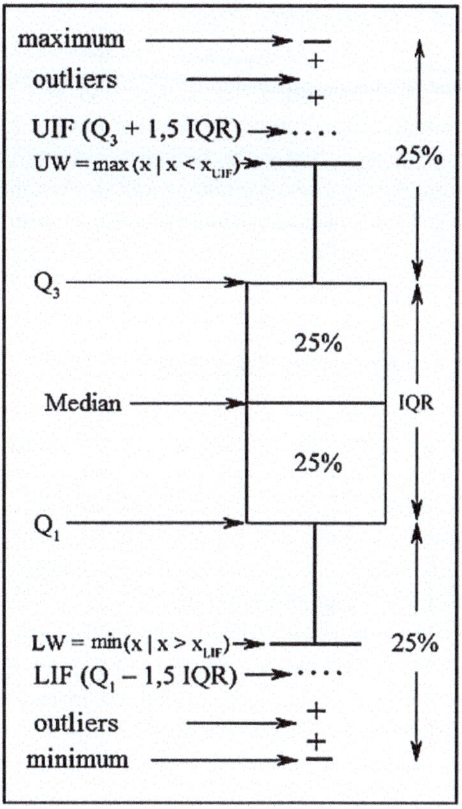

Fig. 1 Boxplot [18]

2.3 *Variance Analysis*

This method is based on the important property that the total observed variance of one variable is equal to the sum of the contributions from each source liable to add variability to that total. There are several types of this test, the most applied in geochemical prospecting to called one-way analysis of variance, which allows the disaggregation of the total variance in its various components, based on a hierarchical sampling structure with essential randomness at each of its levels, which can be addressed only at the top or highest level. It is a very powerful tool that allows you to establish the plan of ideal sampling for prospecting an area; according to the structure designed, soil levels and depths, distances between samples, drainage and granulometric fractions, together with the sampling, preparation, and sample analysis.

2.4 Probability Plots

Before determining the background and threshold values, it is very important to show the distribution graphically and the cumulative frequency diagram, because it can supply a first sight, before any numerical calculations. In Q-normal plots, it is possible to observe changes and deviations, the existence of multiple populations and outliers using normality or log-normality [18]. The background and anomaly separation using probability plot modeling will be discussed later.

2.5 Bivariate Statistics

Some of the processes that are commonly carried out with two variables are not statistical but rather artifices, used to highlight aspects that would not be so clear by examining only the isolated values of these variables. This is the case with certain indexes and normalizations, some classics and others that can be generated empirically, whose results can be treated as new variables or simply displayed on maps or profiles to be interpreted [9–12].

The important and necessary investigation of the reproducibility of the sampling and analytics can be considered as an examination of the relationship between two variables, that is, results obtained in routine samples versus results for the same variable in the respective replicas; the answer can be found in the field of statistics, for analysis of variance, hypothesis tests and formulas such as those presented by Garrett (1969) [21]. Specifically, the level of error and analytical precision are well verified with the use of graphics such as those suggested by Thompson & Howarth (1978), among other types of existing controls (Fig. 2). One of the advantages of graphic monitoring over the simple application of analytical precision formulas is the eventual detection of deviations systematically, allowing the correction of other results by leveling in the function of selected samples [22].

The relationship between two variables, whenever applicable, must be visually verified through the dispersion diagrams built by the arrangement of the values of the two variables in the orthogonal Cartesian axes, the relationship between them being marked by points in the intersections of lines extended from the respective value scales. The exam allows you to check linear trends or not, relative clusters and spreadings and the presence of outliers, information omitted when the relationship is expressed by a simple index numerical [23].

2.6 Correlations and Regression

The relationship between two variables does not necessarily imply cause-effect, and it is said that there is a correlation between them when an increase in the values of

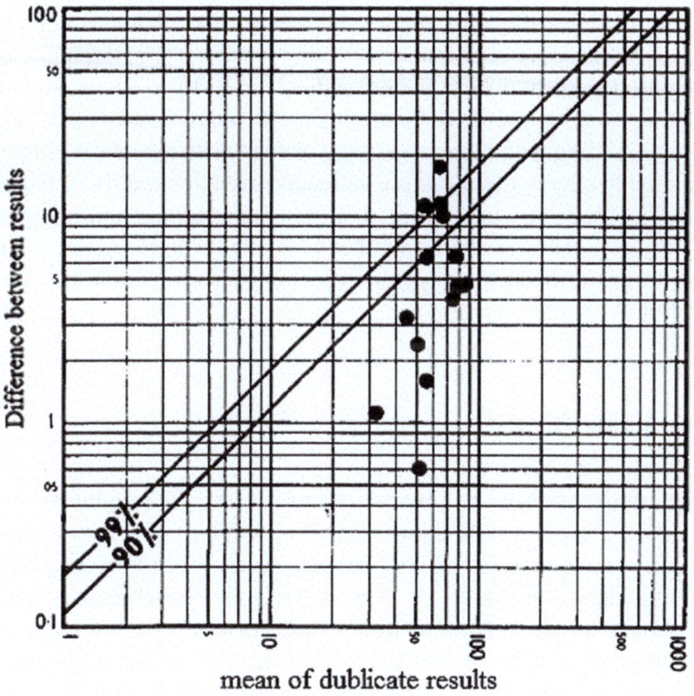

Fig. 2 Accuracy graph [22]

one compares to an increase or decrease in the values of the other. The measure of the intensity of this relationship between the two is the correlation coefficient, which varies from −1 to 1 (perfect inverse correlation to perfect correlation), having a value of 0 when there is no relationship between them.

Two methods are best known for expressing correlation, the Pearson linear product moment (r), with the requirement that the data have a normal distribution, and the Spearman's rank coefficient (rs), which is independent of the underlying theoretical distribution of the data; the latter is similar to the Pearson coefficient, differing in that it does not consider the values of the variables in each sample, but rather the differences in the increasing ordering of these values. Virtually all statistical programs contain correlation modules, saving the manual calculation. The value of r-squared (r^2) constitutes the portion of the total variance that can be interpreted by the relationship between the variables and is called the determination coefficient. The significance of Pearson's correlation is obtained from tables widespread and that of Spearman's correlation can be assessed by an approximation with the Student's t distribution; the fewer samples, the greater the critical value (independent of the sign) below which the correlation should not be considered. The relationship between the points can also be expressed by a straight line that best fits them, constructed using the least squares method. It is known as the regression line of one variable in relation to another. Once the regression line has been calculated and validated by observing the binary diagram, the y values can be calculated from the x results. These regressions are very useful in prospecting when using pathfinder elements to locate the desired element, which is more difficult or costly to detect, and in the case of interpolating values in profiles or sampling mesh, reducing analysis costs for the y element. When dealing with more than two variables, with the relationships between them calculated two by two, a structure called a correlation matrix of the calculated values of the coefficient r can be created. To facilitate reading and interpretation of this apparently confusing matrix of numbers, symbols with ranges of values can be used or even transformed into a type of geochemical association diagram, which greatly facilitates its interpretation.

3 Multivariate Statistics

Multivariate statistical methods provide the possibility of simultaneous statistical analysis of several variables. Problems related to one, two, and even three variables can be imagined or shown graphically, but sometimes in exploratory problems, we are faced with a space of 10, 20, or even 40 variables, which makes it difficult to investigate the relationships between them. In this case, it is necessary to reduce the number of dimensions in the studied space by using multivariate statistical methods, so that the results of these new dimensions (new variables) with a much smaller number than the previous case, can explain most of the variability of the data. For example, in exploratory geochemistry, the simultaneous changeability of several elements (variables) can be studied for more accurate detection of possible anomalies.

3.1 Principal Component Analysis (PCA)

PCA is actually obtaining eigenvalues and vectors from the variance–covariance matrix or the correlation matrix. When the variables, due to different scales of measurements employed, cannot be directly compared, standardization becomes necessary preliminarily, so that the transformed variables have a mean zero and unit variance, which is achieved by the transformation logarithmic. In these cases, with standardized variables, the variance–covariance matrix and the correlation coefficients become identical. As such standardization entails a strong effect on the variance–covariance matrix and, consequently, on the output of the analysis, its use must be judicious, always considering the nature of the geological data under study and the focus it is intended to give.

The first analysis is actually the linear transformation of the initial variables X into the new variables X, as the first new computed variable is responsible for the largest possible variation in the data set, the second by the largest variation possible remainder and so on, until all the variation in the set has been explained.

The analysis of the main components is, therefore, a technique of transformation of variables. If we consider each measured variable as a variability axis that is correlated with other variables. This analysis transmits the data in a way that it reflects the whole variability and does not correlate with other axes. Graphically, this transition can be explained as the rotation of existing points in a multi-dimensional space and the assignment of new axes. New components in a 2D space may reflect the variability necessary to recognize existing patterns [24, 25].

The analysis of the main components begins with the eigenvalue calculation and corresponding eigenvectors of a variance–covariance matrix or correlations between variables and this procedure is known as "R" mode. The first eigenvalue to be determined will correspond to the highest percentage of the total variability present and thus successively. Usually, the first two or three eigenvectors found will explain most of the present variability. When the first eigenvector already explains 90–95% of the variability, this should be looked at carefully and checked if they are not present variables with magnitude values much higher than the others. The eigenvectors correspond to the main components and are the results of loading the variable's originals in each of them. Such shipments can be considered as a measure of the relevant importance of each variable in relation to the main components and the respective signs, whether positive or negative, indicate relations directly and inversely proportional. The load matrix of each variable in the main components, when multiplied by the original data matrix will provide the matrix of scores (scores) for each case in the main components. These values can then be displayed in a diagram dispersion, where the axes are the two most important components, and show the relationship between cases conditioned by the measured variables. The results of principal component analysis are generally shown as biplots (Fig. 3). Figure 3 shows a graphic analysis of the principal component analysis, which includes elements A, B, C, and D, and they are symbols of special elements that are shown with gray dots in the figure. The load vectors are also shown with black arrows in the figure, and four red dotted lines show the connection between these black arrows. The length of the AC line is longer than other connections, which indicates more variance from A to C. The connection length of CD is shorter, which shows a relative relationship between C and D. Also, the small angle between AC and BC shows a correlation between A/C and B/C, while the close angle between C and B shows independence between C/D and B/D. Through this form, the relationship between pairs of elements can be recognized better and more easily.

Below are two examples of principal component analysis on geochemical data. The data used in the first example are related to the analysis of rock samples taken from an exploration region in British Columbia, Canada. Sample location is presented in Fig. 4. Here, the goal is to realize the elements related to the copper element and specifically the subsurface mineralization of copper and reduce the number of variables and problem dimensions for further investigations. Correlation of all measured elements with copper element is presented in Table 1. As shown, Mo, Au, and Ag elements have the highest correlation with copper element, which is expected by geological, petrological, and mineralogy studies conducted in the region [10, 11, 17].

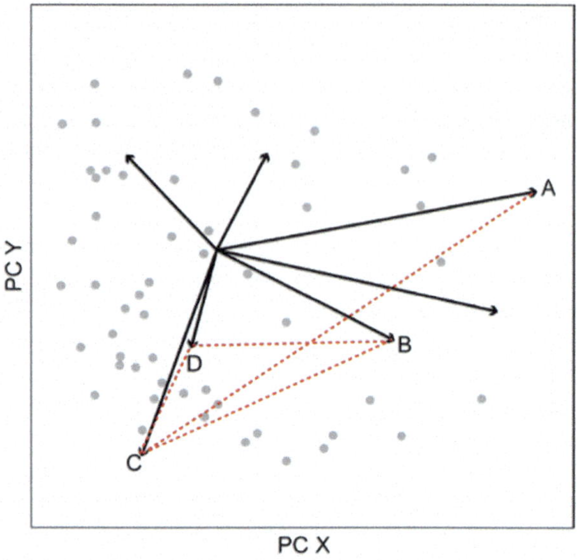

Fig. 3 Example of the center log-ratio transformed biplot. Black arrows of A, B, C, and D, are loading vectors, gray circles are sample scores, and red links plotted along two principal components (X and Y)

After analyzing the main components, the amount of variability provided by the first five components is shown in Table 2. It shows the variable value for each component and its specific value. As it can be seen, the first and second components alone justify 56.8% variability and five components justify 68.4% variability. Regarding the subsequent components, there is no considerable change in justified variability. Therefore, the above components form the new dimensions.

Also, Fig. 5a–c illustrates the coordinates according to components PC1, PC2, and PC3. In Fig. 5a, the elements of Cu, Au, and Mo as the main variables of the deposit, and the elements of As, Ag, Se, Pb, Bi, and S are far away from the center and close to each other. It indicates a good and positive correlation between them. Also, the correlation between elements can be described using Fig. 5b, c so that the correlation and connection between different elements can be better known. On the other hand, the connection between the vectors of two elements can be used to interpret the existing correlation between them. For example, in Fig. 5, the relatively small angle between the two straight lines connecting Fe and Cu and Au and Fe indicates a correlation between the ratio of Cu/Fe and Au/Fe, while the close angle between Cu/Ti and Zn/Ti shows a negative correlation between these elements. Therefore, through such analyses, it is possible to know the relationship between different elements and their relationships with possible mineralization.

The data used in the second example (Case study 2) are related to the analysis of sediment samples taken from an exploration area in the northwest of Iran. The samples location of is presented in Fig. 6. In this example, the goal is to specify the elements related to the arsenic tracer element as a subsurface mineralization tracer and to reduce the variable number and dimensions of the problem for further investigation. The basic statistics of the case study 1 dataset is shown in Table 3.

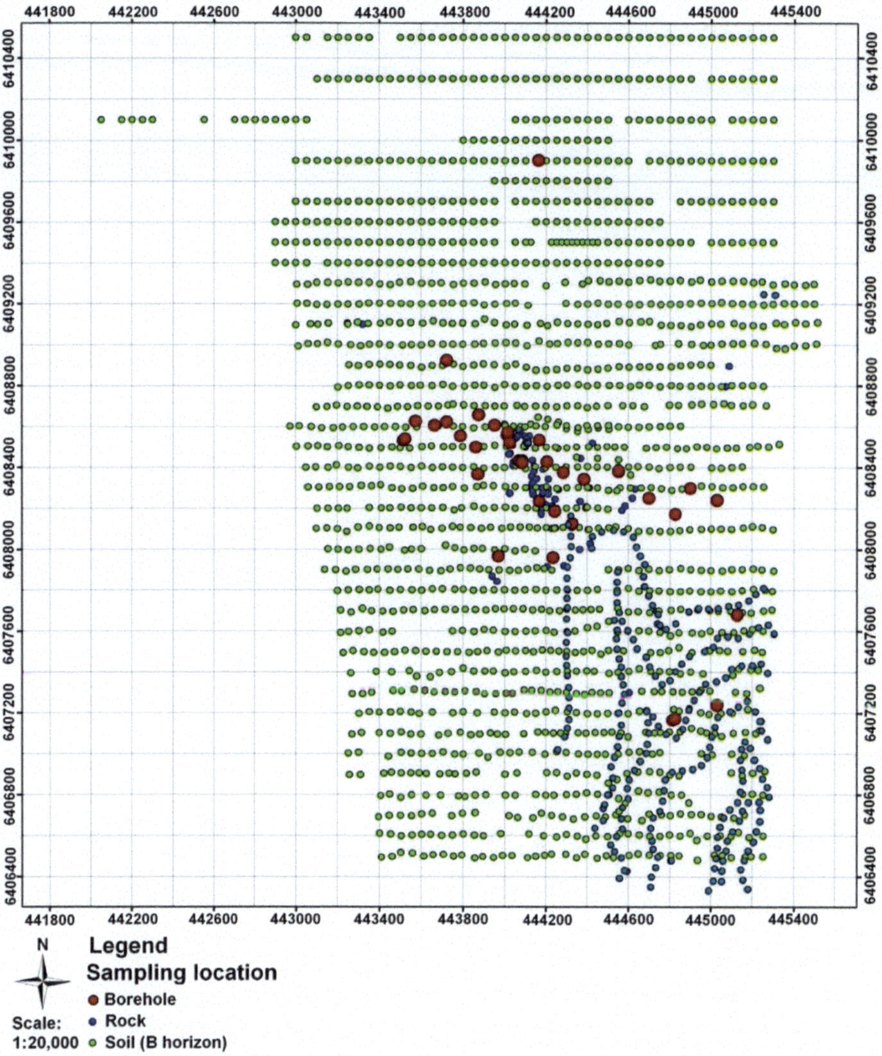

Fig. 4 Location map of the rock samples (blue circles) in British Columbia, Canada exploratory study area (case study 1) [11]

Principal component analysis was done on the data of stream sediments to extract the relationship between the elements. Specific values, justified variability, and the cumulative percentage of justified variability for the first seven components are shown in Table 4. As can be seen, the first two components justify 45.63% of the variability, while this percentage for the first five components is 66.9%. Figure 7 shows that after the first five components, the variability justified by the new components is not so much, and these components can be ignored. [17]

Table 1 Spearman correlation coefficient for Cu elements with other elements

Variables	Cu	Variables	Cu
Cu	**1**	V	0.184
Mo	**0.577**	Ca	**−0.286**
Pb	**0.221**	P	0.054
Zn	0.094	La	**−0.456**
Ag	**0.727**	Cr	0.009
Ni	−0.130	Mg	−0.035
Co	−0.054	Ba	**−0.304**
Mn	−0.172	Ti	0.034
Fe	**0.433**	B	0.043
As	**0.319**	Al	−0.095
Au	**0.805**	Na	−0.054
Th	**−0.342**	K	−0.043
Sr	−0.150	Hg	0.127
Cd	0.029	S	**0.325**
Sb	0.161	Sc	−0.131
Bi	**0.305**	Se	**0.563**
Ga	0.093		

Table 2 The main principal components and their variability

	PC_1	PC_2	PC_3	PC_4	PC_5
Eigenvalue	8.624	6.675	3.435	2.301	1.536
Variability (%)	26.134	20.227	10.409	6.973	4.656
Cumulative (%)	26.134	46.361	56.770	63.743	68.398

Figure 8 shows the relationship and correlation between components 1, 2, and 3 based on the coordinates of the elements. As shown, the loads of component 1 include the elements Ni, Na, Ba, Ti, Mn, Th, Li, Nb, Cr, Sn, Zn, Fe, Co, Cs, Hg, V, Bi, and Tl, and the loads of component 2 includes the elements Rb, Be, K, Mo, Au, As, P, Sb, W, Cu, Ag, Al, Zr, La, Ce, Sc, Pb, and Cd. In addition, the relationship between elements can be explained based on Fig. 8, for example, the most important elements related to (As) as a geothermal tracer are Au, Mo, Sb, W, Cu, Rb, K, Zr, Be, and P. (Fig. 8a). This is due to the small angle between the vectors of each of these elements. On the other hand, due to the small distance of Au, As, Sb, and Mo elements from the center, for a better interpretation of the relationship between these elements, their interaction in the map between other components should also be examined (Fig. 8b, c). On the other side, the relationship between the ratio of elements also provides us with other information for better interpretation, for example, there is no correlation between the ratio of As/Au with As/Cu, but the short length of the connecting line

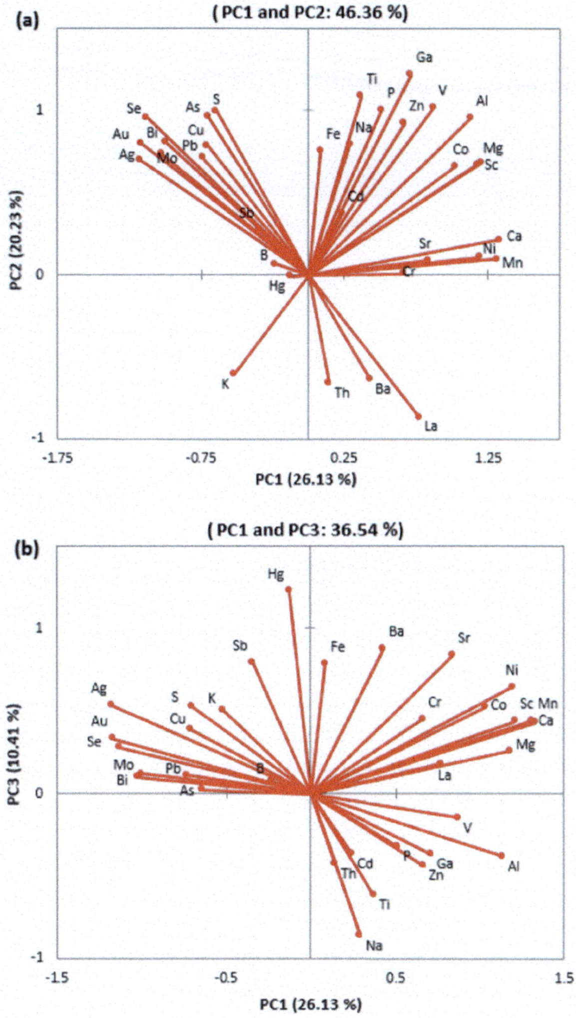

Fig. 5 Variables coordination showing the relationship between variables (elements)

between As and Au and As and Cu, shows It has low variance between each pair of the above elements. For example, for a high correlation between the ratios of two elements, the correlation between the ratio of two elements As/Au and As/Sb can be mentioned. Figure 9a–c shows the coordinates of the samples based on three components 1, 2, and 3.

Fig. 5 (continued)

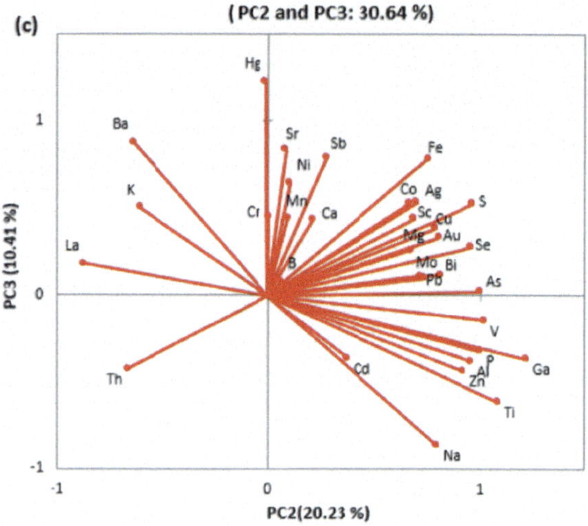

3.2 Clustering Analysis

Cluster analysis is a term used to describe several numerical techniques whose fundamental purpose is to classify the values of a data matrix under study in discrete groups. The multivariate classification technique of clusters can be used when you want to explore the similarities between individuals (Q mode) or between variables (R mode) defining them in groups, considering simultaneously, in the first case, all the variables measured in each individual and, in the second, all the individuals in whom the same measurements were made. The purpose of this method is to seek homogeneous groupings of items represented by points in an n-dimensional space in a convenient number of groups relating them through coefficients of similarities or correspondences. The application of this analysis is controversial among researchers because little is known about the statistical assumptions of their various methods and only limited tests are known to verify the significance of the results. In the introductory analysis of data matrices, however, they become quite efficient for assisting in the formulation of hypotheses regarding the homogeneity or not of these data. According to Davis (2002), different clustering methods can be divided into four categories: approaches with random origin; partition approaches; approaches by hierarchical groupings; and approaches by mutual similarity [26].

The most popular method is hierarchical groupings. The techniques of this method are the most used in geology. From the initial data matrix, a symmetric matrix of similarities and the detection of pairs of cases with the highest similarity, or the lowest distance; for this combination, according to hierarchical levels of similarity, one chooses between the different agglomerative procedures, in such a way that each grouping cycle obeys a successive order in the direction of decreasing similarity. Although several similarity measures have been proposed, only two are generally used: Pearson's correlation coefficient and distance measurement Euclidean. The

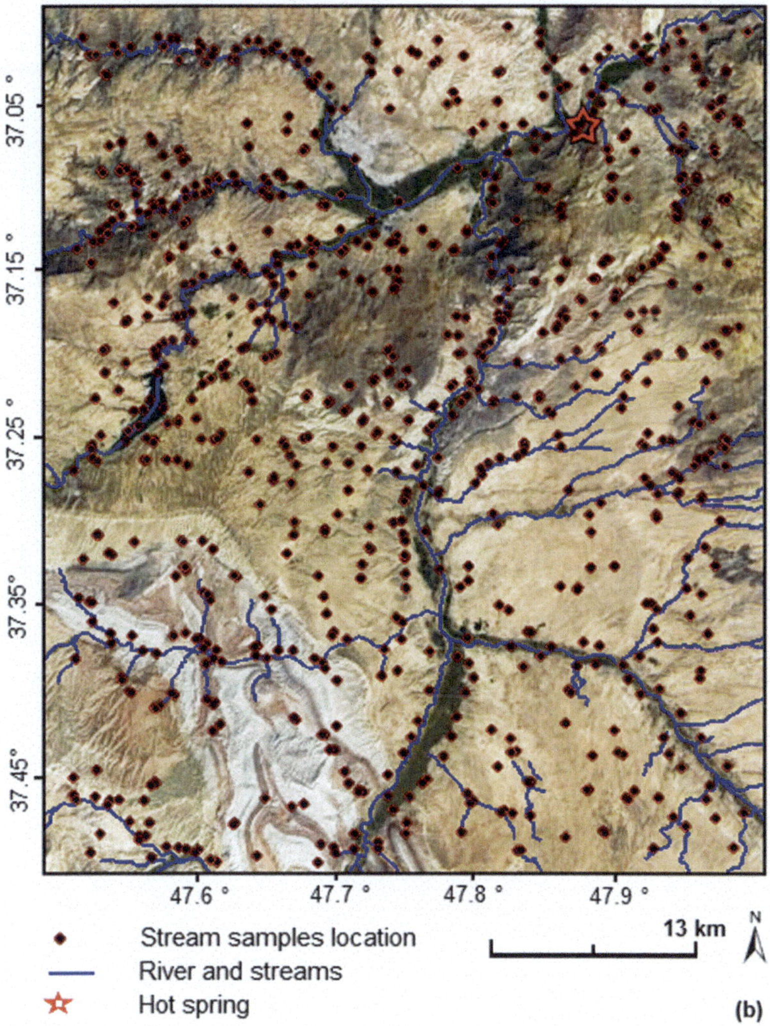

Fig. 6 Location map of the stream sediment samples in Mianeh study area (NW of Iran). (Case study 2) [17]

results when organized in a graph, of the dendrogram type, will show the grouped sample relationships. Various clustering techniques have been proposed, and the most commonly used methods are: simple link (single linkage method or nearest neighbor), complete linkage (complete linkage method or farthest neighbor), proportionally weighted pairwise grouping (weighted pair group method, WPGM); Minimum variance (Ward's method or minimum variance clustering). Here, grouping by minimum variance was used. The most common method of presenting clustering results is in

Table 3 The basic statistics of the case study 1, dataset

Variance	Min	Max	Mean	Standard deviation	Variable	Min	Max	Mean	Standard deviation
As	1.46	174.10	20.069	12.94	Na	1376	73,830	13,422.8	6550.1
Au	1.000	88	2.338	4.18	Nb	4.87	69.67	16.435	5.709
Ag	0.000	2	0.367	0.164	Ni	2.68	74.27	28.366	11.3
Al	23,210	103,300	64,196.7	13,703.5	P	20.14	4321	993.05	758.576
Ba	129.6	2516	675.068	208.582	Pb	9.61	172.3	32.307	13.915
Be	0.64	5.82	1.700	0.594	Rb	12.49	266.9	69.074	30.482
Bi	0.01	12.87	1.501	1.255	S	10	246,100	12,055.0	27,337.8
Ca	11,240	283,200	78,684.4	27,033.9	Sb	0.14	342	2.451	11.79
Cd	0.1	6.48	0.341	0.268	Sc	4.23	25.450	11.781	3.25
Ce	10.94	111	51.739	12.436	Sn	1.8	7.6	3.204	0.676
Co	4.53	57.83	15.712	5.636	Sr	117.8	6124	629.82	584.64
Cr	3.61	227.4	63.184	26.408	Te	0.08	1.0	0.156	0.086
Cs	1.0	27.55	5.889	2.535	Th	3.84	83	14.892	5.887
Cu	9.090	203.5	34.70	20.98	Ti	1363	35,455	5409.1	2383.8
Fe	17,480	335,750	53,608.2	20,787.5	Tl	0.4	2.35	0.918	0.235
Hg	0.05	0.6	0.089	0.026	U	0.94	17.23	4.437	1.554
K	3434	50,800	19,424.5	6558.1	V	41.96	1147	152.98	84.706
La	7.51	75	27.344	7.637	W	0.27	54.3	1.754	2.227
Li	11.33	185.8	28.842	12.291	Zn	24.64	334.1	75.95	29.441
Mg	96.41	32,810	9951.3	4237.1	Zr	75.7	911.2	315.02	121.75
Mn	246.7	4431	903.246	283.6					
Mo	0.010	15.61	1.490	1.338					

Table 4 Main PCs and their eigenvalue and variability

	PC1	PC2	PC3	PC4	PC5	PC6	PC7
Eigenvalue	13.5	5.66	2.73	1.916	1.54	1.42	1.32
Variability (percent)	32.15	13.49	6.51	4.562	3.68	3.37	3.15
Cumulative (percent)	32.15	45.63	52.14	56.705	60.38	63.76	66.9

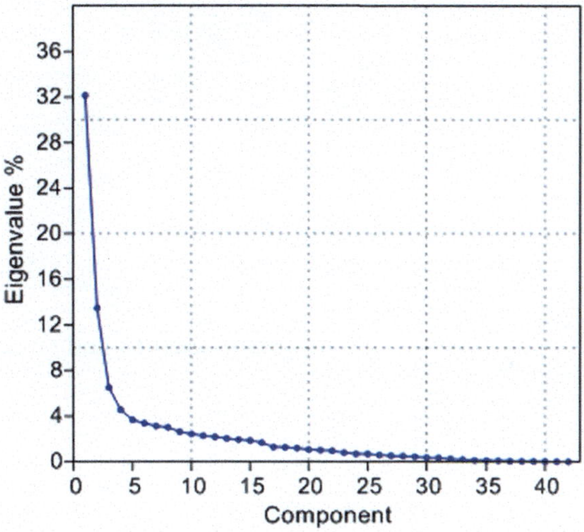

Fig. 7 Variable contribution and the main PCs

the form of a dendogram, which connects the lines based on the similarity between pairs of variables.

Here, the clustering results for the previous two cases (see Figs. 4 and 6) that were discussed in the principal component analysis section are presented. Agglomerative Hierarchical clustering (AHC) method using Ward algorithm was used to explore the relationship between the variables of the two case studies of Figs. 4 and 6. Also, the Euclidean distance method was used to calculate the distance as a dissimilarity criterion which places pairs of variables into groups based on the level of similarity. The results of the AHC method are shown in Figs. 10 and 11. More correlation between variables makes their connection link to be shorter in proportion. As can be seen in Fig. 10, the purple cluster shows the most important related elements to copper (Cu). In Fig. 11, the orange color cluster shows the most important elements related to the arsenic geothermal detector. The mentioned consists of As, W, Au, Zr, Be, K, Mo, Rb, La, Cu, Al, Ag, Ce, P, and Sc, which some of them have a longer distance link and some have a shorter distance. However, the results for the main related elements to Cu and (As) elements of both aforementioned case studies are rather similar to previous results in the PCA section.

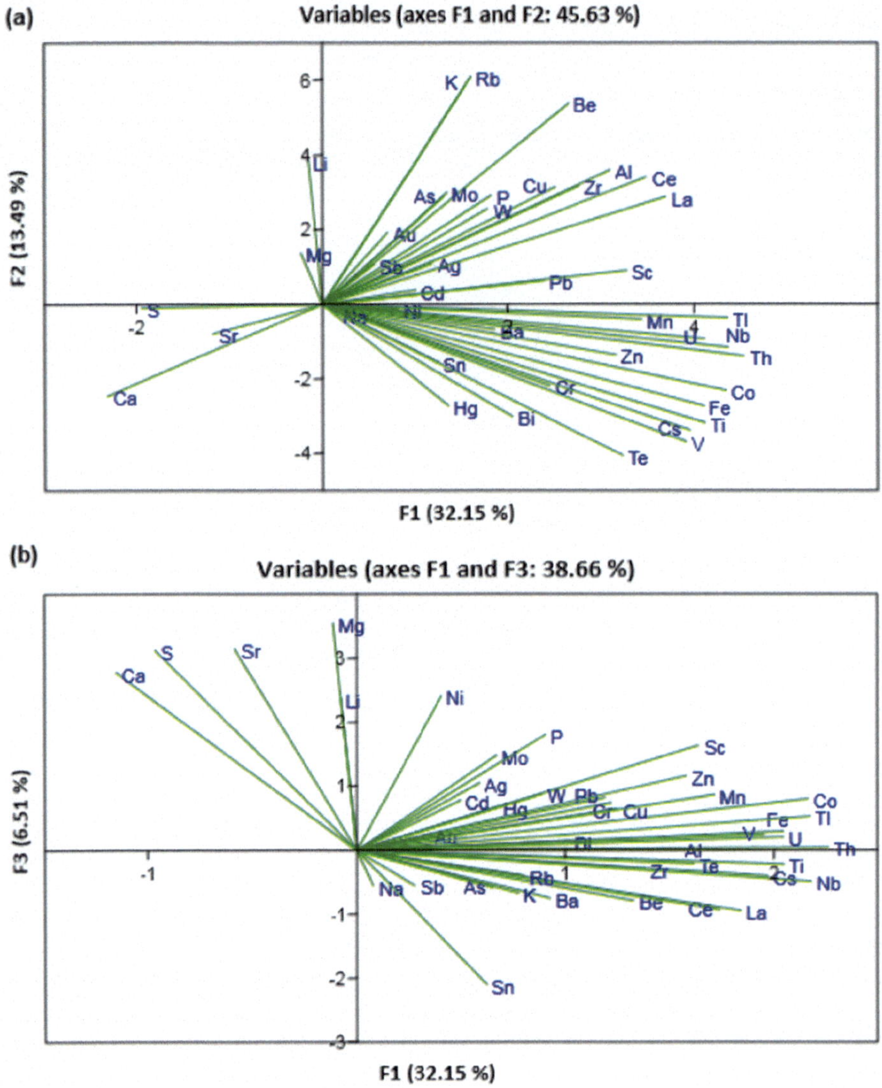

Fig. 8 Coordinates of different variables based on two-by-two combination of components 1, 2, and 3 [10]

3.3 *Discriminant Analysis*

Multivariate statistical analysis using discriminant functions was initially applied to decide which of two groups individuals would belong to on which several identical measurements had been made. The basic idea in this analysis is to replace the set of the various measurements with a single D_i value, defined as a combination linear. To

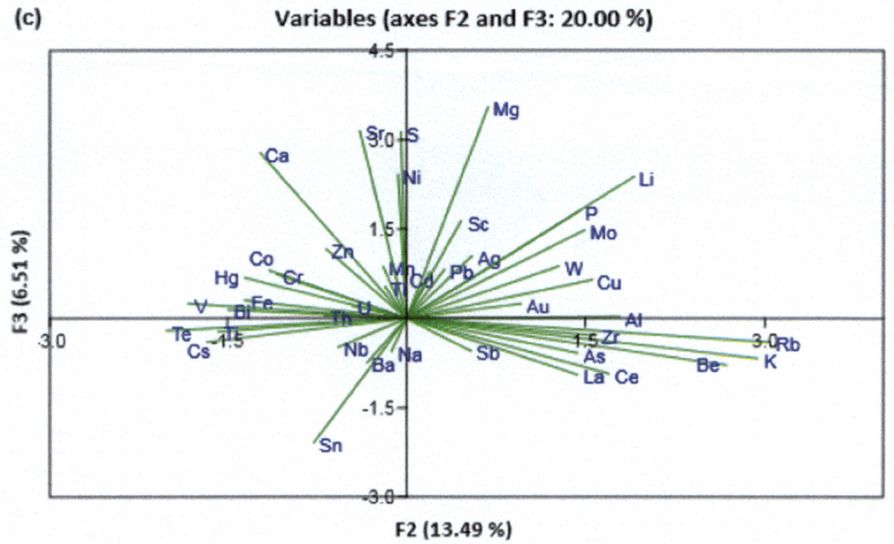

Fig. 8 (continued)

provide a single D_i value, terms are added in a linear function, such that the minimum ratio between the difference between pairs of multivariate means is provided and the multivariate variance within the two groups. Once the D_i value is known, it will be compared with D_0, a value located on the line expressed by the discriminant function halfway between the centers of two groups, to verify which one the individual belongs to. Therefore, the process requires "a priori" knowledge of the relationships between the groups studied. In addition, for the application of tests of significance to discriminant functions, the following assumptions are required:

- That the observations in each group have been chosen at random;
- That the probability that an unknown individual belongs to one of the groups is the same;
- That the variables have a normal distribution;
- That the variance matrices of the compared groups are of the same size;
- That all observations used to calculate the discriminant functions have been classified without error.

To test the significance of the function found, that is, verify that the two groups considered belonging to a single population or to two distinct populations; it is calculated the distance between the two multivariate averages, D_a-D_b. This distance measure is known as *generalized Mehalanobis distance*, or D_2, and measures the separation between the two multivariate means expressed in units of combined variances. The mathematical formulas of multivariate statistical methods are widely disseminated in books such as Davis (2002) [26].

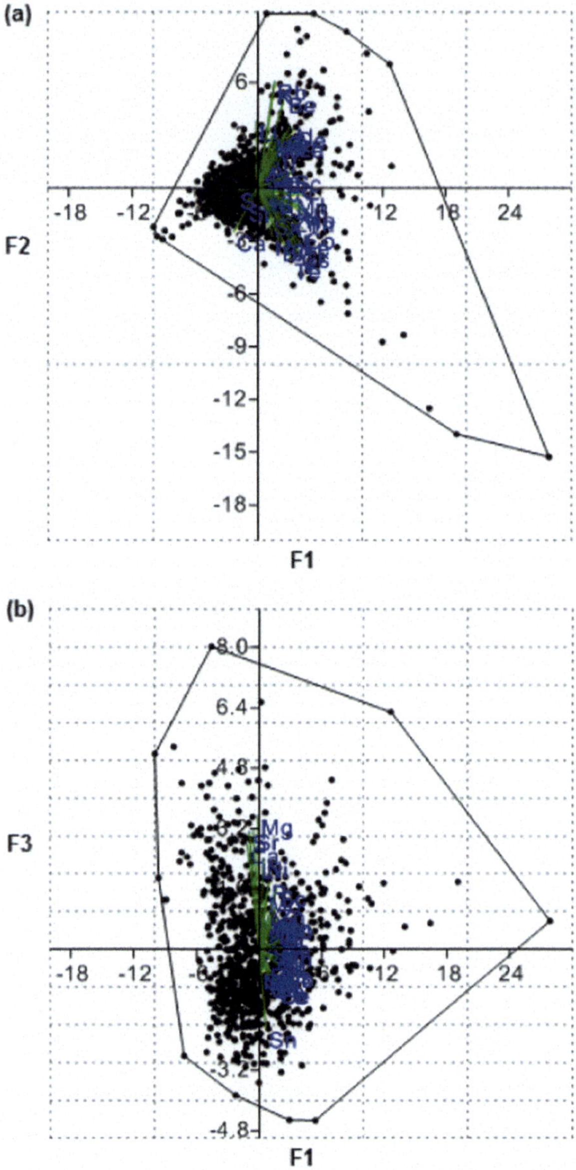

Fig. 9 Coordinates of the samples based on two-by-two combinations of components 1, 2, and 3 [10]

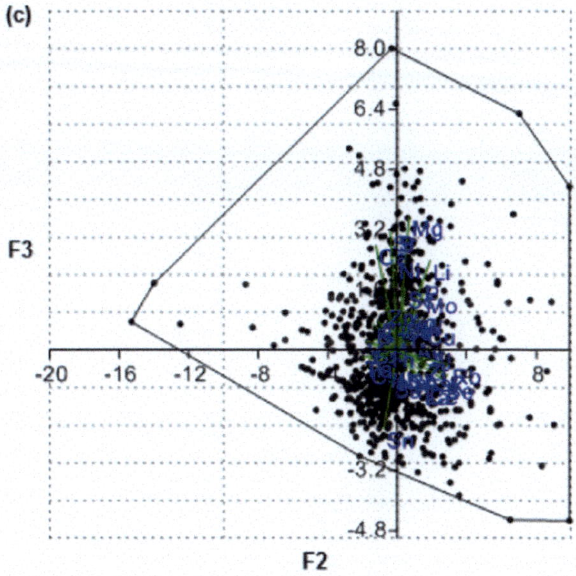

Fig. 9 (continued)

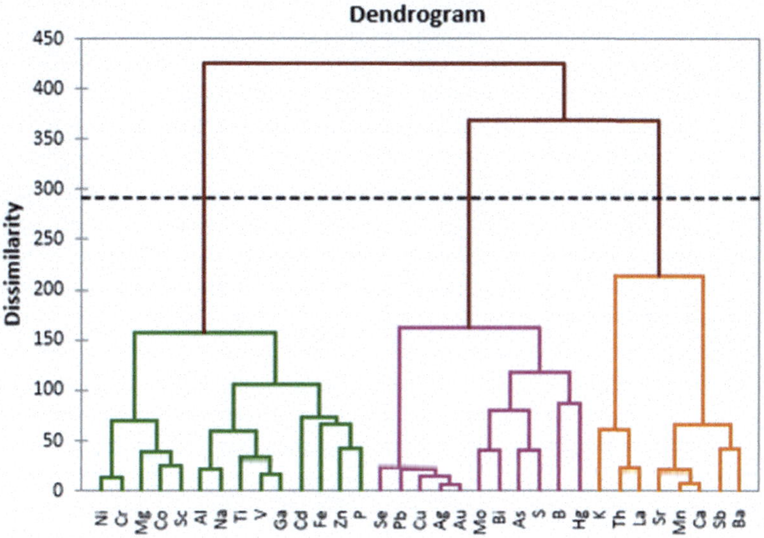

Fig. 10 Dendrogram of distance between different variables [10]

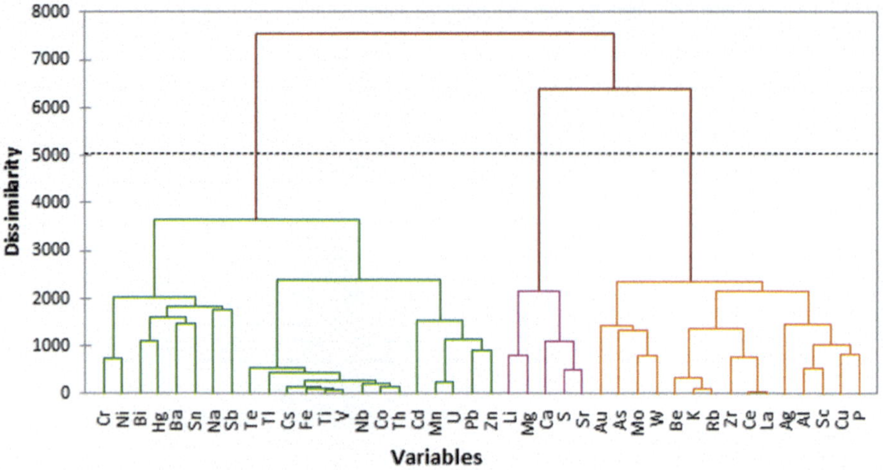

Fig. 11 Dendrogram of distance between different variables [17]

To perform differentiation analysis, the sub-communities in the data were first identified using the probability diagram modeling method (the data from two case studies were mentioned earlier). Figure 12 shows the first stage of the implementation of the PP modeling for the Cu variable to de-convolute the sub-populations. The vertical axis shows the logarithmic values of the concentration of different groups as an empty circle and the horizontal axis is the cumulative percentage. Despite the fact that the mentioned diagram is complicated, but it is possible to see the inflection points, where the background, mixed and anomalous sub-populations can be recognized. As shown in Fig. 12b, the next step is to fit a curve as best as possible to the data so that the existing subpopulations show themselves in the form of straight lines. In the next iterations, we tried to make better fits to the data using the least squares method. Eventually, after successive iterations, the best and most optimal fit to the data was obtained which the threshold values can also be determined (Fig. 12c). Based on the summary information of PP modeling in Table 5, 12.2, 8, and 79.8(%) of the data were designated to the anomaly, background, and mix sub-populations. In addition, it is possible to determine the limits of different sub-communities based on the obtained threshold values as a function of data discrimination.

Discriminant analysis (DA) is carried out to determine whether the different samples were properly discriminated or not. Also, DA also provides the possibility to link related variables to each other. As noted above, in the porphyry deposit of case study 1 (Fig. 4), copper (Cu) is considered as a dependent variable and other variables as independent variables. Therefore, codes 1, 2, and 3 were defined as discrimination functions based on the information obtained from the probability diagram modeling of Cu variable. DA was performed on the data of 87 rock samples with 36 measured elements. According to the table of threshold values, Cu values less than 6 (ppm) were considered as background (code 0), values between 6.6 and 1151 (ppm) as mixed (code 1) and values greater than 1151 (ppm) as anomaly (code 2). The codes

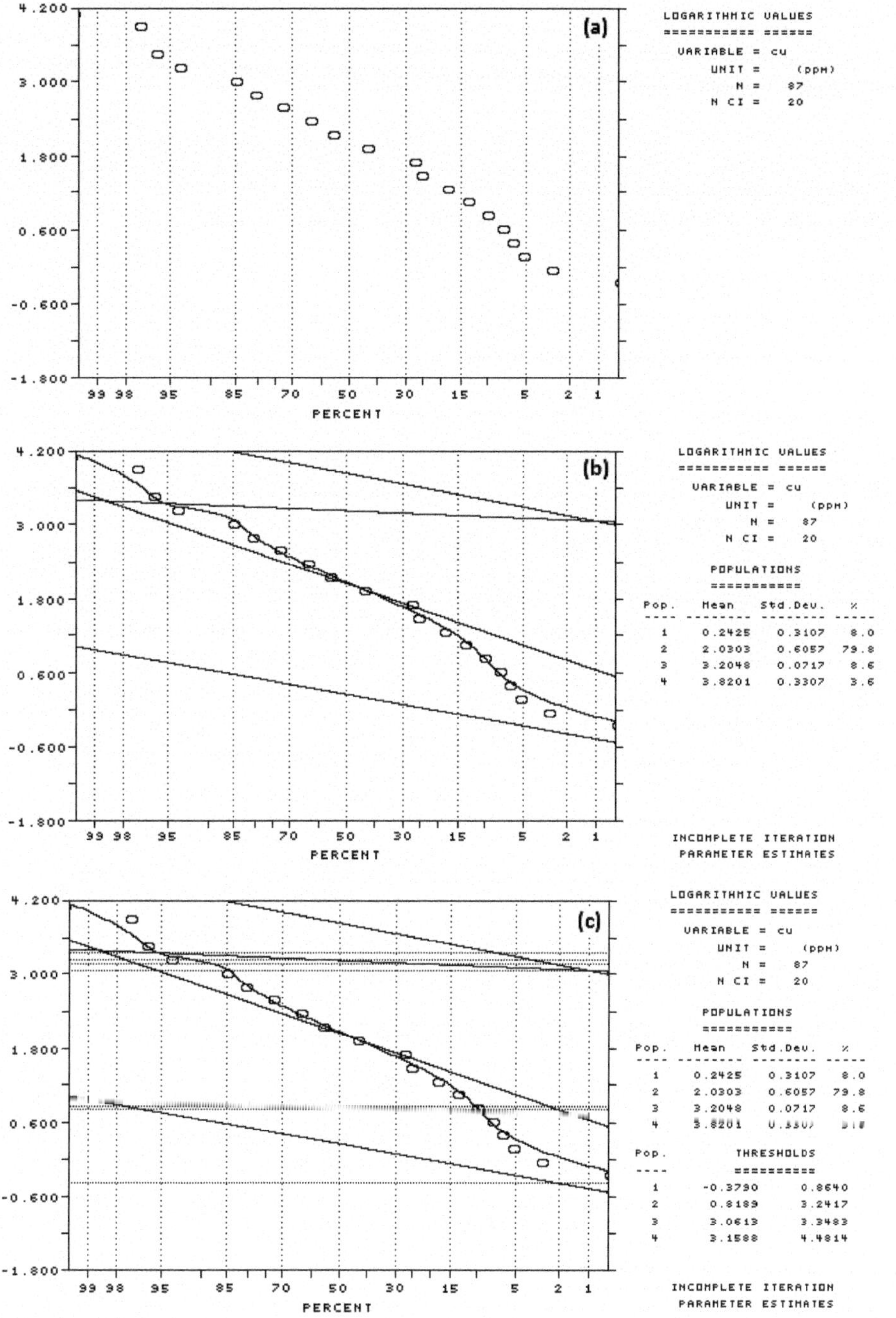

Fig. 12 Different steps of PP modeling. **a** PP of the Cu values (logarithmic). **b** First estimation of the sub-population. **c** Revision of the previous estimation (previous iterations)

Table 5 The details of the PP modeling (the primary threshold values for each subpopulation)

Values (Log.)			
Element: Cu Unit: ppm No. 87			
Sub-populations			
Pop	Mean	Standard Deviation	%
1	0.2425	0.3107	8.0
2	2.0303	0.6057	79.8
3	3.2043	0.0717	8.6
4	3.8201	0.3307	3.6
Pop	*Thresholds*		
1	−0.3790	0.8640	
2	0.8189	3.2417	
3	3.0613	3.3483	
4	3.1588	4.4814	

form the base of the linear function that was applied for DA which assesses the relationship between different variables through the two disputed discriminant functions. Figure 13 presents the results of DA which show the relationship between the Cu and other independent variables.

Also, the results of probability diagram modeling for the case study of Fig. 6 are presented in Fig. 14. The vertical axis shows the logarithmic values of the concentration of different groups as an empty circle and the horizontal axis is the cumulative

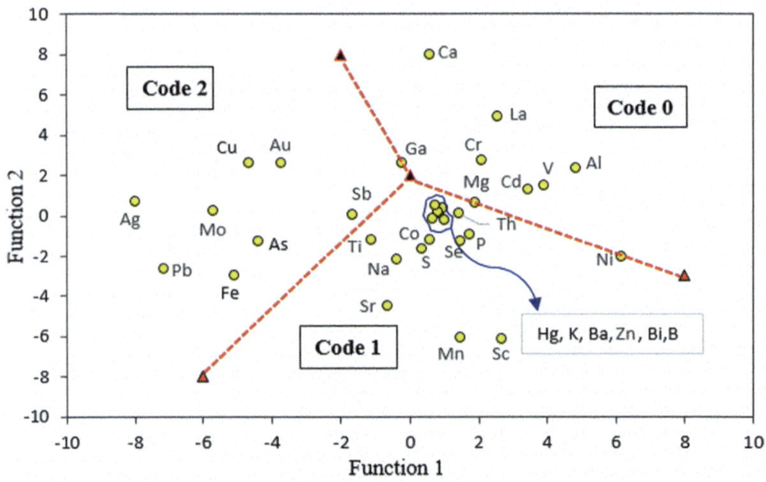

Fig. 13 The results of DA analysis to explore the relationship between Cu and other variables [10]

percentage (Fig. 14a). It is interesting to note that the method of least-squares was applied to fit the best possible model to the data. A solid curve that passed through the empty circles representing different groups of data was the best fit, and the recombination between the data is shown by two sloping straight lines. As shown in Fig. 14b, three sub-populations can be identified. The last step is to improve the prior fitted model that is presented in Fig. 14c, including the final threshold values. The results of PP modeling include the information for each of the sub-populations and the threshold values are presented in Table 6.

According to the results presented in Fig. 14 and Table 6, the values of arsenic (As) element less than 9.21 (ppm), values between 9.21 up to 42.78 (ppm) and values higher than 42.78 (ppm) were considered as background, mixed, and abnormal, respectively.

As noted above, in the geothermal region of case study 2 (Fig. 6), Arsenic (As) is considered as a dependent variable and other variables as independent variables. Therefore, codes 1, 2, and 3 were defined as discrimination functions based on the information obtained from the PP modeling of Arsenic (As). The DA was performed on the matrix obtained from 840 stream-sediment samples and with 42 variables. According to the table of threshold values, AS values less than 9.21 ppm were considered as background (code 0), values between 6.6 and 1151 ppm as mixed (code 1) and values greater than 1151 ppm as anomaly (code 2). The codes form the base of the linear function that was applied for DA, which assesses the relationship between different variables through the two disputed discriminant functions.

As shown in Table 7, two DA functions can be determined. Thus, 90.3% of the variability was described through Function 1, which has a correlation of 84% with the data. Also, 9.7% of the variability with a 45% correlation was described through function 2. As can be seen, the smaller the Wilks' Lambda coefficient, the more suitable the DA function. Also, the greater the Chi-square coefficient causes the function to justify the variability much better.

In continuation, the observations were classified based on the two DA functions. As can be seen in Fig. 15, the classified samples were coded, and various colors were designated for visualization goals. As can be seen in the figure, it indicates that significant discrimination has appeared between the data.

In addition, the discrimination of variables was implemented using the two determined functions that are presented in Fig. 16. Figure 16 presents the fact that represents the separation boundary using (As) variable coded values and the related functions. Also, the other variables' position is the result of the DA considering the (As) variable. Figure 16 shows that As, Mo, Au, K, Th, W, Be, Rb, Fe, Nb, and Ce were situated in the Code 2 area (anomalous) as pathfinder variables for (As) variable and potential mineralization, and Ce, Co, Cs, Ni, P, Li, Zr, and Hg were situated in the mixing area (Code 1).

To assess the results of DA considering the PP modeling and based on the DA functions, 93.2, 86.6, and 81 percent of the observations were accurately classified as codes 0, 1, and 2 respectively, while three distinct groups' memberships were applied (Table 8). Furthermore, the results of Table 8 prove that the used procedure

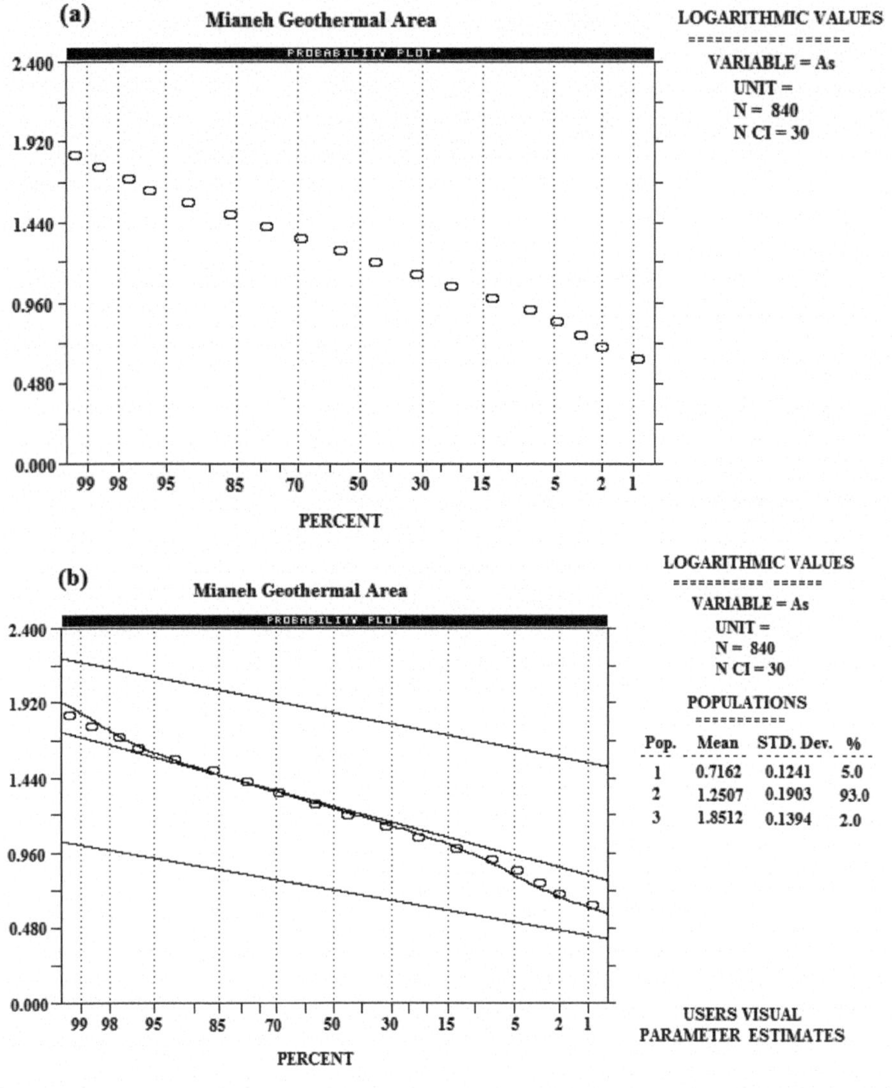

Fig. 14 Steps of PP modeling. **a** PP of the Cu element values (logarithmic). **b** First estimation of three sub-populations. **c** Revision of the previous estimation (previous iterations) [17]

can validate that 92%, 83.5%, and 71.4% of codes 0, 1, and 2, respectively, were classified properly.

Based on the above different methods results (PCA, clustering, and DA), the data were facilitated by realizing seven elements of Mo, Ag, Au, As, Fe, Se, and Pb, as the main related variables to Cu (for the case study of Fig. 4, British Colombia, Canada). For the case study of Fig. 6 (Mianeh, NW of Iran), the variables of Mo, Au, Cu, Be, W, Rb, and K were identified as significant connected elements to the

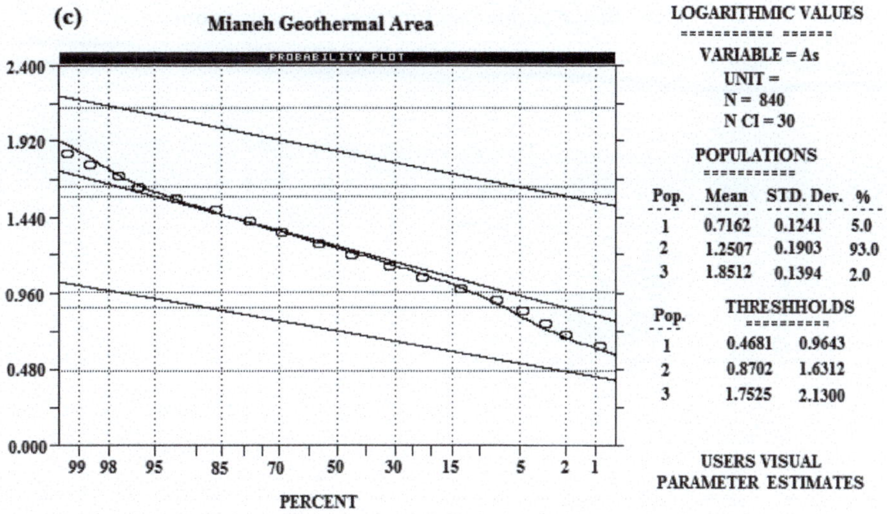

Fig. 14 (continued)

Table 6 The final threshold values for each subpopulation

Elements = As	Unit: (ppm)	No. = 840	Population	Thresholds
Sub-population	Mean	%	1	2.94–9.21
1	5.20	5.0	2	7.42–42.78
2	17.81	93.0	3	37.37–134.9
3	70.99	2.0	**Transform = Logarithmic**	
Populations = 3 **Missing observation = 0**				

Table 7 Computed DA functions and the parameters of validation

Function	Eigenvalue	Variance (%)	Cumulative (%)	Correlation	Wilks' Lambda	(Chi-square)	df	Sig.
1	2.35	90.30	90.30	0.85	0.24	1169.66	84	0.00
2	0.25	9.70	100	0.46	0.79	182.89	41	0.00

(As) variable and using multivariate geochemical analysis (MGA). After specifying the most important variables related to mineralization, the maps of promising copper mineralization areas were presented according to Fig. 17 for the case study of Fig. 4. Also, the prospectivity maps of the main related elements to (As) element are shown in Fig. 18 (Case study 2). It should be added that the explanations and preparations for drawing the map have been neglected here.

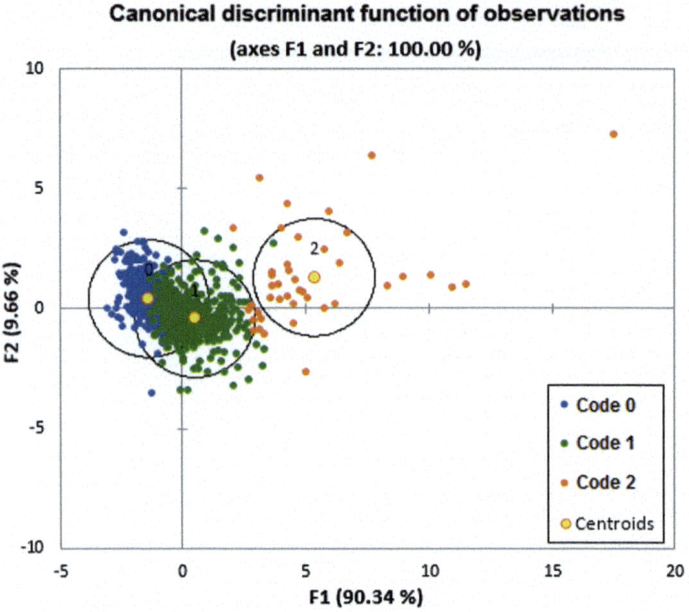

Fig. 15 Observation classification through the discriminant function codes of 0, 1, and 2

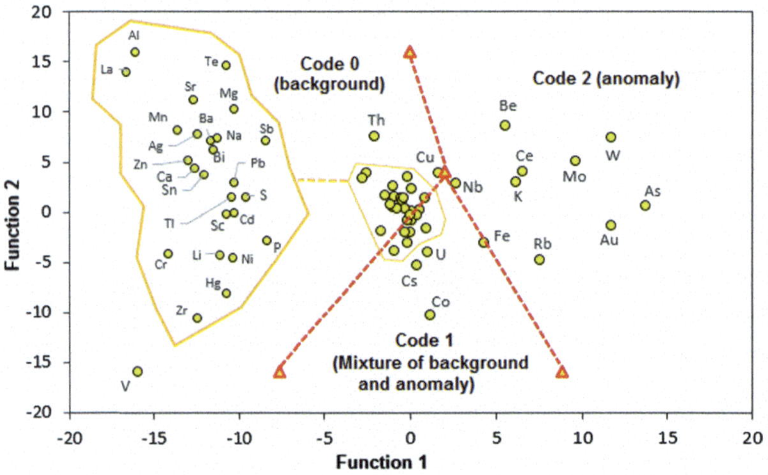

Fig. 16 The results of DA analysis to explore the relationship between (As) variable and other variables [17]

Table 8 DA classification results and their validation

	Code	Membership of predicted group				Total
		0	1	2		
Original	Count	0	314	23	0	337
		1	61	399	1	461
		2	0	8	34	42
	%	0	93.2	6.8	0.0	100.0
		1	13.2	86.6	0.2	100.0
		2	0.0	19.0	81.0	100.0
Cross-validated	Count	0	310	27	0	337
		1	71	385	5	461
		2	0	12	30	42
	%	0	92.0	8.0	0.0	100.0
		1	15.4	83.5	1.1	100.0
		2	0.0	28.6	71.4	100.0

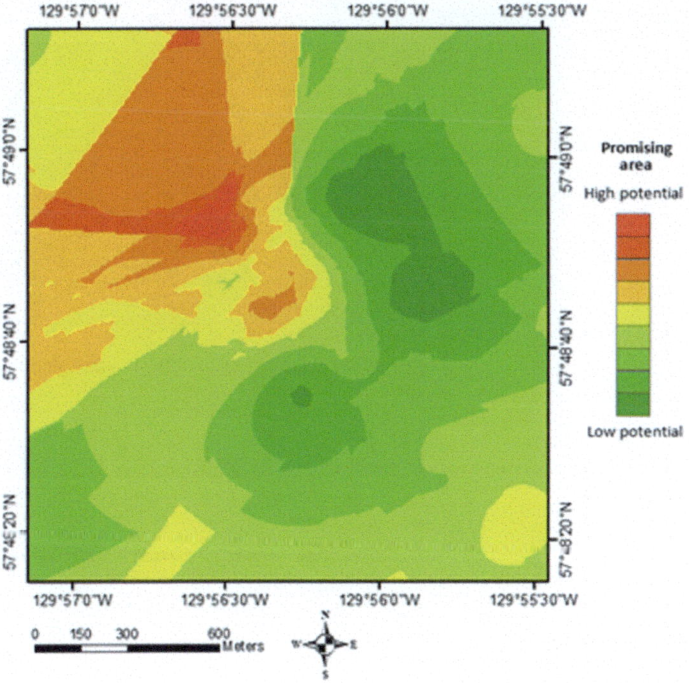

Fig. 17 The prospectivity map results using the integration of all elements related to the Cu elements (case study of Fig. 4) [10]

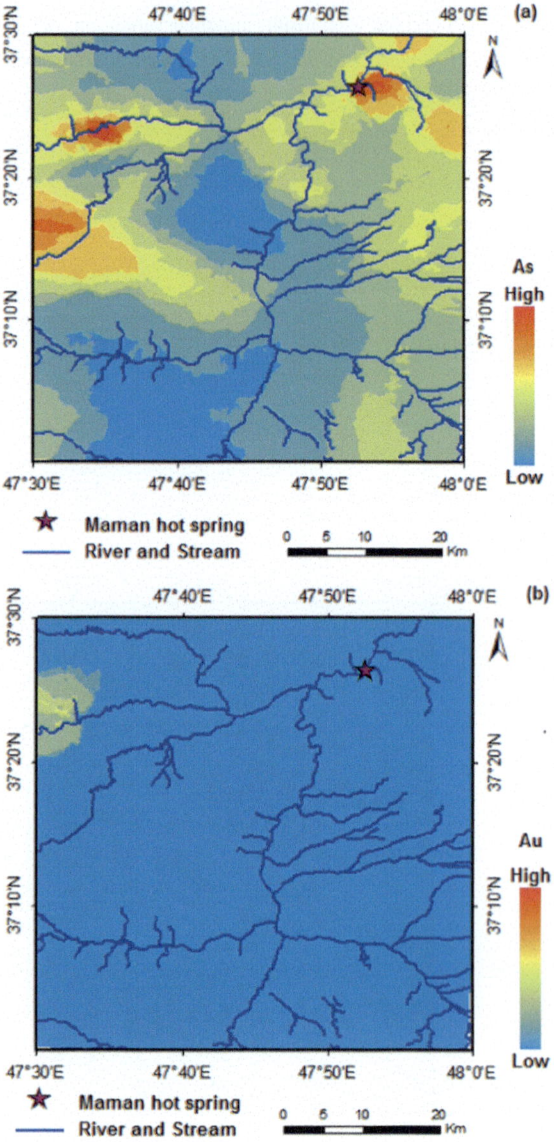

Fig. 18 The promising anomalous regions based on the most important geothermal pathfinders of mineralization [17]

3.4 *Regression Analysis*

Regression is a technique to measure the effects of variables by considering their simultaneous relationships with each other. The literal meaning of regression is to return to the past, and the name of this method is the use of past data to predict the future.

Fig. 18 (continued)

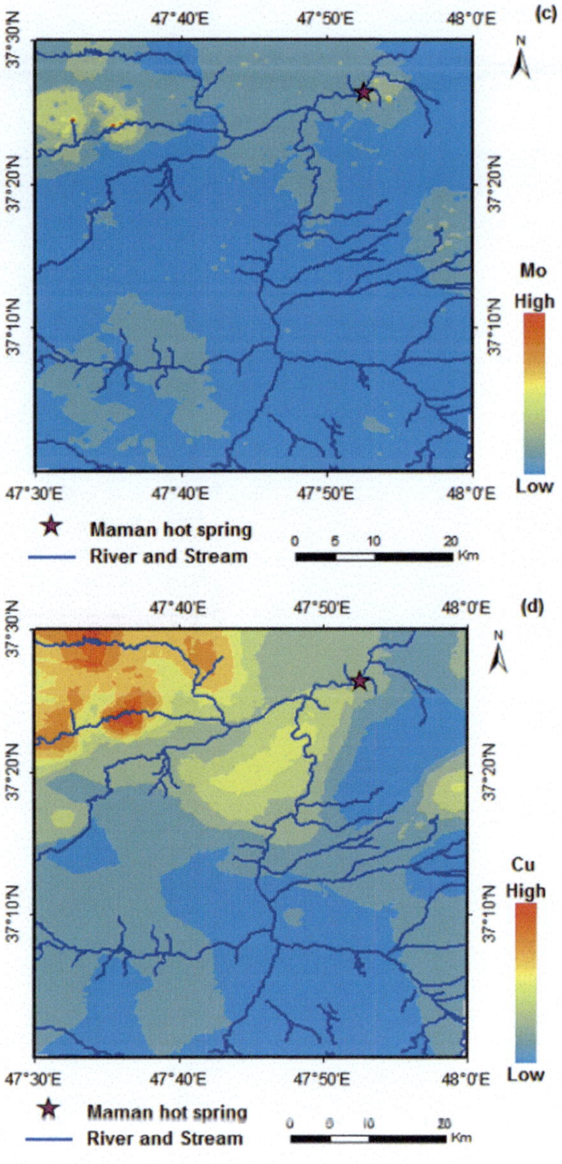

Fig. 18 (continued)

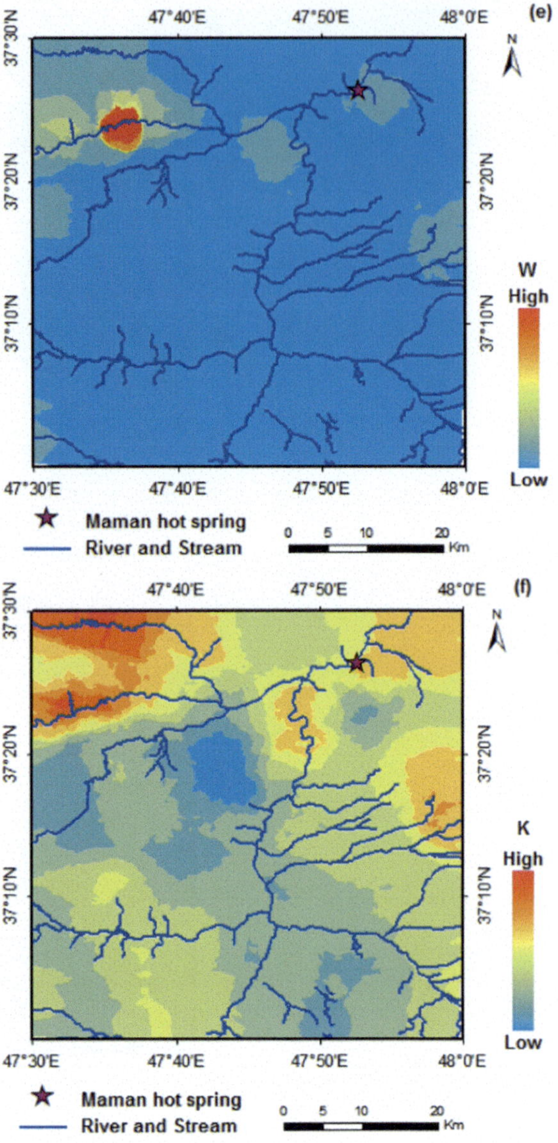

Fig. 18 (continued)

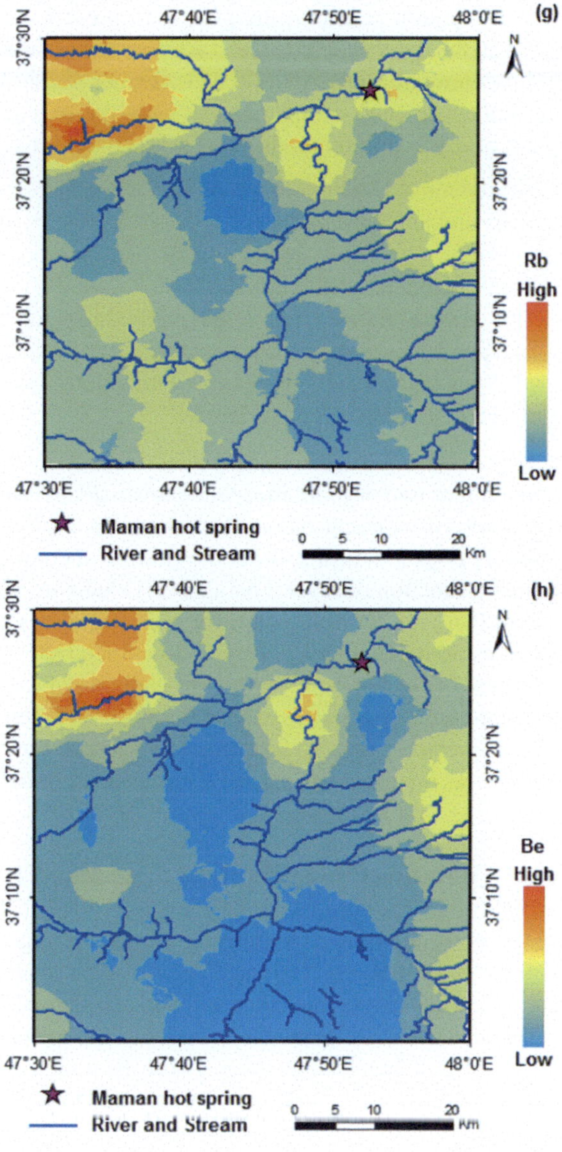

The formation and emergence of regression methods go back to the last decade of the nineteenth century. In 1886, Karl Pearson was able to provide a formula for calculating the correlation coefficient, which, in addition to being innovative, provided a new path in the science of statistics. This formula was the basis of the regression theory. If the Pearson formula shows the correlation between two variables, in the regression theory, the dependent variable variability is shown because of the independent variable variability.

In regression research, the goal is to predict one or more dependent variables based on the independent (predictor) variables. In multiple regressions, the goal is to find predictor variables that predict changes in the dependent variable either alone or jointly. The entry of independent variables in regression is done in several ways. Types of regression methods go as follow:

- *Simple linear regression*: predicting a dependent variable based on one or more independent variables.
- *Multiple regression:* prediction of one or more dependent variables based on multiple independent variables.
- *Uni-variate multiple regression*: predicting a dependent variable based on several independent variables.
- *Multivariate multiple regression*: Prediction of several dependent variables based on several independent variables.

In geochemical exploration projects, in many cases, we deal with variables that are a function of two or more independent variables. For example, gold grade is usually a function of its tracer elements such as arsenic (As) and antimony (Sb). Also, in porphyry copper deposits, the concentration of copper (Cu) element in the samples is usually a function of several paragenesis elements, especially gold (Au) and molybdenum (Mo). In the following, for example of case 1 (Fig. 4), first, the important variables related to copper (Cu) are identified by *F-test feature selection algorithm*, and then, using different regression methods of *linear regression, regression tree, ensemble bagged trees, ensemble boosted trees, three-layered neural network* and *proposed support vector regression* (SVR) models to predict the concentration of Cu elements as the main variable of the subsurface mineralization.

3.4.1 Modeling Procedure of the Proposed SVR

Figure 19 shows the used SVR technique for the target prediction. The method begins with the assumption of various SVR model parameters. The database of dependent and independent variables then implements the method and the SVR model trains and checks with the use of the preparation and testing datasets. The efficiency of the SVR model is then analyzed using statistical metrics. Finally, the prediction model will suggest the final conditions for solving the problem.

The Kernel function used is the Radial Basis Function (RBF) kernel of Eq. 1. The RBF kernel evaluates nonlinearity between the input variables x_i, x_j. Where γ in the function acts as a regularization parameter.

$$K(x_i, x) = e^{-\gamma \left\| x_i - x_j^2 \right\|} \tag{1}$$

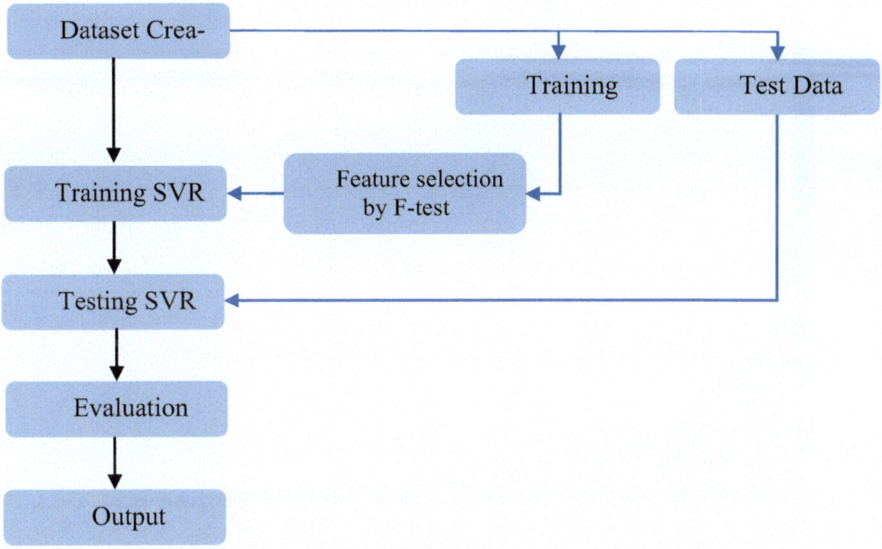

Fig. 19 Modeling flowchart of proposed SVR method

3.4.2 Feature Selection

All those who work in the field of artificial intelligence are faced with the problem of identifying important features and removing unimportant features. Unimportant features do not have a positive effect on the prediction of the known variable but will increase the error in the prediction results. Therefore, feature selection methods play an important role in selecting important features and optimizing the learning method. Thus, the feature selection on the data of case study 1 (Fig. 4) was carried out using F-test feature selection algorithm which is shown in Fig. 20. As can be seen the elements of Ag, Na, Au, Se, and La are the five first important features among all features.

3.4.3 Results of Modeling with Different Results

The data are split into training and test sets using 80–20 split. Also, *k*-fold cross-validation with $k = 5$ is used to validate the efficiency of our proposed SVR model. All these simulations have been implemented using MATLAB® (R2022a).

The results of the different prediction models of Cu on validation data and test data using different regression methods of linear regression model, regression tree model, ensemble bagged trees model, ensemble boosted trees model, three-layered neural network model, and proposed support vector regression (SVR) model are presented in Tables 9 and 10, respectively. The abbreviation of MSE, RMSE, R^2, and MAE

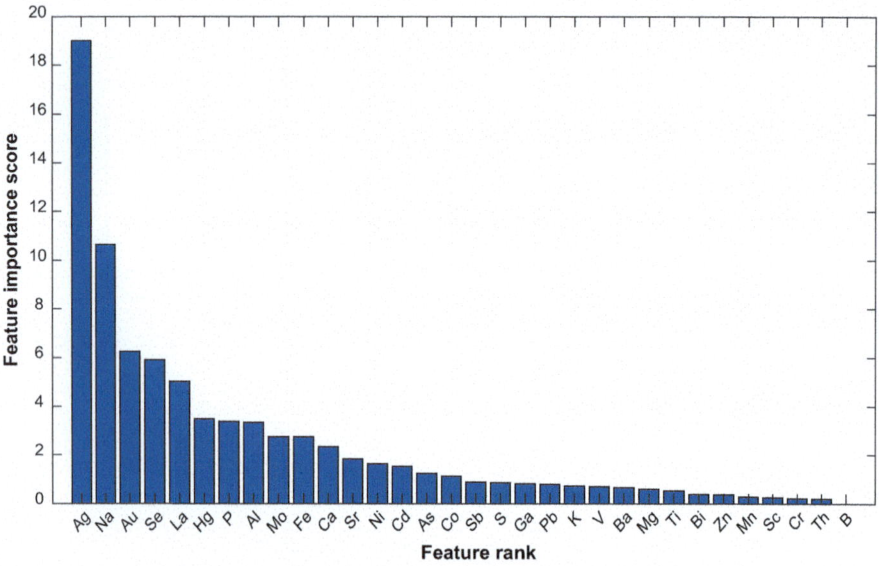

Fig. 20 Input feature importance obtained by F-test feature selection algorithm

in Tables 5 and 6 presents mean squared error, root mean square error, R-squared, mean absolute error for comparison of the results.

Measured and predicted Cu concentration values on validation and test data using disputed regression models are shown in Figs. 21, 22, and 23. The presented results are so understandable and comparable that it does not need much detailed explanation. As can be seen, the efficiency of each of the mentioned regression methods can be easily compared and interpreted using Tables 9 and 10 as well as Figs. 21, 22, and 23.

Table 9 Results of the prediction models of Cu on validation data

Algorithm	MSE	RMSE	R^2	MAE	Training time
Linear regression model	0.0682	0.2612	0.10	0.0804	4.38
Regression tree model	0.0140	0.1182	0.38	0.0538	2.71
Ensemble bagged trees model	0.0141	0.1186	0.37	0.0533	3.58
Ensemble boosted trees model	0.0139	0.1179	0.38	0.0585	4.72
Three-layered neural Network model	0.0142	0.1190	0.37	0.0676	3.67
Proposed SVR model with selected features	0.0077	0.0879	0.66	0.0441	1.34

Table 10 Results of the prediction models of Cu on test data

Algorithm	MSE	RMSE	R^2	MAE
Linear Regression model	0.3205	0.5662	0.08	0.1851
Regression tree model	0.0330	0.1817	0.31	0.0941
Ensemble bagged trees model	0.0310	0.1761	0.35	0.0927
Ensemble boosted trees model	0.0273	0.1652	0.43	0.0986
Three-layered neural network model	0.0898	0.2996	0.31	0.1309
Proposed SVR model with selected features	0.0109	0.1045	0.77	0.0545

3.5 Optimization Methods

The progress of computer systems over the past few decades has led to the development of optimization methods so that different optimization methods have been presented to solve different problems. As mentioned above, the optimization methods are divided into four important categories of *Enumerative method*, *calculation-based method*, and random search-based methods that are *heuristic and Meta-Heuristic methods*. Each of the mentioned categories also has different sub-categories.

In general, if the optimization methods are divided into four categories of *enumerative methods, calculation-based methods, heuristic methods, and meta-heuristic methods* which are described in the next section. This chapter aims to introduce some of the most used techniques of the last group titled *meta-heuristic optimization methods (MHOM)* with its applications and examples in geosciences.

3.5.1 Enumerative Methods

In these methods, only one point from the domain of the objective function is considered in each iteration. Although they need considerable computations, they are easier to implement than other methods. The search domain is very wide and there is no way to reduce it. This method is completely unintelligent, and for this reason, it is rarely used alone today.

3.5.2 Calculation-Based Methods

These methods use all the necessary and sufficient conditions that apply to the solution of the problem. Optimization constraints play a very important role in Calculation-based methods and this is the main reason to divide them into constrained and unconstrained techniques. According to the count of variables, the unconstrained methods try to optimize one or multi-variable functions.

One-variable function optimization methods are categorized into three groups of order zero, one, and two orders. For the zero-order groups, only the objective function

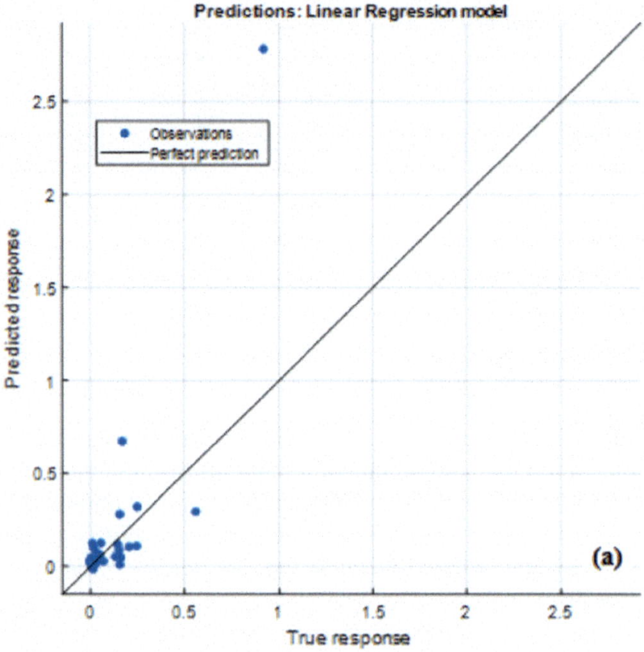

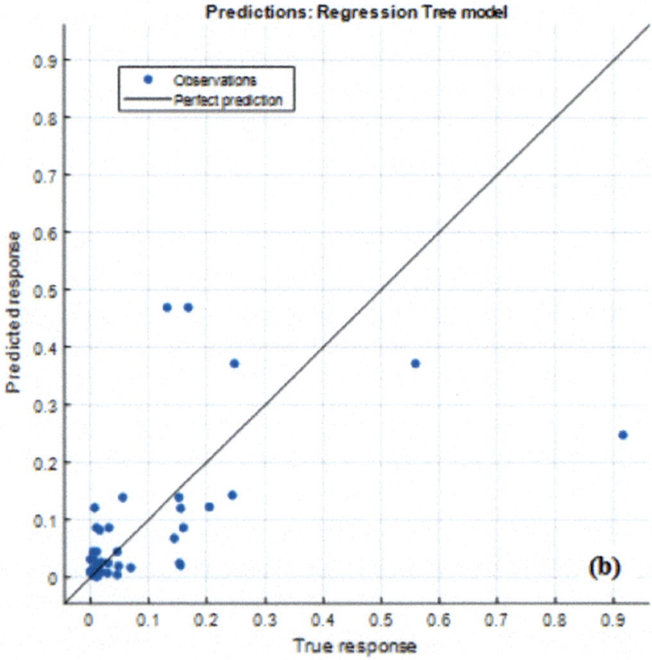

Fig. 21 Measured and predicted Cu concentration values on *validation data* for different models

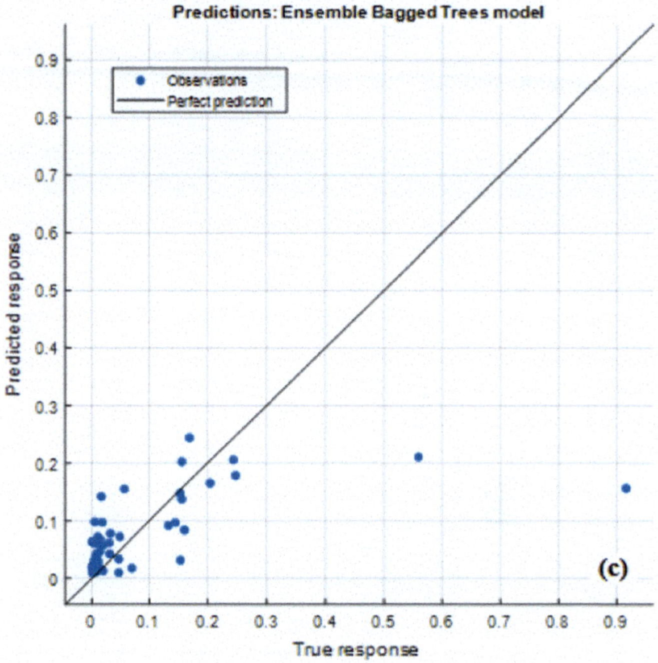

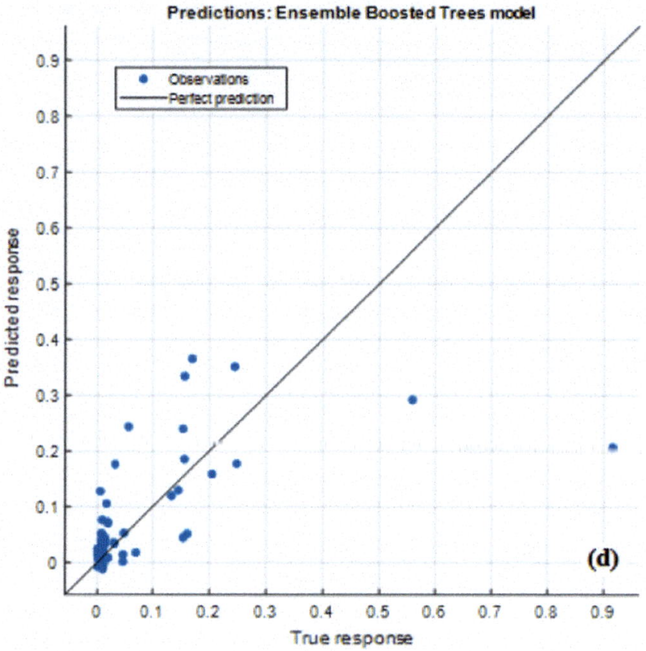

Fig. 21 (continued)

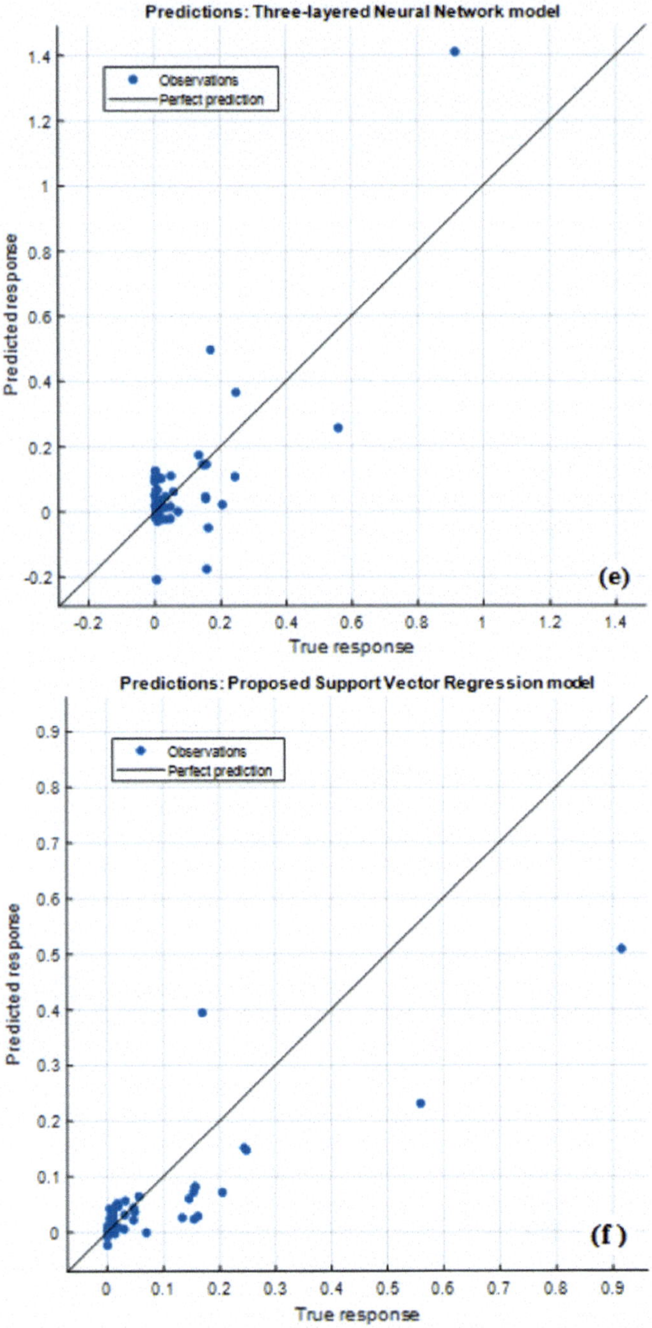

Fig. 21 (continued)

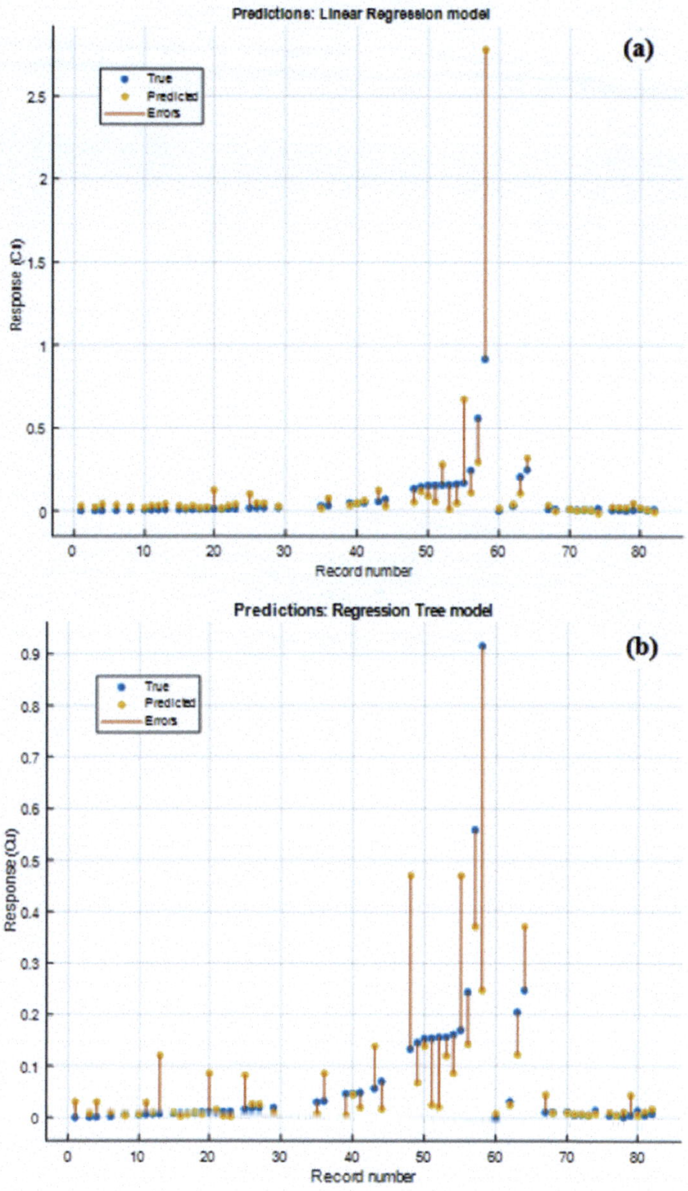

Fig. 22 Error between measured and predicted Cu concentration values on *validation data* for different models

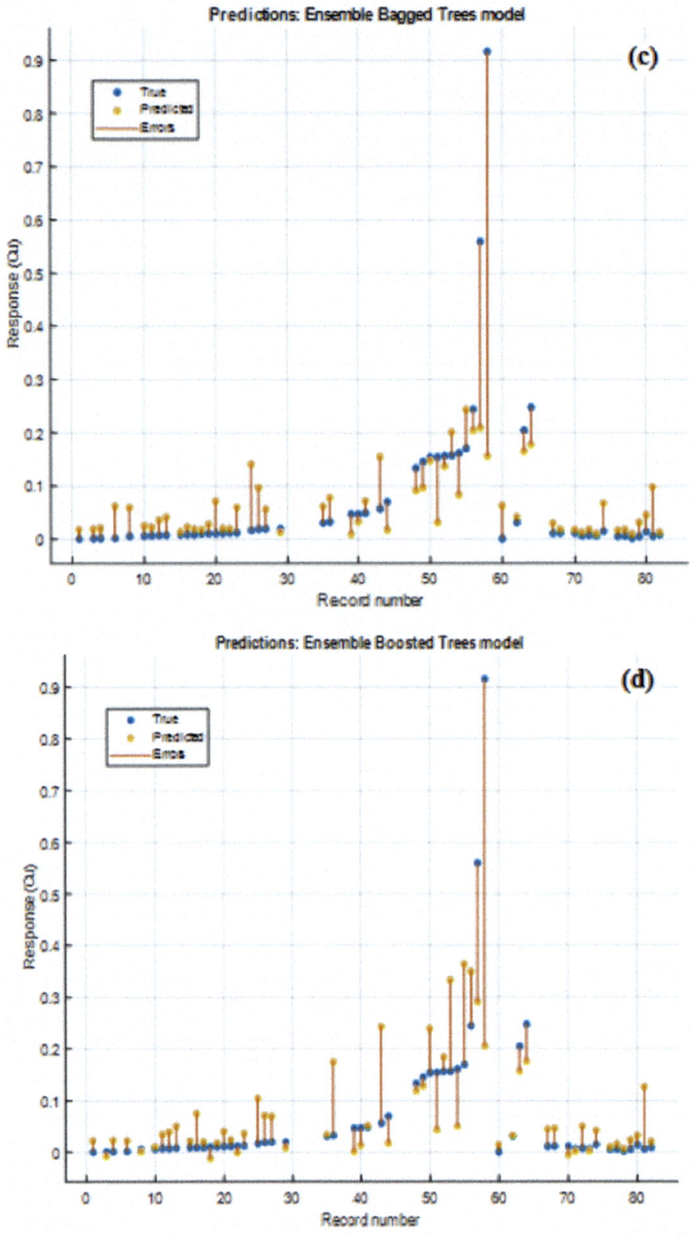

Fig. 22 (continued)

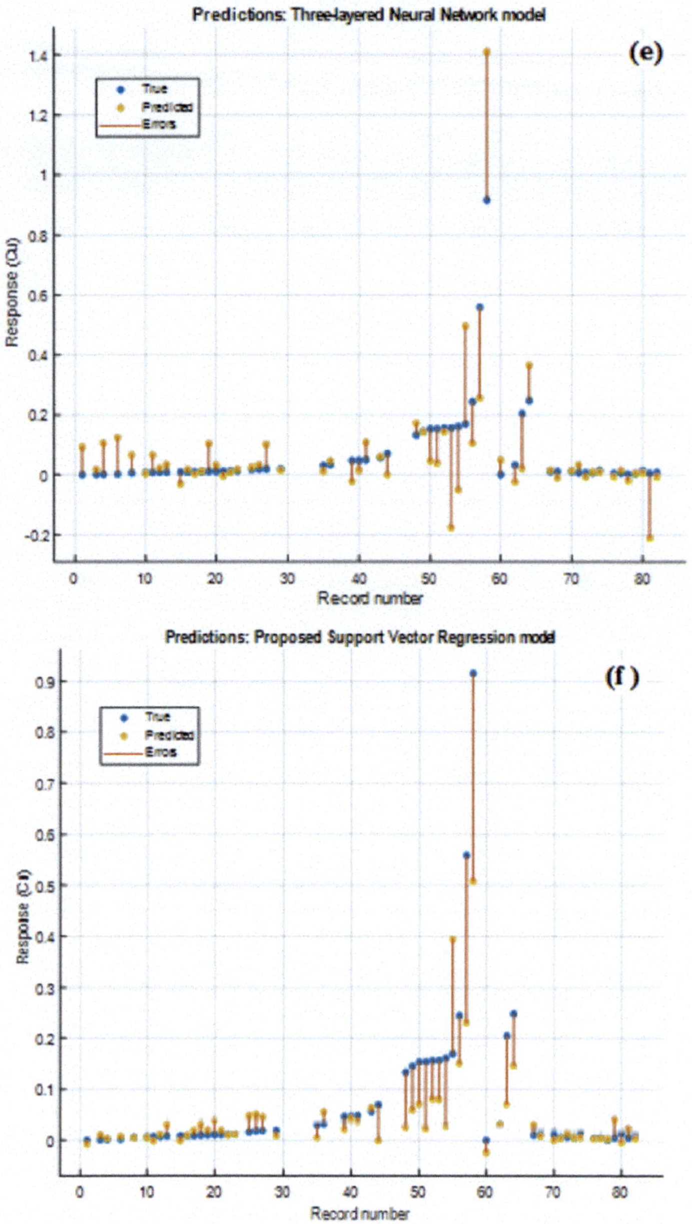

Fig. 22 (continued)

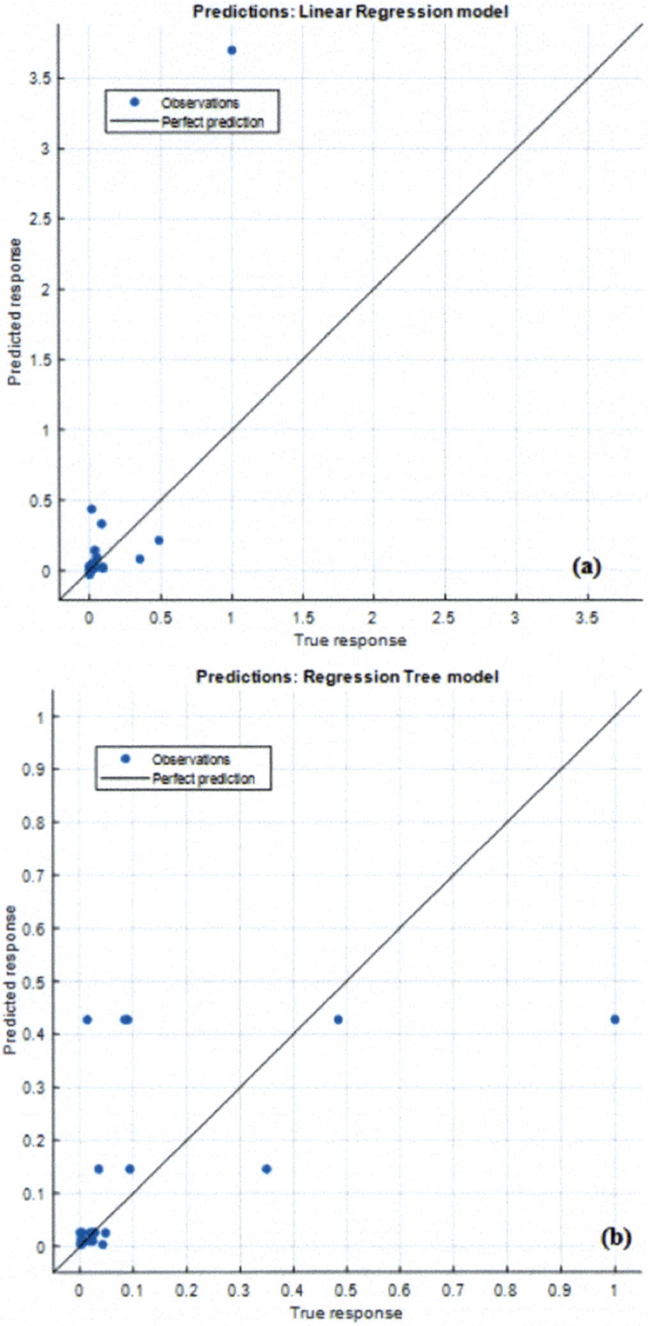

Fig. 23 Measured and predicted Cu concentration values on *test data* for different models

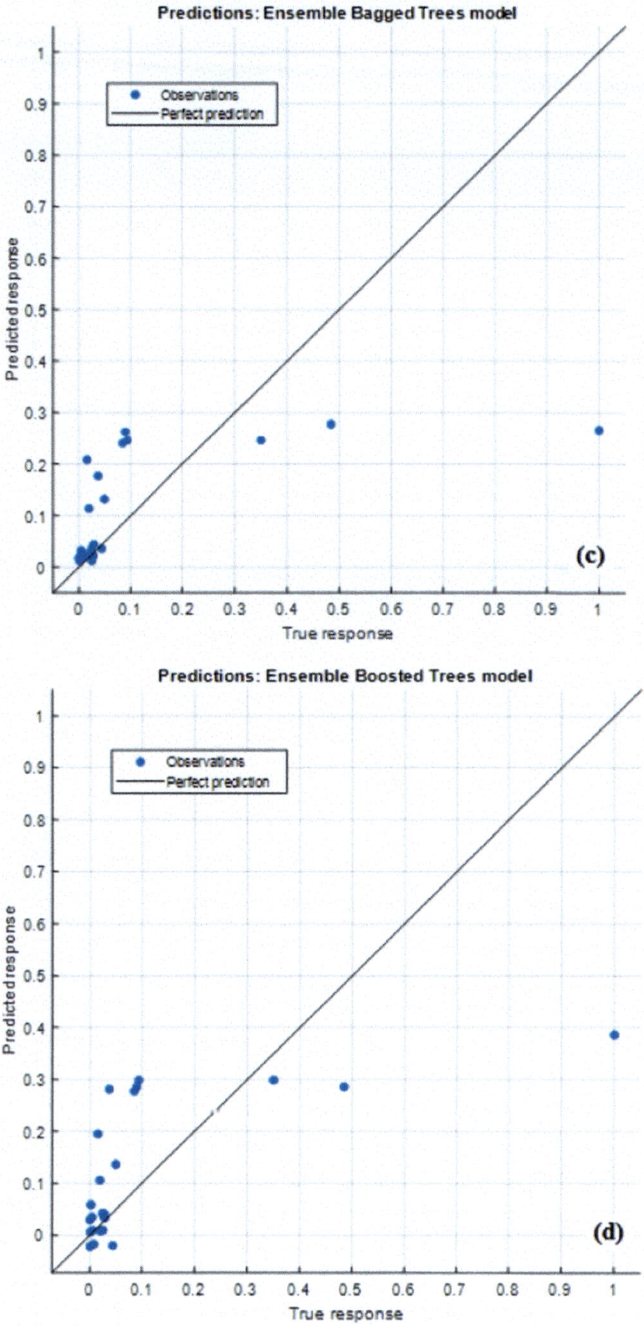

Fig. 23 (continued)

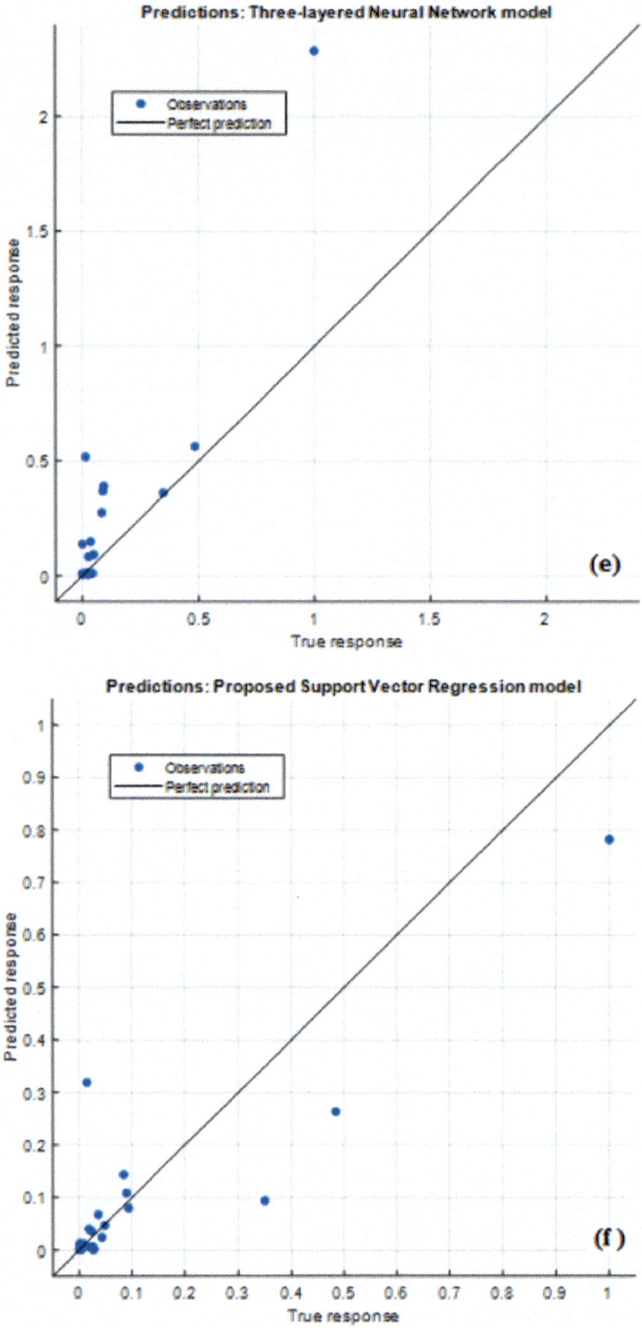

Fig. 23 (continued)

needs to be calculated at different points, for the groups of first-order we need both the objective function and also the related derivative, and for the second-order groups in addition to the objective function and its first derivative, the second derivatives are needed. In the optimization of multi-variable functions, which is widely used in engineering problems, the minimization or maximization of a quantity with many variables is done.

Unconstrained Optimization Problems: Here, the overall goal is to maximize or minimize our objective function, and no constraint is applied on the determined parameters. These problems are less complex and their modeling is also simpler.

Constrained Optimization Problems: In these problems, constraints are applied on some parameters or all parameters. The limitation can be according to our needs or behavioral limitations in the real world and dependent on physics and nature. For example, suppose as the response of an optimization problem, the speed of a car is considered to be 500 km/h. We know that physically and legally this speed is not possible, so a speed limit (for example 100) is applied to the speed variable, or suppose the final cost variable for a simple construction work reaches tens of billions, which seems unreasonable (perhaps reasonable in a few years it seems). Here, we find the final answer by limiting the price on the cost variable. The selection of the algorithm and optimization method is also done according to these limitations.

3.5.3 Heuristic and Meta-Heuristic Methods

Different heuristic and meta-heuristic algorithms are shown in Fig. 24. Heuristic algorithms use a specific domain to reach an acceptable solution in a short time, so they do not guarantee that the given solution is the optimal solution. These algorithms are based on search and try to find the answer by reducing the search space. For very difficult and complex problems, it usually seems logical to use heuristic algorithms; because it may be impossible to find an optimal answer or it may require a long time to find it.

On the other side, meta-heuristic methods are derived from nature; these methods are more similar to the systems governing nature and society. The application of meta-heuristic methods is taken from continuous heuristic techniques which leads such interesting results in solving NP-Hard problems. Between the meta-heuristic methods, the most important goes as follows:

- A real phenomenon of the nature can be modeled.
- They are uninterrupted; they usually present a similar combination (multiple factors) without condition.
- They are adaptable.
- The mentioned features lead to reasonable behavior to supply intelligence.

Meta-heuristic algorithms use a higher-level strategy to search for solutions in optimization problems. These algorithms are usually general-purpose and can be used in many problems. In general, meta-heuristic algorithms use an iterative search

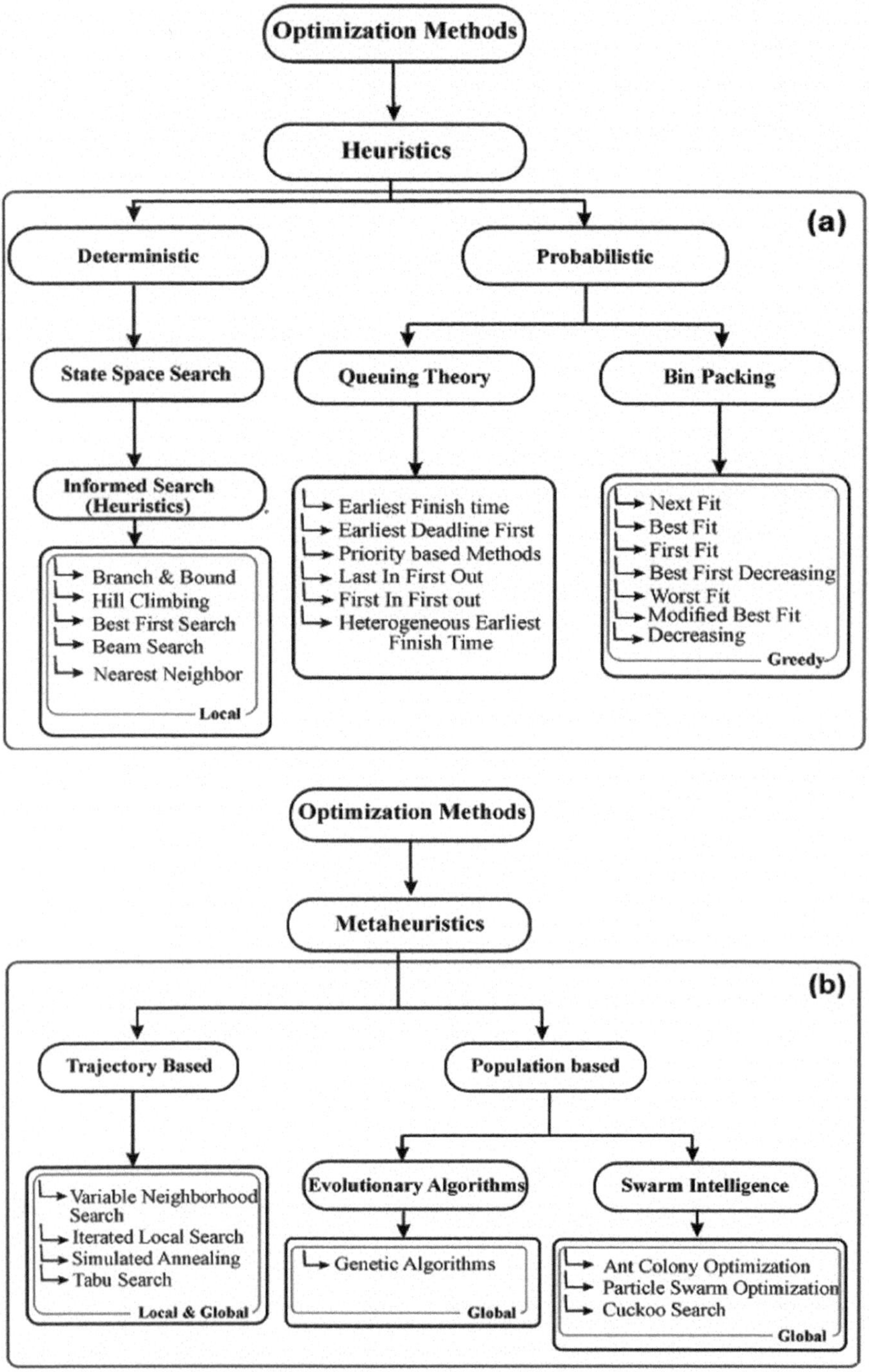

Fig. 24 Heuristic and meta-heuristic methods

routine to intelligently discover the answer instead of relying on a specific domain. Usually, elements of randomness, local search, and global discovery are integrated into these algorithms to reach an acceptable solution. Genetic algorithm and ant colony algorithm are among this category.

3.5.4 Genetic Algorithm

The main idea of evolutionary algorithms was proposed by Rechenberg in 1960. Genetic algorithms, which are one of these types of algorithms, are actually a computer search method based on optimization algorithms, based on the structure of genes and chromosomes, which was proposed by John Holland (1970) at the University of Michigan. One of the best and most comprehensive definitions of genetic algorithms also belongs to Goldberg's book: "Genetic algorithms are a model of machine learning whose behavior is an illustration of evolutionary processes in the natural world." Genetic algorithms are one of the most powerful methods derived from nature, which search the problem space in a randomly guided way, and this search takes place in the form of trying to create better answers in each generation compared to the answers of the previous generation. It provides one of the best forms of numerical optimization in science and engineering problems. Some of the genetic algorithm applications are described in Table 11 [27]. Flowchart of the implementation steps of a simple genetic algorithm is shown in Fig. 25.

Considering that in this section, the main goal is not to introduce the basic concepts and theory of the genetic algorithm method, in the following, an example of the application of this method in solving geochemical problems will be presented. For example, Ziaii et al. (2012) used a neuro-fuzzy method based on a genetic algorithm (GA-based neuro-fuzzy algorithm) to identify geochemical anomalies of a hidden mineralization [28]. They presented genetic algorithm operator which can optimize artificial neural network process. At first, they used the neuro-fuzzy method to identify anomalies, but the modeling results were not compatible with those obtained

Table 11 Applications of genetic algorithm [27]

Field	Application
Control	Gas transmission lines—missile launch—balance systems—interception and…
Designing	Aircraft design—design of VLSI circuits and communication networks and…
Management and planning	Production-scheduling-allocation of resources and…
Robotics	The path of the robot and…
Machine learning	Designing neural networks—classification algorithms and
Signal processing	Filter design and…
Other cases	Art and music—the problem of the traveling salesman—routing in the network, geosciences and…

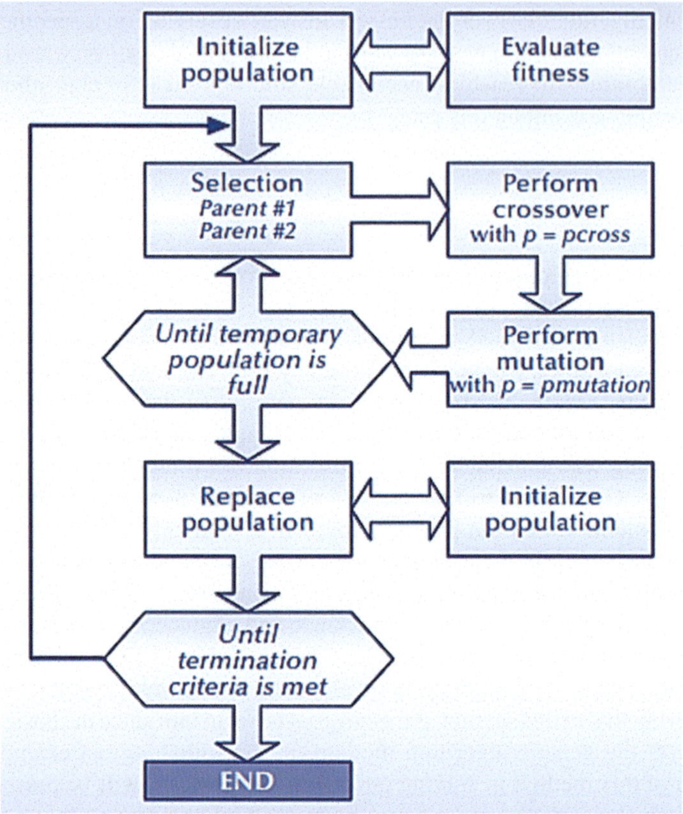

Fig. 25 Flowchart of the implementation steps of a simple genetic algorithm

from conventional methods. Perhaps the most important reason for this problem was the many complications and the impact of factors such as weathering in the environment. They also mentioned the presence of different geochemical landscapes as the reason for the unfavorable results. For this reason, they used a GA-based neuro-fuzzy to remove the effect of geochemical landscapes. In the conducted research, the lithogeochemical data of the Deh-Salm region in eastern Iran were used (Fig. 26). In the study area, two different zones have been studied, and the position of zones A-I and A-II is shown in Fig. 27. Also, the distribution map of subsurface anomalies based on the GA-based neuro-fuzzy method is shown in Fig. 28, and the map of subsurface anomalies of zone A-II is shown in Fig. 29. In Fig. 30, the map of subsurface anomalies of Zone A-II is presented on a larger scale and with more details. It should be mentioned that the anomalies shown have a very high compatibility with the geological and petrological realities of the region which indicates the very good efficiency of the presented GA-based neuro-fuzzy algorithm.

As another example, Liu et al. (2004) used the genetic algorithm optimization method to design and present a classifier using a neural network [29]. They designed

Fig. 26 Location map of the study area [28]

their genetic algorithm operator to be able to optimize the neural network and solve its other parameters such as premature convergence. They used Spot-4 XS image (700 × 900 pixels) of the Jiangning area of Jiangsu province in eastern China for classification to test their algorithm. For this classification, they used three classifiers—maximum likelihood classifier (MLC), back propagation neural network (BP-multilayer perceptron or BP-MLP), and genetic algorithm-based neural network method (GA-MLP). 96 points related to three land cover classes were used for training. They used three MLP layers plus one hidden layer for the genetic algorithm and back-propagation, which is shown in Fig. 31. The classification results of the three mentioned algorithms are presented in Fig. 32, which shows the presented genetic algorithm-based neural network has higher accuracy and better resolution in comparison to the other methods.

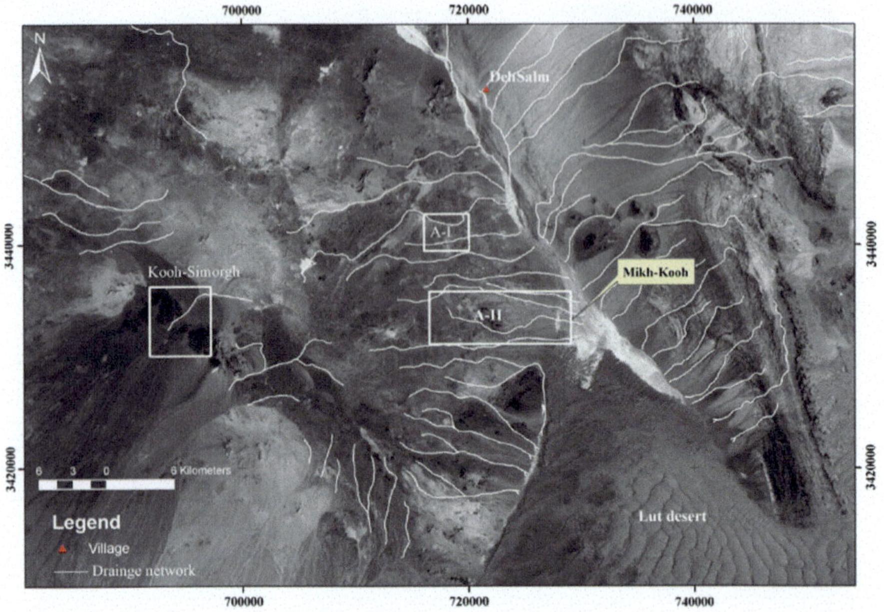

Fig. 27 Satellite image of the study area including the location of two zones of A-I and A-II [28]

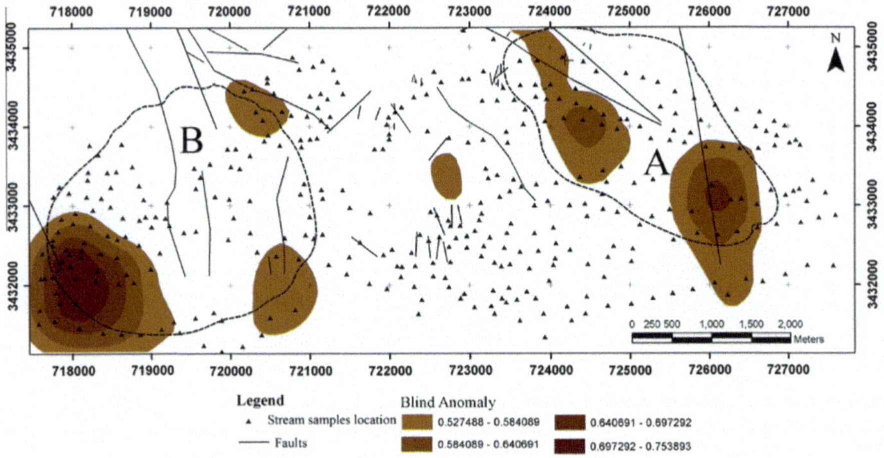

Fig. 28 The detected hidden anomalies of zone A-II which is divided in two anomalies of A and B [28]

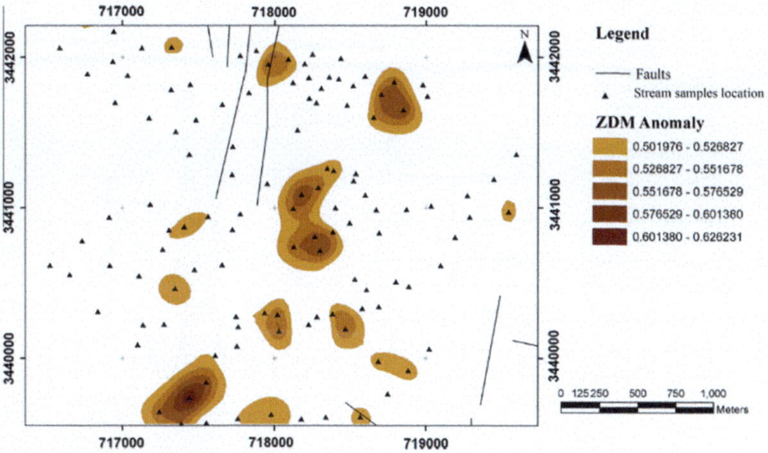

Fig. 29 The detected hidden anomalies of zone A-I [28]

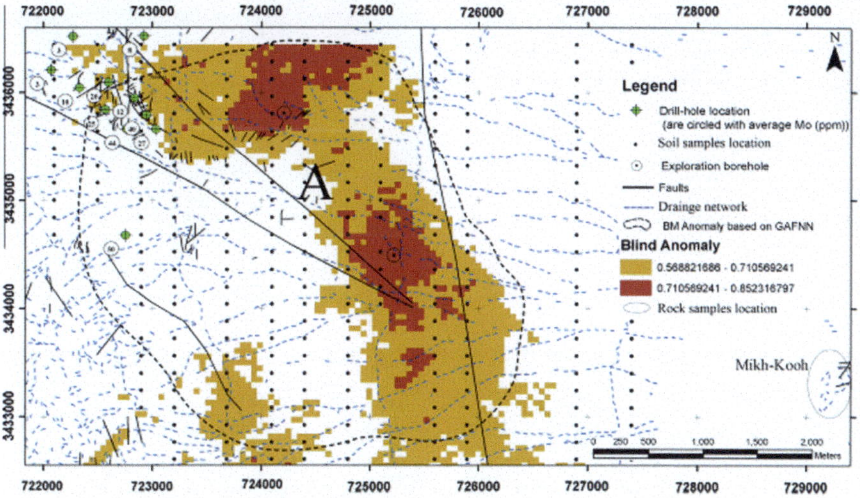

Fig. 30 The detected hidden anomalies of zone A-II on a large scale and in detail [28]

4 Conclusions

Geochemical exploration is very important as one of the steps of the mineral exploration process, which is a step-by-step process. The success of each of these steps will make the project move forward, and the failure will also cause an early stop of the exploration project. Therefore, issues related to cost, time, and risk play an important role in each of these exploration phases and should be managed intelligently. Therefore, the optimization of each of the steps of exploratory operations

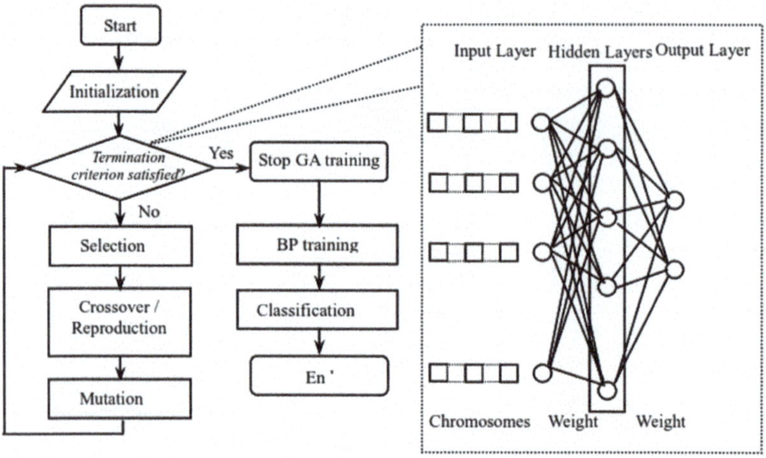

Fig. 31 Flowchart of the genetic algorithm-based neural network (GA-MLP) [29]

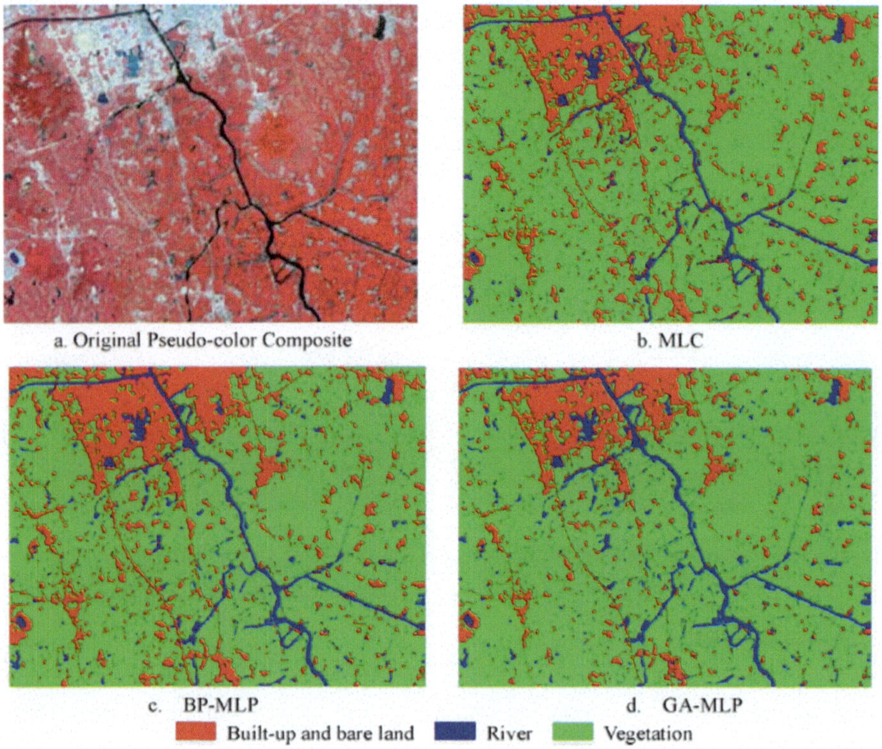

Fig. 32 The results of classification for the two algorithms of MLC, BP-MLP, and the proposed algorithm of GA-MLP [29]

is vital. Considering that geochemical sampling is often done in the preliminary steps of an exploration project, choosing the right sampling environment, correct sampling, and accurate preparation are among the things that should be considered. On the other side, performing one-variable, two-variable, and multi-variable analyses requires accurate analysis by experts. Therefore, each of the above steps needs to be done optimally and correctly. Selecting relevant and important elements from a large number of variables and finding the relationship between them is an optimization problem that should be properly investigated. It is optimal to advance or stop an exploratory project in the group of solving this problem, which is not necessarily a completely mathematical problem, and in addition, the experience of experts is needed. Therefore, in this chapter, some widely used machine learning and optimization methods were introduced in optimizing a geochemical analysis. In addition, it tried to fully explain each of the discussed methods by presenting some practical examples.

References

1. Deb K (2004) Optimization for engineering design: algorithms and examples. [S.l.]: Prentice-Hall of India. ISBN: 9788120309432
2. Gattie DK, Wicklein RC (2007) Curricular value and instructional needs for infusing engineering design into k-12 technology education. J Technol Edu 19(1)
3. Rao SS (2009) Engineering optimization: theory and practice: fourth edition. [S.l.]: John Wiley and Sons. ISBN: 9780470183526
4. Bartholomew-Biggs M (2008) The steepest descent method. In: nonlinear optimization with engineering applications. Boston, MA: Springer US, v. 19, pp 1–8. ISBN: 9780387787237
5. Goldberg DE (1989) Genetic algorithms in search, optimization, and machine learning. 1st. ed. [S.l.]: Addison-Wesley Longman Publishing Co., Inc., ISBN: 9780201157673
6. Cavazzuti M (2012) Optimization methods: from theory to design scientific and technological aspects in mechanics. [S.l.]: Springer Berlin Heidelberg, ISBN: 9783642311864.
7. Dorigo M, Birattari M, Stutzle T (2006) Ant colony optimization. IEEE Computat Intell Mag 1(4):28–39, ISSN: 1556-603X
8. Licht OAB (1998) Geochemical prospecting: principles. Techniques and Methods. CPRM, Rio de Janeiro, p 236p
9. Moradpouri F, Ghavami-Riabi R (2020) A multivariate geochemical investigation of borehole samples for gold deposits exploration. Geochem Int 58(1):40–48. https://doi.org/10.1134/S0016702920010103
10. Moradpouri F, Hayati M (2021) A copper porphyry promising zones mapping based on the exploratory data, multivariate geochemical analysis and GIS integration. Appl Geochem 132:105051. https://doi.org/10.1016/j.apgeochem.2021.105051
11. Moradpouri F, Ahmadi SMH, Ghaedrahmati R, Barani K (2023) Determination of the erosion level of a porphyry copper deposit using soil geochemistry. J South Afr Inst Min Metall 123(2):103–112
12. Moradpouri F (2012) A new approach in reverse time migration for properly imaging of complex geological media. Acta Geophys 69:529–538.
13. Darnley AG, Björklund A, Bölviken B, Gustavsson N, Koval PV, Plant JA, Steenfelt A, Tauchid M, Xie Xuejing X (1995) A global geochemical database for environmental and resource management. Recommendations for international geochemical mapping. Final report of IGCP project 259. UNESCO Publishing

14. He Y, Zhou Y, Wen T, Zhang S, Huang F, Zou X, Ma X, Zhu Y (2023) A review of machine learning in geochemistry and cosmochemistry: method improvements and applications. Appl Geochem 140:105273. https://doi.org/10.1016/j.apgeochem.2022.105273
15. Pan T, Zuo R, Wang Z (2023) Geological mapping via convolutional neural network based on remote sensing and geochemical survey data in vegetation coverage areas. IEEE J Sel Topics Appl Earth Observ Remote Sens 16:3485–3494
16. Ranasinghe PN, Fernando GWAR, Dissanayake CB, Rupasinghe MS, Witter DL (2009) Statistical evaluation of stream sediment geochemistry in interpreting the river catchment of high-grade metamorphic terrains. J Geochem Explor 103:97–114. https://doi.org/10.1016/j.gexplo.2009.07.003
17. Moradpouri F, Sabeti H (2024) Geochemical evaluation of a geothermal region for the trace elements related to the subsurface mineralization using machine learning methods. Jordan J Earth Environ Sci 15(2):94–106
18. Andriotti JLS (2010) Interpretation of geochemical prospecting data with the aid of statistics. Porto Alegre regional superintendence, CPRM, Porto Alegre, 67 p
19. Reimann C, Filzmoser P, Garrett RG (2005) Background and threshold: critical comparison of methods of determination. Sci Total Environ 346:1–16
20. Reimann C, Garret RG (2005) Geochemical background—concept and reality. Sci Total Environ 350:12–27
21. Garrett RG (1969) The determination of sampling and analytical errors in exploration geochemistry. Econ Geol 64:568–569
22. Thompson M, Howarth RJ (1978) A new approach to the estimation of analytical precision. J. Geochemical Exploration 9:23–30
23. Frizzo SJ, Licht OAB (2007) Uni and bivariate statistics applied to geochemical prospecting. In: Licht OAB, Mello CSB, Silva CR (Eds). geochemical prospecting for metallic, non-metallic, oil and gas mineral deposits. SBGq/CPRM. Rio de Janeiro. 778p. Chapter 20, pp 619–660
24. Aitchison J, Greenacre M (2002) Biplots of compositional data. J Roy Stat Soc: Ser C (Appl Stat) 51:375–392
25. Grunsky EC (2010) The interpretation of geochemical survey data. Geochem Exploration. Environ Anal 10:27–74
26. Davis JCS (2002) Statistics and data analysis in geology. 3. ed. Wiley, New York, 656 p
27. Alireza M (2006) An introduction to genetic algorithm and its applications. Naqos Publications, Tehran, p 143p
28. Ziaii M, Pouyan AA, Ziaei M (2009) Neuro-fuzzy modelling in mining geochemistry: identification of geochemical anomalies. J Geochem Explor 100:25–36
29. Liu Z, Liu A, Wang C, Niu Z (2004) Evolving neural network using real coded genetic algorithm (GA) for multispectral image classification. Futur Gener Comput Syst 20(7):1119–1129

Chapter 9
Bat Algorithm for Real-Life Problem-Solving

Swarnali Daw, Dipannita Basu, Abhisek Roy, and Sayan Chakraborty

1 Introduction

Bat Algorithm [1], a type of bioinspired computation, has been used to solve a number of practical issues. The BAT algorithm is a bioinspired optimization tool based on bat echolocation behavior. It navigates and locates prey via sound waves, and it mimics this approach while solving complex tasks. Introduced in 2010, the method [1, 2] combines local search capabilities with global exploration, making it useful for a variety of applications such as image processing, medical data classification, and workflow scheduling. Its innovative method of balancing exploration and exploitation has resulted in significant advances in optimization issues across multiple disciplines. This technique, which is based on echolocation calls made by bats for hunting and navigation, is extensively used in many different domains to address real-world optimization issues. Route optimization via the bat method has been applied to operations research and logistics. This involves, but is not restricted to, determining the best delivery routes, reducing expenses, and increasing logistical efficacy. The algorithm's greatest utility to all of us is that it computes and displays the shortest path, saving time and minimizing [2] traffic bottlenecks while lowering fuel usage in an environmentally sustainable manner. Like bat algorithms are employed successfully in portfolio management, they are also used in financial engineering,

S. Daw
Deparment of CSE, Narula Institute of Technology, Kolkata, West Bengal, India

D. Basu
Deparment of Data Science, Haldia Institute of Technology, Kolkata, West Bengal, India

A. Roy
Budge Budge Institute of Technology, Kolkata, West Bengal, India

S. Chakraborty (✉)
Deparment of CST, JIS College of Engineering, Kalyani, West Bengal, India
e-mail: sayan.cb@gmail.com

N. Dey (ed.), *Optimizing Solutions for Real-Life Problems*, Springer Tracts in Nature-Inspired Computing, https://doi.org/10.1007/978-981-96-3216-9_9

which helps investors choose the best way to divide the assets in a portfolio to minimize risk and increase rewards. Investors can enhance their financial performance by making more thoughtful investment decisions via the bat algorithm [3]. The bat algorithm has proven to be invaluable for engineering design and control. It is useful in determining the ideal settings for production processes or in optimizing designs to satisfy certain requirements, including maximizing performance or using the least amount of material. Better products and more affordable manufacturing techniques are the outcomes of this. The bat algorithm provides a unique way to solve problems in the real world because of its echolocation-inspired [4, 5] methodology. Its numerous uses in engineering, finance, and logistics highlight how valuable it is at providing the best answers to challenging issues, making it an effective tool across a range of problem-solving fields. Yang created the bat algorithm [6, 7] in 2010 as a nature-inspired algorithm based on the echolocation habits of microbats. The algorithm is a component of a larger class of bioinspired computational algorithms, which include particle swarm optimization and ant colony optimization. These algorithms use real-world phenomena' behaviors to address intricate computer problems.

The bat algorithm [8] makes use of a virtual population of bats, each of which has unique frequency and loudness characteristics in addition to position and velocity. It imitates how bats use echolocation to locate their prey by using these features to explore the issue space iteratively. To discover the global optimum, the algorithm in [9] strikes a compromise between exploration, which involves examining the larger problem space, and exploitation, which entails fine-tuning the existing best solutions. Optimization problems represent one of the bat algorithm's most important applications. These kinds of issues, where the optimal answer must be chosen from a plethora of options, are prevalent in numerous professions. Because the bat algorithm can thoroughly explore and utilize the problem space, it is very useful for handling discrete and continuous optimization problems. The bat algorithm can be used in logistics and transportation to optimize truck delivery routes, reduce fuel consumption and trip time while increasing customer satisfaction. Determining the quickest and least crowded routes improves service efficiency. The field of finance also uses algorithms [10], especially for portfolio management, which help determine the best way to divide the various assets in an investment portfolio to maximize profits and minimize risk at the same time. Investors can make better decisions and improve portfolio performance by using the bat algorithm. The bat method has been applied to design optimization and control in the realm of engineering. For example, it has been applied to optimize beam and truss designs, reducing material consumption without sacrificing structural integrity. It has also been used to find the ideal manufacturing process parameters, enhancing both operational effectiveness and product quality. It is important to consider the application of the bat algorithm in power systems. It has been applied to solve the economic dispatch problem, which involves determining the best generation schedule for power plants to meet demand and adhere to operational constraints while minimizing overall operating costs. Moreover, the bat method has shown encouraging results in the field of data and picture processing. It has been applied to image segmentation, clustering, and feature selection in machine learning—all crucial problems in the big data and artificial intelligence eras. The bat algorithm is a strong and adaptable computational tool with a wide range of important applications. Its distinctive method, which draws inspiration from bats' echolocation behavior, has been successful in resolving a variety of optimization issues in the fields of data processing, engineering, finance, logistics, and power systems. The bat algorithm continues to offer creative and practical answers as our computer problems become more complicated, highlighting the enormous potential of bioinspired computational algorithms. The rest of the chapter is organized as follows: Sect. 2 discusses the related literature, the bat algorithm is thoroughly discussed in Sects. 3 and 4 discuss step-by-step bat algorithms that can solve real-life challenges. Section 5 discusses different case studies where real-life problems are solved via the bat algorithm. The advantages of the bat algorithm over similar existing algorithms are discussed in Sect. 6, and the advantages and disadvantages are discussed in Sect. 7. Future directions and possible research opportunities are discussed in Sect. 8, and the chapter concludes in Sect. 9.

2 Literature Review

The BAT algorithm [11] is a bioinspired algorithm that works on echolocation to identify prey. Echolocation is the type of sonar seen in the BAT. Echolocation is the process of using sound waves and the echoes of the target object to identify the location of objects in space. In the dark, bats utilize echolocation to find their way and find food. Through their mouths or noses, they release sound waves that aid with object detection in their surroundings. These sound waves cause echoes that return to the bats' ears when they strike an item. Bats can identify the location, size, and shape of an object by listening to these echoes. In the field of medicine, the bat algorithm has significantly contributed to advancements in various ways. In [12], the context of picture thresholding was discussed, and the method was used to maximize entropies, such as Kapur's and Tsallis' entropies. This method is used to distinguish between different tissue components, such as gray matter (GM), white matter (WM), and cerebrospinal fluid (CSF). The Dice coefficient (Dice) is used to assess the thresholding technique's efficacy. According to the findings, Tsallis' entropy performs better at segmenting CSF and WM than Kapur's entropy does. Furthermore, the BAT algorithm's entropy maximization yields a higher Dice value than does particle swarm optimization.

Binu et al. [13] demonstrated how to use the Bat method to create a fuzzy classifier for the classification of medical data. This classifier uses fuzzy logic, but it also incorporates the Bat algorithm in the fuzzy rule generation phase. To improve speed, this hybrid system combines the optimization powers of the Bat algorithm with fuzzy logic principles. This technique, which is called the bat fuzzy classification process, consists of three main steps: creating rule bases, automatically designing fuzzy membership functions, and classifying via the fuzzy system. Crucial information can be found scattered throughout different parts of the MRI scans utilized for brain imaging. Consequently, segmenting these images is crucial for efficiently focusing on regions of interest. Akhtar et al. [14] utilized the Bat algorithm in image and video processing. They used the bat algorithm to address the task of tracking human body movement from captured video segments. By combining the powerful local search capabilities of Harmony search with the exploratory capabilities of particle swarm optimization (PSO), the Bat algorithm achieves a low computing cost. The parameters of the normal Bat Algorithm were modified to meet the unique needs of this issue. Analyses of performance utilizing fidelity criteria revealed that the Bat algorithm performed noticeably better than alternative optimization methods did. Suárez et al. [15] used a two-stage procedure to apply the Bat Algorithm in swarm robotics. They created and built a working prototype of a true robotic system in the first phase. They created a robotic system simulation framework in the second phase. The goal of locating a certain location in three-dimensional surroundings was used to test the simulation. The findings of the investigation demonstrated a high degree of similarity between the simulated and real robotic swarms. This study's main contribution is that, by using these simulations, it offers an ideal way to pinpoint the Bat Algorithm's shortcomings for a range of practical uses.

Tamiru and Fakhruldin in 2011, examined energy losses [16] in gas turbines via a fuzzy system in addition to the Bat algorithm. They first created a fuzzy model, which was further refined by training it and modifying its parameters via a modified bat algorithm. Raghavan et al. presented a BAT algorithm for scheduling workflow [17] applications in cloud computing. These programs attempt to minimize workflow expenses by allocating workloads to available resources in the best possible way. The modified version, BBA, was utilized for feature selection, whereas the original BA was used to determine a minimal cost. The outcomes demonstrated that the BA outperformed the BRS algorithm in terms of cost. The authors of [18] used a combination of the BAT algorithm and the interval type 2 fuzzy C-means (IT2FCM) algorithm, which was thoroughly examined and evaluated to segregate tumor locations in different brain MRI scan sequences. The computational efficiency and performance of this novel algorithm outperform those of existing approaches, such as the cuckoo-based IT2FCM. It is a potentially useful technique for precise brain tumor identification since it efficiently detects different types of tumors and segregates tissue sections with high specificity.

Despite the progress made with the BAT algorithm in numerous sectors, there are still substantial gaps in its use and optimization. Current research is mostly focused on specialized applications such as medical imaging and fuzzy categorization, but there is little investigation into its adaptability to dynamic contexts and real-time processing. Furthermore, while algorithm improvements have increased performance, there has been a paucity of comprehensive research comparing its usefulness to a larger spectrum of optimization strategies in varied settings. Further research into integrating the BAT algorithm with future technologies, such as deep learning, may improve its capabilities and widen its applicability.

3 Overview of the Bat Algorithm

The bat algorithm (BA) [1] is a nature-inspired optimization algorithm developed by Xin-She Yang in 2010. The basic Bat algorithm is bioinspired. It is based on the characteristics of bat biosonars or echolocation. Bats in the wild emit ultrasonic waves into the environment to aid in hunting or navigation. Following the emission of these waves, it receives the echoes of the waves. It locates itself and identifies obstacles by using the received echoes. It is designed to solve optimization problems, particularly those involving nonlinear and complex functions. Here, we provide an overview of how the Bat Algorithm works (Fig. 1).

3.1 Inspiration from Bats

Bats use echolocation to detect prey and obstacles. They emit ultrasonic pulses and listen to the echoes to determine the distance, direction, and nature of the objects around them. This process involves emitting sound pulses, adjusting them on the basis of the echoes received, and homing to targets.

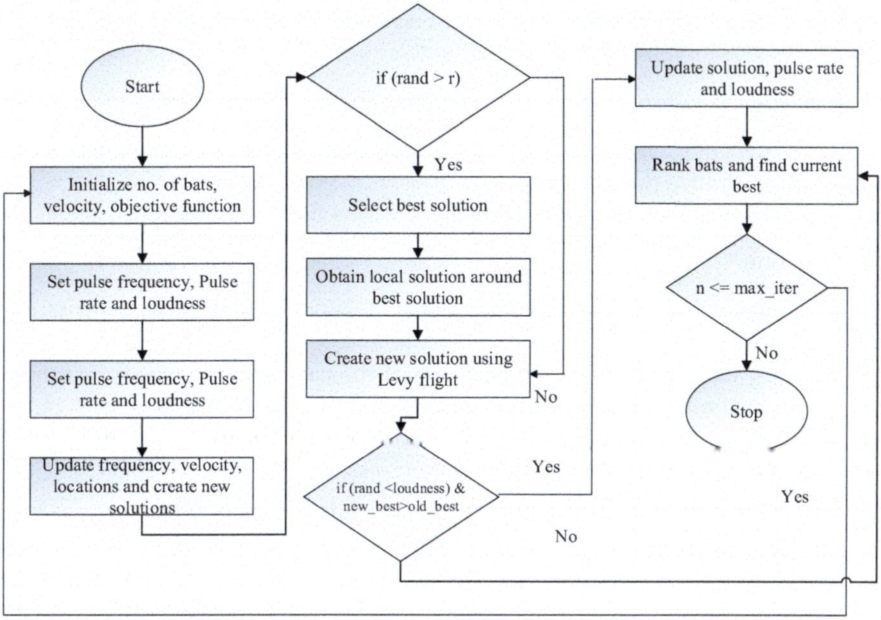

Fig. 1 Bat Algorithm Block diagram

3.2 *Algorithm Basics*

(a) **Initialization**:

   i. Initialize a population of bats randomly within the search space.
   ii. The initial velocity and position for each bat are assigned.
   iii. The initial frequency and loudness parameters are set.

(b) **Echolocation and Movement**:

   i. Each bat emits pulses (or signals) with a specific frequency.
   ii. The emitted pulse represents a potential solution (candidate solution) in the search space.
   iii. Bats adjust their frequency (rate of pulse emission) and loudness on the basis of their proximity to the optimal solution.
   iv. The loudness of the emitted pulse typically decreases with time, simulating a decrease in energy expenditure.

(c) **Movement Towards Optimal Solutions**:

   i. Bats move towards the best solutions found thus far (either globally or locally).
   ii. The movement is guided by the loudness and frequency of emitted pulses.
   iii. This process balances exploration (searching for new areas) and exploitation (refining known good solutions).

(d) **Updating Positions**:

   i. Update the positions of bats on the basis of the frequencies and loudness of their emitted pulses.
   ii. Randomness is incorporated to allow exploration of the search space.

(e) **Iteration**:

   i. The process is repeated for a number of iterations or until a stopping criterion is met.
   ii. Each iteration updates the positions of the bats and evaluates their fitness on the basis of the objective function.

4 Steps Involved in Solving Real-Life Challenges Via the Bat Algorithm

On the basis of the principle of echolocation behavior, Yang developed the bat algorithm [19] in 2010 to address various optimization problems. The process of using the Bat algorithm is as follows:

Step 1: Problem Formulation:

The first step is to define the problem clearly where the objectives, constraints, and desired outcomes should be mentioned properly. This entails comprehending the real-world problem and converting it into an algorithm-manageable format.

Step 2: Initialization:

First, a virtual bat population is created, with each bat representing a possible fix. For every bat, the starting characteristics, including position, velocity, frequency, loudness, and pulse emission rate, are established.

Step 3: Bat Movement:

The movement of bats essentially depends on several components, such as frequency tuning, loudness, and pulse rate.

- Frequency Tuning: The frequency of motion of a bat is changed to regulate the size of each step.
- Loudness and pulse rate update: Depending on how close each bat is to the ideal solution, each bat's pulse emission rate and volume are adjusted. Typically, when bats approach the ideal solution, they become quieter, and their heart rate increases.

Step 4: Updating Bat Positions

Update the velocity and position of each bat via the adjusted frequency. By going through the movement and update processes again, constant updates and improvements to the solutions can be made. To enable a complete search of the solution space and prevent premature convergence, this entails striking a balance between local exploitation and global exploration.

Step 5: Evaluating Fitness

Each bat's fitness is determined by comparing it to the goal function. This function calculates how well a bat's location (or solution) satisfies the problem's restrictions and desired result.

Step 6: Termination Criterion

The method converges by observing whether the number of iterations has been reached or if the improvement in solutions has stopped. The algorithm's best result is extracted. This bat best fits the goal and restrictions of the challenge and has the highest fitness value. The viability and efficacy of the extracted solution were verified in an actual setting. The solution should be put into practice, and its effectiveness should be considered to address any possible modifications or enhancements.

5 Solving Real-Life Challenges with the Bat Algorithm

The bat algorithm (BA) is a bioinspired optimization algorithm that is based on the echolocation behavior of bats. It is primarily used for solving optimization problems, especially those that involve finding optimal solutions in complex search spaces (Fig. 2).

5.1 Optimization of Wireless Sensor Networks (WSNs)

WSNs often require optimization of sensor placement, energy consumption, and data routing to maximize coverage and efficiency. Researchers have applied the bat algorithm to optimize the deployment of sensors in WSNs, ensuring effective coverage with minimal energy consumption. An enhanced bat algorithm-based [20] energy-efficient congestion control strategy for wireless sensor networks was suggested by Manshahia et al. in 2016. In this work, an enhanced bat algorithm inspired by nature was implemented at the transport layer to control congestion in wireless sensor networks. To obtain the best answer, the algorithm was run on the fitness function. The outcomes of the simulation demonstrate improvements in parameters such as throughput and network longevity compared with the CODA (congestion detection and avoidance), PSO (particle swarm optimization), and ACO (ant colony optimization) algorithms. Mohar et al. (2022) suggested employing the bat optimization algorithm with enhanced exploration [21] and exploitation characteristics to localize sensor nodes in wireless sensor networks. Two variations of the bat optimization

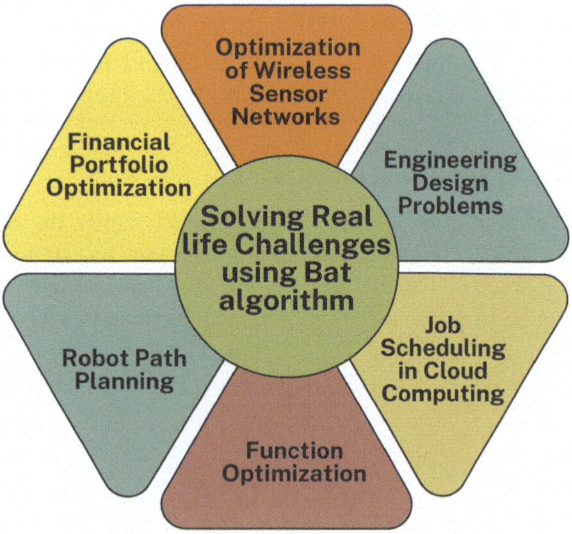

Fig. 2 Different Domains that the Bat algorithm has intercepted in

algorithm (BOA) are presented in this study to solve the limitations of the original BOA, namely, being caught in the local optimal solution and locating the sensor nodes more effectively. The suggested BOA variants 1 and 2 use enhanced global and local search strategies to modify the exploration and exploitation features of the original BOA. A number of simulations have been run for varying numbers of target and anchor nodes to verify the effectiveness of the proposed BOA versions 1 and 2. The outcomes are compared with those of the original BOA and other optimization methods that have been applied to the node localization problem in the past.

5.1.1 Engineering Design Problems

In various engineering fields, ranging from mechanical to civil engineering, design optimization problems in which the parameters need to be adjusted to achieve the desired performance metrics exist. The Bat algorithm has been utilized to optimize the design parameters of structures, mechanical components, and systems, such as minimizing weight while maintaining structural integrity or maximizing efficiency. Vu-Huu et al. discussed an enhanced bat method [22] for truss structure optimization design in 2023. To enter the conventional bat algorithm (BA) performance for optimization engineering problems and the optimization design of truss structures, they used a novel improved metaheuristic in this work. Like many other metaheuristic algorithms, the BA is easy to implement and may be used to solve a wide range of problems with a great deal of flexibility. However, it may stagnate after an early stage if its exploitation stage is altered too soon by abruptly altering the loudness and pulse emission rate. This can lead to a low chance of successfully locating the true prey. Consequently, a novel loudness function is suggested to adjust the random walk's step size effectively. Furthermore, instead of requiring numerous calculations, this work suggests a push-process technique that can produce an intervention in the BA algorithm to reveal the genuine global ideal in a few generations.

5.1.2 Job Scheduling in Cloud Computing

Cloud computing environments require efficient job scheduling to maximize resource utilization and minimize processing time. The bat algorithm [23, 24] has been applied to optimize job scheduling algorithms in cloud computing, ensuring that tasks are allocated to resources in a way that reduces the overall completion time and resource waste. Jacob worked on the bat algorithm for cloud computing resource allocation [24] in 2014. His architecture schedules resources in the cloud environment via the bat algorithm. Compared with other optimization methods, this method offers high efficiency and accuracy. Raghavan et al. employed the bat algorithm in 2015 to schedule cloud-based workflow applications [23]. Their method schedules processes in a cloud by using the bat algorithm in conjunction with the binary bat algorithm. In particular, this approach was used to map out tasks and resources. The best available resources were used to minimize the workflow's overall cost.

5.1.3 Robot Path Planning

Path planning for robots involves finding the optimal path from a start to a goal location while avoiding obstacles and minimizing travel time. Researchers have applied the bat algorithm to optimize robot path planning algorithms, ensuring that robots can navigate efficiently in complex environments. On the basis of an enhanced bat algorithm, Yuan et al. presented path planning for mobile robots in 2021 [25]. They created a better bat method in this work that uses a Cauchy disturbance and a logarithmic decreasing technique. They used the bat algorithm (BA) and the dynamic window approach (DWA) to meet the requirements of global optimal and dynamic obstacle avoidance in path planning for a mobile robot. Virtual points were set up to create an undirected weighted graph that gives the robot path switching methods. On the basis of the simulation findings, the enhanced bat algorithm outperforms both the original bat algorithm and the particle swarm optimization (PSO) method in terms of finding the ideal solution. When hybrid path planning techniques are used instead of the dynamic window approach, the path length can be considerably reduced. Their simulations demonstrated the effectiveness of the path switch method.

5.1.4 Financial Portfolio Optimization

In finance, portfolio optimization aims to select the best mix of assets to maximize returns while minimizing risk. BA has been used to optimize investment portfolios by adjusting asset allocations on the basis of historical data and risk-return profiles. In each of these cases, the Bat algorithm demonstrates its versatility and effectiveness in solving complex optimization problems across various domains. Its ability to balance exploration and exploitation, mimicking the echolocation and foraging behavior of bats, makes it a valuable tool for tackling real-life challenges where traditional optimization methods may fall short. Using the bat method, Kamili and Riffi discussed portfolio optimization [26]. The portfolio selection problem was solved in this paper by using the bat algorithm. Using four indices—the Hang Seng in Hong Kong, the DAX 100 in Germany, the FTSE 100 in the USA, and the NIKKEI in Japan—this method was modified to fit the cardinality constrained efficient frontier model (CCEF). The findings were compared to those of the unconstrained efficient frontier model (UEF). An ideal approximation between the restricted and unconstrained models was demonstrated via the bat approach.

5.2 Case Study 1: Optimization of Antenna Array Design

Researchers have applied the bat algorithm to optimize the geometry and configuration of antenna arrays. The goal was to achieve specific radiation patterns (such as directive gain and beam width) while minimizing side lobes and maintaining impedance matching across different frequencies. The BA was able to efficiently search the design space, resulting in antenna arrays that exhibited improved performance metrics compared with traditional design methods. The optimized arrays demonstrated enhanced signal reception and transmission capabilities, which are crucial for applications in radar systems, satellite communications, and wireless networks. An optimization of a rectangular patch antenna at 5 GHz via the bat search technique was proposed by Kumar et al. [27]. They used MATLAB R2009b to evaluate their optimization framework with programs based on the bat algorithm, and CST Studio Suite 2011 was used to build the antenna on the basis of the results. After the antenna's transient solver stage is finished, the S-parameter plot is examined.

5.3 Case Study 2: Parameter Optimization in Biomedical Signal Processing

Tuning parameters for signal processing algorithms used in biomedical applications, such as EEG signal classification. The bat algorithm was employed to optimize the parameters of the machine learning classifiers and signal processing filters. This included optimizing features extracted from EEG signals to increase accuracy in detecting neurological conditions or abnormalities. By leveraging BA, researchers have achieved significant improvements in the classification accuracy and robustness of EEG signal analysis. The optimized parameters led to more reliable diagnostic tools, aiding healthcare professionals in making informed decisions on the basis of EEG data. Using cutting-edge bioinspired algorithms, Jeevitha (2018) explored innovative optimization parameters of power quality disturbances [28]. Through their study, a novel method for the automatic classification of single and hybrid PQDs was introduced. The discrete wavelet transform (DWT) and probabilistic neural network-based BAT (PNN-BAT) optimal feature selection of PQDs comprise the suggested approach. The DWT is utilized in conjunction with multiresolution analysis (MRA) to extract disturbance features. Signals such as voltage sag, voltage swell, voltage transients, flickers, voltage imbalance, and harmonics are examples of power quality disturbances. Owing to their high sampling rate, these disturbance signals generate megabytes of data and cover a wide frequency band, necessitating a large amount of storage capacity. A discrete wavelet transform was employed in their study to analyze the signals of power quality disturbances and minimize the amount of storage space needed. The PNN classifier proved to be an efficient method for classifying the PQDs.

5.4 Case Study 3: Optimization of Supply Chain Management

BA was applied to optimize the logistics and distribution networks within supply chains. This included determining optimal warehouse locations, inventory levels, and transportation routes to minimize costs associated with inventory holding, transportation, and order fulfillment time. The Bat algorithm facilitated the discovery of near-optimal solutions that balanced cost reduction with improved service levels. Companies implementing these optimized supply chain strategies experienced reduced operational costs, improved delivery times, and increased customer satisfaction through more efficient management of inventory and logistics. In each of these case studies, the Bat algorithm demonstrated its effectiveness in tackling complex optimization problems across diverse fields. By leveraging the BA's ability to explore and exploit search spaces effectively, researchers and practitioners were able to achieve superior results compared with traditional methods, thereby demonstrating the practical applicability and versatility of bioinspired algorithms in solving real-world challenges. In 2014, Chaoran et al. examined how to optimize supply and demand scheduling via the bat algorithm within a cloud manufacturing platform. To limit the creation of the initial population and resolve the batch scheduling issue, they employed an improved bat algorithm (IBA) [29]. Consequently, it decreased the number of irrelevant searches and decreased the time needed to identify the best work strategy. The efficiency of IBA was demonstrated through experimental comparison. IBA steers the idle resource and resource overload phenomena, which encourages the use of CMfg in the supply chain.

6 Comparison of the Bat Algorithm with Other Metaheuristic Algorithms

The collaborative combination of the BA with other metaheuristic algorithms, as discussed by Puchinger and Raidl [30], highlights a synergistic approach where information exchange enhances the overall performance and efficiency of the BA in solving complex optimization problems. In a collaborative approach, the BA exchanges information with other algorithms but remains independent in its operation. This means that the BA and another algorithm work alongside each other, potentially sharing information such as solutions or fitness evaluations.

An example of collaborative combination could involve using solutions generated by another metaheuristic algorithm (such as genetic algorithms or particle swarm optimization) as an initial population for the BA. This initialization can help the BA [31] start with diverse and potentially promising solutions from the outset, enhancing its ability to find high-quality solutions quickly. In contrast, integrative combinations involve embedding the BA as a submodule within another metaheuristic or exact algorithm. Here, the BA becomes an integral part of the larger algorithmic framework, contributing to its overall performance. For example, the BA might be integrated into a hybrid algorithm where it handles specific aspects of the optimization process, such as exploration or fine-tuning solutions found by another method. Other metaheuristic algorithms can be utilized to initialize the population in the BA. This approach leverages the strengths of multiple algorithms by starting the BA with potentially diverse and well-performing solutions from the outset. The initialization phase is crucial in optimization algorithms because it can significantly influence the exploration and exploitation capabilities of the algorithm.

Reference [32] serves as a comprehensive survey that highlights the potential and effectiveness of hybrid approaches that combine metaheuristics such as the Bat Algorithm with exact algorithms in combinatorial optimization. This study provides valuable insights into how these hybridization strategies can be applied to improve the efficiency and effectiveness of optimization algorithms in practical applications. The BA represents a synthesis of the PSO, HS, and SA techniques, incorporating their advantageous features while introducing its own mechanisms inspired by bat echolocation behavior. The parameter α in the BA [33] acts analogously to a cooling factor in the SA, illustrating how different optimization concepts [19] can be combined synergistically to create a robust metaheuristic algorithm. Compared with other metaheuristic algorithms such as simulated annealing (SA), genetic algorithms (GA), differential evolution (DE), and particle swarm optimization (PSO), the bat algorithm (BA) indeed has distinct characteristics and performance, as discussed in [34, 35]. BA stands out among metaheuristic algorithms because of its unique characteristics, such as population diversity, dynamic frequency tuning, and echolocation-inspired movements. It has demonstrated competitive or superior performance compared with the GA, DE, and PSO in many studies, making it a valuable tool in optimization research and practical applications. The Bat Algorithm (BA) offers several distinct advantages over other metaheuristic algorithms, as highlighted by studies [36–38]. The dynamic parameter adjustment capability of the BA, along with its competitive performance across diverse combinatorial optimization problems, makes it a promising choice for researchers and practitioners seeking efficient and effective solutions. Its flexibility and adaptability contribute to its widespread application and favorable performance compared with other established metaheuristics in the field.

7 Advantages and Limitations of the Bat Algorithm

The Bat Algorithm indeed harnesses the natural behavior of bats to strike a balance between exploration and exploitation in optimization problems. The ability of the Bat Algorithm to emulate the echolocation behavior of bats enables it to explore and exploit the search space efficiently, making it particularly effective for tackling optimization problems where a balance between exploration and exploitation is crucial for finding high-quality solutions [39]. BA, as a scheduling tool, excels in optimizing the sequence of operations across multiple stages and machines, offering efficiency, adaptability, and robust performance in complex manufacturing environments. This makes it a valuable asset for industries seeking to streamline their production processes and enhance overall productivity [37]. Sabba and Chikhi's [40] binary bat algorithm with sigmoid function transformation offers a structured approach for solving the multidimensional knapsack problem, effectively balancing exploration and exploitation while ensuring that solutions adhere to the problem's constraints through binary representation. Talafuse and Pohl [1] applied the Bat algorithm to the redundancy allocation problem in series–parallel systems with component mixing and demonstrated the effectiveness of the BA in optimizing [41] the allocation of redundancies to enhance system reliability and efficiency while considering practical constraints and system configurations. Dapa et al. [42] demonstrated the effectiveness of the Bat algorithm in optimizing machine layout design to minimize material handling distances in manufacturing settings. This application highlights the BA's ability to address complex spatial optimization problems while considering practical constraints and operational objectives. Frequency tuning, automatic zooming, and parameter control are the three features that collectively contribute to the BA's effectiveness [43] in optimizing complex problems across diverse domains. By dynamically adjusting frequencies, automatically zooming in on promising solutions, and controlling parameters to adapt to specific tasks, the BA efficiently balances exploration of new areas and exploitation of current knowledge to achieve optimal solutions. This makes BA a versatile and powerful tool in the toolkit of optimization algorithms [44].

The advancements introduced in the enhanced version of the BA, as applied to UCAV problems in reference [45], have resulted in notable improvements in accuracy, feasibility, and the ability to locate multiple solutions. These advancements significantly contribute to optimizing UCAV operations and enhancing decision-making capabilities in the defense and aerospace domains. The technique proposed by Mallick et al. for a smart post buckled beam actuator (PBA), as detailed in reference, leverages innovative design principles and optimization methods to enhance the performance of piezoceramic stack actuators in driving buckled beam actuators. Mallick et al. utilized piezoceramic stack actuators as prime movers for buckled beam actuators. These actuators are chosen for their characteristics, such as small displacement, high force output, high bandwidth, and high energy density, which are advantageous for driving precision mechanisms such as the PBA. A key innovation introduced in their work was the design of a motion amplification device, specifically a linear-to-linear motion converter. This device addresses the challenge of small displacements typical of piezoceramic actuators by amplifying the motion, thereby enhancing the actuation capability of the system. The Mathieu–Hill equation is employed to model the dynamics of the post buckled beam actuator, capturing its nonlinear behavior under actuation. The BA is utilized as an optimization tool to formally optimize [46] the design parameters of the PBA. BA's ability to explore diverse solution spaces and dynamically adjust parameters aids in finding configurations that maximize performance metrics such as large end rotations (indicative of greater actuation range) and reduced weights. Mallick et al.'s approach demonstrates how the integration of innovative design concepts with advanced optimization techniques such as the Bat algorithm can lead to significant advancements in smart actuator systems. The optimized PBA design not only enhances the performance parameters but also expands the applicability of piezoceramic actuators in demanding aerospace and mechanical applications. The study by Rauf et al. introduced an optimized approach using long short-term memory (LSTM) networks for predicting COVID-19 cases. They integrate the bat algorithm (BA) to increase the performance of LSTM, addressing challenges such as local minima and premature convergence, which are typical in optimization processes. BA optimizes the LSTM parameters to improve its predictive accuracy for COVID-19 cases. Parameters such as learning rates, weights, and biases within LSTM are tuned via BA to minimize the prediction error [47] effectively. LSTM is designed to preserve optimal solutions by continuously comparing personal best solutions with the swarm's local best solutions. Gaussian walk further aids in fine-tuning the LSTM parameters to improve model performance. Rauf et al. evaluated their optimized LSTM approach against several benchmarks, including nonoptimized LSTM, gated recurrent units (GRUs), recurrent neural networks (RNNs), and other contemporary algorithms used for COVID-19 case prediction. Rauf et al.'s study highlights the effectiveness of integrating BA with LSTM for optimizing COVID-19 case prediction models. By addressing optimization challenges through an enhanced BA variant and leveraging LSTM's capabilities, they achieve notable improvements in prediction accuracy, highlighting the potential of hybrid approaches in enhancing the performance of neural network-based predictive models in epidemiological forecasting [20]. The modification proposed by Cai et al. to enhance the bat algorithm (BA), as detailed in [21], focuses on improving its exploitation capabilities, particularly in the context of solving the coverage problem in wireless sensor networks (WSNs). By enhancing the exploitation capabilities of the BA, the modification of Cai et al. contribute to making the BA more competitive in solving complex optimization problems, particularly those requiring fine-tuned local search strategies such as the WSN coverage problem. The results highlight the potential of tailoring modifications to metaheuristic algorithms such as the BA to adapt to specific application requirements and improve the overall solution quality and efficiency. Cai et al. modified the BA to illustrate how targeted enhancements can optimize its performance in tackling real-world optimization challenges such as the WSN coverage problem. By refining the algorithm's exploitation capabilities, they achieve improved efficiency and effectiveness compared with the original BA, demonstrating the value of adaptive approaches in metaheuristic algorithm design.

8 Future Directions and Research Opportunities

In recent years, the bat algorithm (BA) has emerged as a powerful optimization technique inspired by the echolocation [48, 49] behavior of bats. This section explores the future directions and research opportunities of the BA in solving real-world problems. By understanding its capabilities and potential applications, we can unlock new possibilities for problem-solving across various domains.

(a) **Optimization in Engineering and Design**: The BA has shown promising results in solving complex optimization problems in engineering and design. Future research opportunities include enhancing the algorithm's efficiency, scalability, and adaptability to different problem domains. Exploring its applications in areas such as structural engineering, mechanical system design, and process optimization can lead to improved designs, reduced costs, and better resource allocation.

(b) **Energy Systems and Renewable Resources**: The transition towards sustainable energy systems and the efficient use of renewable resources pose significant challenges. The BA has the potential for optimizing energy production and distribution systems, improving energy efficiency, and facilitating the integration of renewable energy sources into existing grids. Future research in this area can focus on addressing the dynamic nature of energy systems and optimizing their operation under uncertain conditions.

(c) **Transportation and Logistics**: Efficient transportation and logistics systems are critical for economic growth and sustainability. The BA can be applied to optimize vehicle routing, scheduling, and supply chain management. Future research directions include exploring its potential in solving real-time transportation problems, optimizing multimodal transportation networks, and addressing the challenges of urban logistics. By leveraging the BA, we can improve efficiency, reduce costs, and minimize environmental impacts in these domains.

(d) **Healthcare and medical applications**: The healthcare industry faces numerous complex challenges, including resource allocation, scheduling, and optimization of medical treatments. The BA can be utilized to solve these problems, optimizing patient scheduling, healthcare facility layout, and resource allocation in hospitals. Future research opportunities involve adapting the BA to address specific challenges in healthcare, such as optimizing cancer treatment plans, healthcare data analytics, and personalized medicine.

(e) **Financial Optimization and Portfolio Management**: The BA has demonstrated potential in financial optimization and portfolio management, assisting investors in making informed decisions. Future research can explore its capabilities in portfolio optimization under uncertainty, risk management, and algorithmic trading. By integrating the BA with machine learning and predictive modeling techniques, we can enhance its performance and enable more accurate financial decision-making.

(f) **Environmental Management and Conservation**: Preserving our environment and natural resources is of paramount importance. The BA can contribute to environmental management by optimizing water resource management, land use planning, and ecosystem conservation efforts. Future research can focus on leveraging the BA to address the challenges of climate change adaptation, biodiversity conservation, and sustainable land development. By doing so, we can ensure the responsible use of resources and protect fragile ecosystems.

(g) **Social Sciences and Policy-Making**: Understanding and addressing societal challenges requires interdisciplinary research. The BA can be applied in the social sciences to aid in policy-making, urban planning, and social network analysis. Future research opportunities involve exploring the BA's potential in modeling social dynamics, analyzing public sentiment, and optimizing social welfare programs. By integrating the BA with social science methodologies, we can gain valuable insights and inform evidence-based policy-making processes.

The Bat Algorithm presents a promising avenue for solving complex real-world problems across a wide range of domains. By focusing on optimization in engineering and design, energy systems, transportation, healthcare, finance, environmental management, and social sciences, we can unlock its full potential. Further research and development efforts must be directed towards enhancing its efficiency, scalability, and adaptability to address specific challenges in these domains. By embracing the opportunities offered by the BA, we can pave the way for more effective problem-solving, sustainable development, and societal progress.

9 Conclusions

The current chapter discusses the importance of the bat algorithm (BA) in resolving real-world problems in a variety of disciplines in this 1000-word conclusion. The BA has demonstrated enormous promise in optimization and problem-solving, drawing inspiration from the echolocation activity of bats. We can solve challenging real-world problems in engineering, energy systems, transportation, healthcare, finance, environmental management, and the social sciences by utilizing the power of nature-inspired algorithms. The effectiveness of the BA in streamlining engineering and design procedures has been shown, leading to better designs, lower costs, and more efficient use of resources. It may be applied in structural engineering, mechanical system design, and process optimization because of its versatility in solving many problem domains. Improving the algorithm's scalability, flexibility, and efficiency will open new avenues for optimization and design research in the future. The integration of renewable energy sources, enhancing energy efficiency, and optimizing energy production and distribution are all made possible by the BA in the context of energy systems and renewable resources. Its ability to optimize energy system operation in unpredictable circumstances and address their dynamic nature is potentially useful. We may strive for more sustainable energy systems and effective utilization of renewable resources by utilizing the BA.

Systems for logistics and transportation are essential to sustainability and economic progress. By using the BA to optimize supply chain management, truck scheduling, and routing, expenses can be decreased, and environmental effects can be minimized while increasing efficiency. The issues of urban logistics, multimodal transportation network optimization, and real-time transportation concerns are areas of future research potential. Transportation and logistics operations can be revolutionized by utilizing the power of the BA. The healthcare sector faces several intricate issues, such as optimizing medical treatments and allocating resources. Hospital resource allocation, healthcare facility layout, and patient scheduling can all be improved with the use of the BA. Potential avenues for further research entail tailoring the BA to particular healthcare problems, such as customized medicine, healthcare data analytics, and cancer treatment plan optimization. The BA has the power to improve patient care and transform healthcare systems.

By minimizing risks, facilitating algorithmic trading, and optimizing portfolios under uncertainty, the BA helps investors make well-informed decisions in the areas of financial optimization and portfolio management. We can improve the BA's performance and make financial decision-making more precise by combining it with machine learning and predictive modeling approaches. Both investors and financial experts can benefit greatly from the BA. To protect our delicate ecosystems and natural resources, environmental management and conservation are essential. By maximizing land use planning, ecosystem preservation, and water resource management, BA may support these initiatives. Subsequent studies can concentrate on adapting to climate change, conserving biodiversity, and utilizing the BA in sustainable land development. By doing so, we can ensure that resources are used responsibly and save the environment for coming generations. The BA has much to offer the social sciences and the field of policy-making. A bachelor's degree can help with social network analysis, urban planning, and policy-making by illuminating social dynamics and enhancing social welfare initiatives. Prospective avenues for research include social dynamics modeling, sentiment analysis, and providing guidance for evidence-based policy-making procedures. We can learn a great deal and build more just and inclusive societies by combining BA with social scientific methods. The Bat Algorithm offers a strong and adaptable method for resolving problems in a variety of real-world settings. Its capacity to optimize difficult issues leads to better designs, lower costs, and better decision-making. It is critical that we concentrate on improving the BA's effectiveness, scalability, and adaptability for certain issue domains, as we look into potential future research directions. Through the utilization of nature-inspired optimization, we can open novel avenues, develop enduring resolutions, and address the intricate problems of our day. With the help of the Bat algorithm, we can move towards a more robust and promising future.

References

1. Yang XS, He X (2013) Bat algorithm: literature review and applications. Int J Bioinspired Comput 5(3):141–149
2. Mirjalili S, Mirjalili SM, Yang XS (2014) Binary bat algorithm. Neural Comput Appl 25:663–681
3. Al-Betar MA, Awadallah MA (2018) Island bat algorithm for optimization. Expert Syst Appl 107:126–145
4. Yılmaz S, Küçüksille EU (2015) A new modification approach on bat algorithm for solving optimization problems. Appl Soft Comput 28:259–275
5. Agarwal T, Kumar V (2022) A systematic review on bat algorithm: Theoretical foundation, variants, and applications. Arch Comput Methods Eng, 1–30
6. Nakamura RY, Pereira LA, Costa KA, Rodrigues D, Papa JP, Yang XS (2012) BBA: a binary bat algorithm for feature selection. In 2012 25th SIBGRAPI conference on graphics, patterns and images, pp 291–297. IEEE
7. Chakri A, Ragueb H, Yang XS (2018) Bat algorithm and directional bat algorithm with case studies. Nature-inspired algorithms and applied optimization, 189–216
8. Chakri A, Khelif R, Benouaret M, Yang XS (2017) New directional bat algorithm for continuous optimization problems. Expert Syst Appl 69:159–175

9. Komarasamy G, Wahi A (2012) An optimized K-means clustering technique using bat algorithm. Eur J Sci Res 84(2):263–273
10. Fister Jr, I, Fister D, Yang XS (2013) A hybrid bat algorithm. arXiv preprint arXiv:1303.6310.
11. Dey N, Ashour AS, Bhattacharyya S (2019) Applied nature-inspired computing: algorithms and case studies. Springer tracts in nature-inspired computing
12. Kaur T, Saini BS, Gupta S (2019) Optimization techniques for the multilevel thresholding of the medical images. In: Medical data security for bioengineers. IGI Global
13. Binu D, Selvi M (2015) BFC: bat algorithm based fuzzy classifier for medical data classification. J Med Imaging Health Inform 5:599–606
14. Akhtar S, Ahmad A, Abdel-Rahman EM (2012) A metaheuristic bat-inspired algorithm for full body human pose estimation. In: 2012 Ninth conference on computer and robot vision. IEEE, pp 369–375
15. Suárez P, Iglesias A, Gálvez A (2019) Make robots be bats: specializing robotic swarms to the bat algorithm. Swarm Evol Comput 44:113–129
16. Lemma TA, Hashim FBM (2011) Use of fuzzy systems and bat algorithm for exergy modelling in a gas turbine generator. In: 2011 IEEE Colloquium on humanities, science and engineering. IEEE, pp 305–310
17. Raghavan S, Sarwesh P, Marimuthu C, Chandrasekaran K (2015) Bat algorithm for scheduling workflow applications in cloud. In: 2015 International conference on electronic design, computer networks & automated verification (EDCAV). IEEE, pp 139–144
18. Kotwal A, Bharti R, Pandya M, Jhaveri H, Mangrulkar R (2021) Application of BAT Algorithm for detecting malignant brain tumors. Applications of bat algorithm and its variants, 119–132
19. Yang XS (2010) A new metaheuristic bat-inspired algorithm. In Nature inspired cooperative strategies for optimization (NICSO 2010), pp 65–74. Berlin, Heidelberg: Springer, Berlin, Heidelberg
20. Manshahia M, Dave M, Singh S (2016) Improved bat algorithm based energy efficient congestion control scheme for wireless sensor networks. Wirel Sens Netw 8:229–241. https://doi.org/10.4236/wsn.2016.811018
21. Mohar SS, Goyal S, Kaur R (2022) Localization of sensor nodes in wireless sensor networks using bat optimization algorithm with enhanced exploration and exploitation characteristics. J Supercomput 78:11975–12023. https://doi.org/10.1007/s11227-022-04320-x
22. Vu-Huu T, Pham Van S, Pham Q, Thanh-Cuong l (2022) An improved bat algorithms for optimization design of truss structures. Structures 47:2240–2258.https://doi.org/10.1016/j.istruc.2022.12.033
23. Raghavan S, Sarwesh P, Marimuthu C, Chandrasekaran K (2015) Bat algorithm for scheduling workflow applications in cloud. 2015 International conference on electronic design, computer networks & automated verification (EDCAV), Shillong, India, pp 139–144, https://doi.org/10.1109/EDCAV.2015.7060555
24. Jacob L (2014) Bat algorithm for resource scheduling in cloud computing. https://www.semanticscholar.org/paper/Bat-Algorithm-for-Resource-Scheduling-In-Cloud-Jacob/72511f74f5675dd230fcd07d305714757c9ace31
25. Yuan X, Yuan X, Wang X (2021) Path planning for mobile robot based on improved bat algorithm. Sensors 21:4389. https://doi.org/10.3390/s21134389
26. Kamili H, Riffi M (2016) Portfolio optimization using the bat algorithm. Int Rev Comput Software (IRECOS) 11:277. https://doi.org/10.15866/irecos.v11i3.8154
27. Kumar M, Sahoo A, Sao R, Mangaraj B (2015) Optimization of rectangular patch antenna at 5 GHz using bat search algorithm. https://doi.org/10.1109/CSNT.2015.47
28. Jeevitha SRS, Mabel MC (2018) Novel optimization parameters of power quality disturbances using novel bioinspired algorithms: a comparative approach. Biomed Signal Process Control 42:253–266. https://doi.org/10.1016/j.bspc.2018.02.003
29. Chaoran W, Li F, Jiang H (2014) Time optimization of supply and demand scheduling based on bat algorithm under the cloud manufacturing platform. J Modern Inform 34(10):35–40. https://doi.org/10.3969/j.issn.1008-0821.2014.10.007

30. Puchinger J, Raidl GR (2005) Combining metaheuristics and exact algorithms in combinatorial optimization: A survey and classification. In International work-conference on the interplay between natural and artificial computation, pp 41–53. Berlin, Heidelberg: Springer, Berlin, Heidelberg
31. Umar SU, Rashid TA, Ahmed AM, Hassan BA, Baker MR (2024) Modified Bat Algorithm: a newly proposed approach for solving complex and real-world problems. Soft Comput, 1–16
32. Sadeghi J, Mousavi SM, Niaki STA, Sadeghi S (2014) Optimizing a biobjective inventory model of a three-echelon supply chain using a tuned hybrid bat algorithm. Transp Res Part E Logist Transp Rev 70:274–292
33. Dey N, Rajinikanth V (Eds) (2021) Applications of bat algorithm and its variants. Springer
34. Yang XS, Hossein Gandomi A (2012) Bat algorithm: a novel approach for global engineering optimization. Eng Comput 29(5):464–483
35. Marichelvam MK, Prabaharam T (2012) A bat algorithm for realistic hybrid flowshop scheduling problems to minimize makespan and mean flow time. ICTACT J Soft Comput 3(1):428–433
36. Marichelvam MK, Prabaharan T, Yang XS, Geetha M (2013) Solving hybrid flow shop scheduling problems using bat algorithm. Int J Logist Econ Globalization 5(1):15–29
37. Talafuse TP, Pohl EA (2016) A bat algorithm for the redundancy allocation problem. Eng Optim 48(5):900–910
38. Musikapun P, Pongcharoen P (2012) Solving multistage multimachine multiproduct scheduling problem using bat algorithm. In: 2nd international conference on management and artificial intelligence (Vol 35, pp 98–102). IACSIT Press Singapore
39. Sabba S, Chikhi S (2014) A discrete binary version of bat algorithm for multidimensional knapsack problem. Int J Bio-Inspired Comput 6(2):140–152
40. Dapa K, Loreungthup P, Vitayasak S, Pongcharoen P (2013) Bat algorithm, genetic algorithm and shuffled frog leaping algorithm for designing machine layout. In: Multidisciplinary Trends in artificial intelligence: 7th international workshop, MIWAI 2013, Krabi, Thailand, December 9–11, 2013. Proceedings 7, pp 59–68. Springer, Berlin, Heidelberg
41. Dey N, Rajinikanth V, Lin H, Shi F (2021) A study on the bat algorithm technique to evaluate the skin melanoma images. Applications of bat algorithm and its variants, 45–60
42. Li YG, Peng JP (2014) An improved bat algorithm and its application in multiple ucavs. Appl Mech Mater 442:282–286
43. Janakiraman S (2023) Improved bat optimization algorithm and enhanced artificial bee colony-based cluster routing scheme for extending network lifetime in wireless sensor networks. Int J Commun Syst 36(5):e5428
44. Rauf HT, Gao J, Almadhor A, Arif M, Nafis MT (2023) Retraction Note: Enhanced bat algorithm for COVID-19 short-term forecasting using optimized LSTM
45. Cai X, Wang L, Kang Q, Wu Q (2015) Adaptive bat algorithm for coverage of wireless sensor network. Int J Wirel Mobile Comput 8(3):271–276
46. Rajinikanth V, Dey N, Kavitha S (2021) Multi-thresholding with Kapur's entropy—a study using bat algorithm with different search operators. Applications of bat algorithm and its variants, 61–78
47. Satapathy SC, Sri Madhava Raja N, Rajinikanth V, Ashour AS, Dey N (2018) Multi-level image thresholding using Otsu and chaotic bat algorithm. Neural Comput Appl 29:1285–1307
48. Dey N (Ed) (2017) Advancements in applied metaheuristic computing. IGI global
49. de Jesús Rubio J (2023) Bat algorithm based control to decrease the control energy consumption and modified bat algorithm based control to increase the trajectory tracking accuracy in robots. Neural Netw 161:437–448

Chapter 10
Enhancing Feature Selection Using Multi-objective Optimization Concept

Hetavi Mehta, Nirali Parekh, Ramchandra Mangrulkar, Aruna Gawade, Pallavi Chavan, and Nitin Goje

1 Introduction

1.1 Introduction to Multi-objective Optimization

Vilfredo Pareto proposed multi-objective optimization as a mathematical field which can be applied in the realm of decision-making. It focuses on solving problems that require concurrent optimization of two or more objective functions. Generally, these problems concern optimizing objectives that exhibit a conflicting nature, whereby improvement to one objective implies the expense of another. Such problems do not provide a single best solution for all objectives, rather a set of optimal solutions are obtained which represent the most efficient trade-offs possible among the

H. Mehta (✉) · R. Mangrulkar · A. Gawade
Department of Computer Engineering, Dwarkadas J. Sanghvi College of Engineering, Mumbai, India
e-mail: hetavidm22@gmail.com

R. Mangrulkar
e-mail: Ramchandra.mangrulkar@djsce.ac.in

A. Gawade
e-mail: aruna.gawade@djsce.ac.in

N. Parekh
Department of Management Science and Engineering, Standford University, Stanford, CA, USA
e-mail: nirali25@stanford.edu

P. Chavan
Department of Information Technology, Ramrao Adik Institute of Technology, DY Patil's Deemed to be University, Navi Mumbai, India

N. Goje
Webster University, Tashkent, Uzbekistan
e-mail: nitingoje@webster.edu

N. Dey (ed.), *Optimizing Solutions for Real-Life Problems*, Springer Tracts in Nature-Inspired Computing, https://doi.org/10.1007/978-981-96-3216-9_10

conflicting objectives that are considered. This type of optimization problem can be used by professionals in various fields, such as mathematics, engineering, economics, aviation, automotive, and many others.

1.1.1 Elements of Multi-objective Optimization Problem

The components of a multi-objective optimization (MOO) problem serve to collectively define the problem and direct the exploration for Pareto-optimal solutions. These components are comprised of the following:

1. Decision Variables:

They refer to the parameters that are manipulated by a multi-objective optimization algorithm in order to seek optimal solutions. Decision variables often have a physical interpretation in the context of the problem. For example, in a manufacturing optimization problem, decision variables could represent production quantities, energy utilization, or material allocation. More variables may result in a larger search space, which may increase the computational complexity associated with the optimization process. A careful selection of decision-making factors is crucial as they have a direct impact on the final outcome of the problem. Thorough deliberation and precise definition of decision variables are essential for formulating an accurate and effective multi-objective optimization problem.

2. Objective Functions:

In the context of multi-objective optimization (MOO), multiple objective functions are defined, each representing a different criterion or aspect of the given problem. The optimization process is guided by the quantification of objectives in a mathematical manner, taking into account the choice variables and problem context. This quantification facilitates the search for solutions that achieve the best possible balance among them. Objectives can be assigned weights to accurately represent their relative influence on the problem in question. Each objective function pertains to a direction of optimization, which is either minimization or maximization. For example, in the context of a manufacturing optimization problem, an objective function would aim to maximize production quantities while another would be focused on minimizing energy utilization. In multi-objective optimization, the primary challenge is to find a set of optimal solutions that consider the trade-offs between conflicting objectives, while ensuring that no single solution dominates the others.

3. Constraints:

Constraints are mathematical expressions that involve decision variables and are used to describe the constraints, boundaries, and conditions that solutions must adhere to. Constraints are used to define the feasible region within the decision space. This region represents an array of valid solutions that satisfy the requirements of the problem. Any solution that fails to meet one or more constraints is deemed infeasible and is hence eliminated from further consideration. Constraints often have a physical interpretation. In the context of the manufacturing optimization problem, constraints can be defined involving many factors such as the availability of materials, production capacity, and equipment efficiency. Constraints are of utmost importance in multi-objective optimization (MOO) since they are crucial in ensuring the practicality and feasibility of solutions. These constraints interact with the objective functions to establish the feasible region, where solutions must strike a balance among many conflicting objectives.

4. Feasible Region:

The feasible region defines the space within which the decision variables must lie to be regarded as feasible solutions. Infeasible solutions, which are beyond the feasible region, are excluded from consideration throughout the optimization process. The feasible area is defined as the set of points that satisfy all constraints, including both equality and inequality constraints. The feasible region can introduce trade-offs between constraints and objectives. By relaxing a single constraint, it is possible to get enhancements in specific objectives; nevertheless, this action may have adverse effects on other objectives. The concept of feasibility refers to the set of limitations that solutions must comply to in order to be deemed viable. Feasibility plays a crucial role in directing the optimization process toward identifying Pareto-optimal solutions that satisfy the constraints associated with the problem.

1.1.2 Formulating a Multi-objective Optimization Problem

Formulating a multi-objective optimization (MOO) problem entails defining the key components (decision variables, objective functions, and constraints) in a way that captures the trade-offs among multiple conflicting objectives [1].

1. Identify the decision variables: This includes assigning symbols to represent each decision variable and determining its type (continuous, discrete, or mixed).
2. Define objective functions: Identify the objectives that need to be optimized, even though they are conflicting. Formulate a mathematical expression for each objective function and specify the direction of optimization, i.e., minimization or maximization.
3. Specify constraints: Formulate mathematical expressions—equality or inequality—for each constraint, based on the decision variables.
4. Combine objectives and constraints: Combining the objective functions and constraints is a requisite step while formulating a multi-objective optimization problem. The general form is represented in (see Eq. 1).

$$\text{Minimize/Maximize}\,F = \{f_1(x), f2(x), \ldots, f_n(x)\} \quad (1)$$

Subject to:

$g_1(x) \leq 0; g_2(x) \leq 0; \ldots; g_m(x) \leq 0.$

$h_1(x) = 0; h_2(x) = 0; \ldots; h_p(x) = 0.$

where x is the vector of decision variables, F is the set of all objective functions $f_k(x)$, $g_i(x)$ denote the inequality constraints, while $h_j(x)$ correspond to the equality constraints.

1. Assign weights: If some objectives are more important than others, assign weights to the objective functions to reflect their relative priorities.
2. Define feasible region: This establishes the boundaries within which the optimization will occur.

Now, the formulated problem can be solved for set of optimal solutions, which can be achieved with the help of any multi-objective optimization algorithm.

1.2 Pareto Optimality

The theory of Pareto optimality was initially introduced by Vilfredo Pareto within the field of economics. However, the concept exhibits widespread relevance across a variety of different fields, encompassing engineering, economics, finance, healthcare, and various other domains, owing to its foundational role in the sphere of multi-objective optimization. The framework offers a systematic approach for managing complex decision problems that involve competing objectives. This enables decision-makers to analyze and comprehend the compromises involved, facilitating informed decision-making based on their preferences.

Pareto optimality (or Pareto efficiency) refers to a condition in which resource allocation is most effective, so that enhancing one criterion does not result in a deterioration of other criteria [2]. Pareto optimality aptly captures the inherent trade-offs that arise from addressing many objectives simultaneously. A solution qualifies as Pareto-optimal when it is not possible to enhance any single objective without adversely impacting another objective. This suggests that a Pareto-optimal solution exhibits the quality of not being superior to any other solution in the given set, rendering it a distinct and valuable point of compromise. The set of Pareto-optimal solutions constitutes a non-dominated front within the objective space. Solutions present on this front represent potential trade-offs between competing objectives. Convexity of the Pareto front signifies that no solution within the front is superlative among all solutions across the entire objective space. The decision-maker possesses the power to select the best alternative from the Pareto front, considering the demands and priorities of the problem at hand.

Various techniques, such as Multi-Objective Differential Evolution (MODE), Multi-Objective Particle Swarm Optimization (MOPSO), and Non-dominated

Sorting Genetic Algorithms (NSGA), have been specifically developed to identify Pareto-optimal solutions for problems that require multi-objective optimization. As optimization algorithms evolve, they aim to enrich the quality and expand the coverage of the Pareto front. The aforementioned algorithms explore the decision space to determine solutions that are not dominated by others and gradually build the Pareto front.

2 Related Works

2.1 *Single-Objective Feature Selection*

The recent years have witnessed a substantial rise in the focus on feature selection as a vital step in pre-processing data used for machine learning and data analysis. This section provides an exhaustive review of the existing literary works associated with single-objective feature selection. It seeks to explain the evolution of techniques in this field and their impact in enhancing both the performance and interpretability of models.

2.1.1 Filter Methods

Filter methods represent an initial approach to feature selection, wherein the evaluation of features is conducted independently of the learning process. The concept of utilizing correlation and statistical tests for feature ranking was first established in the seminal dissertations of Kohavi and John [3]. Furthermore, Guyon and Elisseeff [4] conducted a comprehensive assessment on different filter algorithms, highlighting their computational efficiency and model-agnostic characteristics. Filter methods are commonly used as a fundamental approach to assess the significance of features. However, it is necessary to note that these methods may not thoroughly capture the interactions between features.

2.1.2 Wrapper Methods

The concept of wrapper techniques was first presented by Kohavi and John [3]. These methods involve treating feature selection as an optimization problem as well as incorporating the performance of the chosen machine learning algorithm as the fitness function for selecting features. Kohavi and John [3] underscored the significance of considering the correlation between feature selection and the learning strategy opted. However, it also highlighted some of the limitations, such as computational cost and the risk of overfitting to a particular learning algorithm. Following this, Brownlee [5] introduced Clever Algorithm, a parallel particle swarm optimization toolkit, which enabled the application of wrapper methods on various datasets. Wrapper methods provide an excellent technique for evaluating the impact of feature subsets on model performance.

2.1.3 Recent Advancements and Challenges

Most recent research studies have proposed extended feature selection techniques wherein domain-specific knowledge has been integrated, ensemble methods have been explored, and high-dimensional datasets have been addressed. In their study, Saeys et al. [6] conducted a thorough examination of feature selection methods in the sector of bioinformatics. The significance of these techniques in enhancing the performance and interpretability of models when dealing with intricate biological data is very well highlighted in the paper. Nevertheless, there still persist issues that need to be addressed in this domain. These challenges include the curse of dimensionality, which refers to the difficulty of analyzing and interpreting data in high-dimensional spaces, as well as the necessity of efficiently managing noisy or redundant attributes in order to provide accurate and reliable results. The studies discussed above have made substantial contributions to the field of feature selection, with a primary emphasis on optimizing a single objective.

2.2 Multi-objective Optimization Methods

This section offers a comprehensive overview of laudable developments and contemporary advancements that have influenced multi-objective optimization. The article emphasizes notable contributions within the discipline and presents research publications that have had a substantial impact on its progress.

2.2.1 Multi-objective Genetic Algorithms

Genetic Algorithms (GA) are computational algorithms that draw inspiration from the principles of natural selection. These algorithms aim to solve complex problems by iteratively evolving viable solutions through time, with the ultimate objective of identifying the optimal solution. Goldberg [7] offers a thorough overview of genetic algorithms, elucidating the fundamental concepts of natural selection and evolution, and their use in the context of optimization problems. The proposed utilization of crossover and mutation operators in genetic algorithms enables quick traversal of the search space and ensures preservation of diversity within the population. The NSGA-II algorithm, which was proposed by Kalyanmoy Deb [8] in a seminal publication, represents a significant contribution to the domain of multi-objective optimization.

It addresses the challenges of optimizing problems with multiple conflicting objectives. It also introduces an innovative, non-dominated sorting algorithm that effectively classifies solutions into distinct Pareto fronts, preserving the best solutions while maintaining diversity. The NSGA-II algorithm provides a reliable and scalable methodology for addressing complex multi-objective problems.

2.2.2 Multi-objective Particle Swarm Optimization

Coello et al. [9] laid the foundation for Multi-Objective Particle Swarm Optimization (MOPSO), emphasizing the handling of multiple conflicting objectives in particle swarm optimization. Li and Zhang [10] extended the concept of MOPSO by employing a decomposition approach to break down multi-objective problems into various single-objective sub-problems. The proposed method leverages decomposition-based hyper-heuristics in order to optimize multiple conflicting objectives simultaneously. Kennedy and Eberhart [11] proposed particle swarm optimization as a technique that is especially applicative for feature selection tasks, owing to its competency to effectively traverse extensive search spaces.

2.2.3 Multi-objective Optimization in Feature Selection

Li and Zhang's paper [10] centers around the topic of multi-objective feature selection (MOFS) using decomposition-based hyper-heuristics. Instead of a single objective (e.g., accuracy), it considers multiple objectives, such as accuracy, dimensionality reduction, and robustness. The authors apply the principles of balancing trade-offs between conflicting objectives to feature selection, highlighting the potential benefits of employing multi-objective optimization techniques in order to obtain feature subsets that are both resilient and interpretable. Deb et al. [12] discuss the hybrid approach of integrating surrogate modeling with multi-objective optimization. The deployment of surrogate models for efficient approximation of the Pareto front is introduced, hence enhancing the computational feasibility of MOFS.

2.2.4 Challenges and Emerging Trends

Al-Tashi et al. [13] present an extensive review of the potential challenges associated with the multi-objective feature selection problem. Additionally, it offers a critical analysis of the many strategies employed to address this problem. A study conducted by Li [14] focuses on the issue of scalability in MOFS. The authors propose an algorithm that is specifically designed to handle scalability concerns, using dominance and roughness-based selection strategies. Yuan et al. [15] underscore the substantial impact of noise in data on the performance of feature selection algorithms. Recent research by Zhou [16] studies multi-objective optimization techniques to select pertinent features within deep learning models, addressing the challenge of feature selection in complex deep neural networks.

3 Importance Multi-objective Optimization for Feature Selection

3.1 Limitations of Single-objective Feature Selection

Single-objective feature selection (SOFS) methodologies enable the identification of relevant features from large datasets, enhancing the interpretability and efficiency of models. Nevertheless, these methodologies possess inherent constraints that could potentially hinder their applicability in specific situations.

The SOFS approach may fail to consider those subsets that have the potential to enhance several aspects of a model simultaneously. Consequently, useful solutions that optimize multiple objectives may be inadvertently omitted. The choice of the single objective is often context-dependent and subjective, making SOFS inflexible and dependent on problem context. Additionally, these methodologies evaluate the relevance of each feature independently, disregarding the potential interdependencies across features, which are frequently observed in real-world datasets. SOFS algorithms typically employ heuristic search methods to find the best feature subset. Nevertheless, if the initial selection fails to yield substantial enhancements in the selected objective, these algorithms may become bound to local optima.

Another significant drawback of employing algorithms that optimize a single objective, often pertaining to model performance metrics, is the risk of overfitting [17]. The selected features may exhibit a high degree of specificity to the training set by capturing noise and idiosyncrasies rather than the underlying patterns. Overfitting occurs when a model is excessively sensitive to variations in the training data or outliers, resulting in a disproportionate influence on the model's predictions. This could lead to the production of unreliable and unstable results, undermining the model's practicality.

The scalability of SOFS algorithms may be compromised when confronted with high-dimensional datasets that encompass numerous features [18]. The pursuit for the most optimal feature subset can become computationally intensive as the number of features increases. Such methods that require evaluation of each subset might result in exponential time complexity. They also demand substantial computer resources, encompassing both memory capacity and processing capabilities. Moreover, it is essential to note that the feature subset selected by SOFS for a particular model may not exhibit strong adaptability when implemented in other machine learning techniques. Consequently, the selected features may lack compatibility, limiting their broader applicability.

3.2 Advantages of Multi-objective Feature Selection

In a medical diagnosis application, focusing exclusively on optimal classification accuracy might lead to the selection of a feature subset that is difficult for medical practitioners to interpret. A better solution might involve a compromise between accuracy and interpretability. As we know, optimizing a single objective may lead to suboptimal feature subsets in multi-criteria scenarios due to its inability to effectively handle trade-offs between different objectives.

In contrast, multi-objective optimization aims to find results that balance multiple, and often competing, objectives simultaneously. An objective function is formulated in order to optimize a system by integrating various objectives into a unified composite objective function. A variety of techniques have been suggested for addressing composite objective functions. Each optimization algorithm requires thorough exploration of the search space to identify a set of suitable solutions. This set of solutions allows practitioners to select a solution from the set that aligns with their specific priorities or preferences, effectively balancing objectives according to the problem context.

Multi-objective feature selection (MOFS) can identify feature subsets that are robust to noisy data [19]. The feature selection process aims to identify and prioritize those features that have a noteworthy influence on the desired objectives, while simultaneously reducing the impact of irrelevant or noisy features. Robust models built with MOFS are less sensitive to outliers or small fluctuations in the data. They offer stable predictions, making them accurate and reliable machine learning models. In addition, the inclusion of feature selection techniques in machine learning models helps mitigate the danger of overfitting by excluding noisy characteristics that may contribute to the development of overfitted models.

As opposed to single-objective feature selection, MOFS techniques often explicitly consider feature dependencies. While assessing feature subsets, they consider feature interactions and redundancy. Consequently, MOFS tends to select feature subsets that accurately capture the complex relationships within the data more accurately, thereby resulting in feature subsets that are more informative and effective. Feature interactions are frequently observed in applications of natural language processing (NLP) as feature interactions are common because words or phrases may have different meanings when combined. MOFS can enhance the performance of sentiment analysis [20] or language translation models [21] by facilitating the selection of feature subsets that effectively capture these intricate relationships.

Evidently, MOFS is adaptable to evolving objectives. As objectives undergo adaptations, MOFS can provide novel Pareto-optimal feature subsets that are in accordance with the updated criteria. Moreover, MOFS method employs a ranking system to determine the significance of features in the optimization of several objectives. This approach provides valuable insights into the relevance of individual characteristics within the chosen subsets.

4 Algorithms and Approaches for Multi-objective Feature Selection

4.1 Multi-objective Wrapper Methods

Multi-objective wrapper methods primarily aim to determine the feature subset that simultaneously optimizes numerous objectives that reflect different aspects of model performance (accuracy, precision, recall, etc.). A specific supervised machine learning model or algorithm is employed to evaluate the quality of feature subsets [22]. The process of feature selection involves evaluating different subsets of features by training and testing the chosen model on each subset. Wrapper methods can capture complex feature interactions, as they assess the impact of every feature on model performance in combination with other features. They start with an initial feature subset (empty or the full feature set) and then iteratively add or remove features based on the model's performance. This process continues until certain stopping criteria are met. To avoid overfitting and ensure robustness, wrapper methods employ cross-validation approaches, which involve splitting the data into multiple subsets (e.g., training and validation sets) to evaluate the model's ability to generalize. A Pareto front is generated, encompassing solutions that represent trade-offs between conflicting objectives. Decision-makers have the liberty to choose a solution from the Pareto front that aligns with their specific priorities or preferences.

4.1.1 Strengths

Multi-objective wrapper methods are particularly valuable when there are trade-offs between different aspects of model performance, such as accuracy, interpretability, and computational efficiency. These strategies aid in the identification of feature subsets that strike the right balance according to the problem context. Global search strategy, which is adopted by multi-objective wrapper methods, helps identify feature subsets that might be missed by other feature selection methods [23].

Practitioners can define and prioritize objectives based on the specific requirements of their application. This adaptability allows for tailoring feature selection to meet specified objectives and constraints. The methods employed yield a Pareto front that offers a transparent representation of the trade-offs between objectives. Allowing practitioners to opt a solution from the Pareto front that aligns with their priorities empowers them to make informed decisions based on the inherent trade-offs in the problem.

Multi-objective wrapper methods are model-agnostic, allowing practitioners to select the best appropriate model for their particular problem while still benefiting from the advantages of multi-objective feature selection [23]. Feature subsets selected by multi-objective wrapper methods are often robust to changes in data distribution or shifts in problem requirements. Since they consider multiple aspects of performance, they tend to produce solutions that are less sensitive to variations in data. In situations where the relationships between features are intricate and nonlinear, these methods can identify feature subsets that account for these interactions, leading to improved model performance.

4.1.2 Limitations

The effectiveness of multi-objective wrapper methods heavily depends on how well the objectives are defined to represent the problem's goals. Along with this, correctly determining their relative weights can be challenging. Poorly chosen objectives can lead to suboptimal results. Apart from this, the selection of a solution from the Pareto front is often subjective and depends on the practitioner's preferences and priorities. This subjectivity can introduce bias into the feature selection process.

Multi-objective wrapper methods, while theoretically powerful, may have limited real-world applications due to their computational demands and complexity. Generating a Pareto front and depicting probable non-dominated (Pareto-optimal) solutions, can be computationally intensive. This complexity grows significantly with the scope of the feature space, number of objectives, and complexity of the machine learning model. Consequently, multi-objective wrapper methods may not be feasible for large datasets or complex models [23]. The advanced optimization techniques employed by multi-objective wrapper methods demand considerable computational resources, including memory and processing power.

The multi-objective optimization process in feature selection may be more challenging when dealing with categorical data. Most multi-objective wrapper methods are designed for continuous feature spaces. Adapting them to handle categorical or mixed-type data can be non-trivial and may require additional pre-processing steps. One-hot encoding is a common approach to handle categorical data, wherein each category is converted into a binary feature (0 or 1). However, this can substantially expand the dimensionality of the dataset. The nonlinear relationships and non-convex spaces that can arise from one-hot encoding make it harder to explore the objective space effectively.

4.1.3 Case Study: Multi-objective Wrapper Methods for Credit Scoring in Banking

Background:

Credit scoring is a critical task in the banking industry, as it involves making loan decisions that must effectively consider a range of objectives [24]. Banks aim to minimize credit risk while maximizing profitability and ensuring regulatory compliance. Traditional credit scoring models often rely on a single objective, such as minimizing default rates, which can lead to overly conservative lending practices. To address these challenges, the bank decides to utilize multi-objective wrapper approaches for feature selection as a means to optimize its credit scoring model [25].

Data and Features:

The bank collects historical data on loan applicants, including their financial information, credit history, employment status, and demographic data. Features include both numerical variables (e.g., income, credit score) and categorical variables (e.g., employment type, education level).

Multi-Objective Feature Selection:

1. Objective Definition: The bank formulates its objectives as follows:

   Objective 1 (Default Risk Minimization): Minimize the number of false negatives (applicants wrongly classified as low-risk but default) to accurately identify high-risk applicants.

   Objective 2 (Profit Maximization): Maximize the net profit, which is the difference between interest income and expected default losses.

   Objective 3 (Fair Lending Practices): Ensure that the model does not discriminate against any protected attribute while maintaining objectives 1 and 2.
2. Feature Subset Evaluation: Multi-objective wrapper methods involve systematic assessment of various feature subsets by training and testing a machine learning model (e.g., a gradient boosting classifier) on each subset. Each subset is evaluated based on the predefined objectives.
3. Pareto Front Search: A multi-objective optimization algorithm (e.g., NSGA-II [26]) is employed to search for the Pareto front, which represents feature subsets that balance the three objectives. It identifies solutions that provide trade-offs between objectives, such as including demographic features to ensure fairness yet preserving default risk and profitability.
4. Solution Selection: The bank selects a solution from the Pareto front that aligns with its particular priorities. For instance, it may choose a feature subset that balances fairness with default risk and profitability, ensuring that loan decisions are both equitable and financially sound.

Challenges [27]:

1. Data Privacy: Ensuring impartiality of the model while adhering to privacy regulations, such as GDPR, requires careful handling of sensitive information.
2. Model Interpretability: Maintaining model transparency and interpretability while optimizing for multiple objectives remains a challenge, especially when complex machine learning models are used.
3. Model Validation: Rigorous model validation and continuous monitoring are needed to assess the model's performance against the defined objectives in real-world scenarios.

4.2 *Multi-objective Genetic Algorithm Based Approaches*

Multi-Objective Genetic Algorithm (MOGA) [28] strategies leverage genetic algorithm, a search and optimization heuristic inspired by natural selection, to address multiple conflicting objectives in feature selection. Like other multiple objective optimization methods, MOGA starts by establishing multiple objectives that capture different aspects of model performance and feature subset quality. These objectives often include model complexity, classification accuracy, robustness, and interpretability. An initial population of feature subsets needs to be generated either randomly or through heuristic methods. The quality of each feature subset is calculated based on the defined objectives by training and testing machine learning models. The process of selecting feature subsets to create a new generation of solutions is accomplished through the utilization of mechanisms that draw inspiration from natural selection. These mechanisms favor solutions that perform well across multiple objectives. Genetic operators (crossover and mutation) are applied to produce the next generation of solutions from the selected feature subsets. These operations introduce diversity into the population and enable exploration of the search space. A set of non-dominated solutions are generated along the Pareto front and are retained across generations. The algorithm continues to evolve the population for a predefined number of generations or until termination criteria are met. Typical termination criteria include a limit on the number of generations, convergence of the Pareto front, or a user-defined stopping point.

4.2.1 Strengths

Multi-objective genetic algorithms are model-agnostic, meaning they can be applied in conjunction with a wide spectrum of machine learning algorithms. This adaptability allows practitioners to opt for the most suitable model for their specific problem while still benefiting from their advantages. Also, they are highly flexible and adaptable to different feature selection scenarios. Practitioners can define and prioritize objectives based on the specific requirements of their application. This adaptability allows for customization of feature selection to meet diverse goals and constraints [29].

Genetic algorithm based feature selection typically results in feature subsets that exhibit resilience to variations in data distribution. This approach employs global search strategies inspired by natural evolution, such as crossover and mutation. This enables them to explore the feature space more effectively and discover feature subsets that might be missed by other feature selection methods, particularly in high-dimensional spaces. The utilization of genetic operators in these approaches facilitates the exploration of feature subsets that encompass a broad range of potential solutions, hence ensuring diversity in the set of solutions generated, as indicated by the Pareto front [26]. By offering a set of Pareto-optimal solutions, MOGA enhances decision support in complex feature selection tasks. Practitioners can explore the trade-offs and select the most suitable solution for their individual requirements, leading to better-informed decision-making.

Moreover, the transparency of the trade-offs between different objectives offered by the Pareto front aids the practitioners to make informed decisions by selecting the most suitable solution for their specific application.

4.2.2 Limitations

MOGA faces the curse of dimensionality. Exploring a vast search space becomes increasingly challenging and may necessitate an impractically high number of generations and assessments, rendering the optimization process infeasible. The iterative progression of evolving a population over multiple generations, evaluating solutions for multiple objectives, and maintaining a Pareto front proves to be very time-consuming, thus increasing the computational complexity.

The efficacy of this strategy in properly exploring the objective space may be variable. The algorithm's convergence behavior and the quality of the Pareto front can vary depending on factors like population size, mutation rates, and crossover operators. Finding an optimal configuration for these parameters can be a non-trivial task. Furthermore, the parameters (e.g., population size, mutation rate, selection mechanism) of genetic algorithms require parameter fine-tuning for optimal performance. Determining the right parameter settings can be time-consuming and may require expert knowledge.

The effectiveness of MOGA depends on the precise design of genetic operators (crossover and mutation) [30]. Poorly chosen operators can lead to premature convergence, suboptimal solutions, or computational inefficiency. Additionally, this approach may struggle with scalability when dealing with a large number of objectives. The Pareto front may become overcrowded with solutions, making it challenging in discerning distinct trade-offs and choosing the most suitable option.

4.2.3 Case Study: Multi-objective Genetic Algorithm for Feature Selection in Medical Diagnosis

Background:

Medical diagnosis is a critical domain where the selection of relevant features for predictive models is crucial. Inaccurate or overly complex models can lead to misdiagnosis, increased healthcare costs, and patient discomfort. To address these challenges, a healthcare institution decides to leverage multi-objective genetic algorithm—based (MOGAs) feature selection for development of a diagnostic model for a specific disease [31]. The proposed diagnostic model will adopt a patient-centric approach, prioritizing the safety and well-being of individuals.

Data and Features:

The institution gathers a comprehensive dataset that includes patient demographics, medical history, and results from various medical tests. This dataset contains mainly numerical and categorical features.

Multi-Objective Feature Selection:

1. Objective Definition: The institution formulates objectives as follows:

   Objective 1 (Diagnostic Accuracy): Maximize the classification accuracy of the diagnostic model.

   Objective 2 (Model Simplicity): Minimize the number of selected features to create a model that is simple, interpretable, and requires fewer resources.

   Objective 3 (Patient Safety): Prioritize features that do not require invasive or expensive tests to minimize healthcare cost and patient discomfort.
2. Population Initialization: An initial population of feature subsets is generated using random sampling or a heuristic method. Each feature subset represents a potential solution.
3. Fitness Evaluation: The quality of each feature subset is determined based on the defined objectives. A machine learning model (e.g., a support vector machine [31]or decision tree) is trained and tested using the selected feature subset, and performance is assessed across multiple objectives.
4. Selection: Feature subsets are selected to form a new generation of solutions, favoring those that perform well across multiple objectives while balancing diagnostic accuracy, simplicity, and patient safety.
5. Crossover and Mutation: The genetic operators are applied to produce offspring solutions. These operations introduce diversity into the population and enable exploration of the feature space.
6. Pareto Front Maintenance: A set of Pareto-optimal solutions are mapped on the Pareto front. Each solution on the Pareto front represents compromises between different objectives. The institution selects a solution from the Pareto front that aligns with its specific priorities.

Challenges [32]:

1. Data Quality: Ensuring data quality and reliability is crucial, as inaccurate or biased data can lead to suboptimal feature selection and model performance.
2. Ethical Considerations: The institution must carefully consider ethical implications, especially when dealing with patient data, to ensure privacy and compliance with regulations like HIPAA.
3. Validation: Rigorous validation and testing are necessary to assess the generalization of the diagnostic model to new patient cases and different healthcare settings.

4.3 Multi-objective Particle Swarm Optimization

Particle swarm optimization (PSO) is a very effective algorithm for addressing continuous optimization problems, drawing inspiration from the collective behavior observed in bird flocking or fish schooling. In PSO, each potential solution to an optimization problem is depicted as a particle in a multidimensional search space. All particles have a velocity that determines their traversal inside the search space. This algorithm employs an iterative approach to continuously update the positions and velocities of all particles, enabling them to explore the search space and ultimately discover solutions that are either optimal or near-optima. The velocity of each particle is updated based on its best-known position (personal best) and the best-known position among all particles in the swarm (global best). A fitness function assesses the quality of each particle's solution in the search space. It quantifies how well the solution satisfies the optimization objectives.

Similarly, each particle in the MOPSO swarm represents a feature subset [33]. The presence or absence of features in the subset is encoded in the particle's position, while the velocity vector denotes the speed and direction of the particle's movement in the search space. Particles keep a track of their personal as well as global best-known positions and utilize this information to guide their movement through the search space. Multiple objective functions are defined to evaluate the fitness of each particle based on its performance across all defined objectives. Given its iterative nature, this methodology involves updating the velocities and positions of each particle with the aim of exploration (moving toward global best-known position) and exploitation (following personal best-known position). Exploration aims to find diverse solutions across the entire search space, which prevents the algorithm from getting trapped in local optima and ensures that it has a chance to discover potentially better solutions. On the contrary, exploitation focuses on refining and improving the solutions that are already known. It aims to converge toward the best-known solutions and enhance them further.

Maintaining diversity within the population of particles in a MOPSO swarm is essential to prevent premature convergence and facilitate effective exploration of the entire Pareto front effectively. Crowding distance is a common diversity maintenance metric used in MOPSO [34]. It quantifies how densely particles are distributed in the objective space. Crowding distance is used to prioritize particles for selection and removal. Some MOPSO algorithms employ the utilization of multiple sub-swarms that operate independently and periodically exchange information to introduce diversity. This multi-swarm approach allows for diverse exploration and can help avoid premature convergence.

4.3.1 Strengths

The strength of PSO lies in its competence to efficiently explore the search space by combining both local search (personal best) and global search (global best) strategies. It's a population-based optimization technique that does not require gradient information, making it suitable for solving problems where gradients are unknown or discontinuous.

The MOPSO algorithm indirectly addresses the issue of feature dependencies by promoting the selection of relevant feature subsets that balance objectives and eliminate redundancy. The optimization procedure of MOPSO does not make any assumptions about linearity. Instead, it allows for the discovery of nonlinear relationships between attributes and the target variable. The reason for this is that the optimization process is designed to identify subsets of features that optimize multiple objectives, and these objectives can capture nonlinear feature importance. This is valuable for capturing complex patterns that might be missed by linear feature selection methods.

MOPSO algorithm incorporates methods of preserving diversity within the population of solutions. This prevents premature convergence to a single solution and ensures that a wide range of trade-off solutions is available. One of these mechanisms is elitism [35], which involves preserving the best solutions (non-dominated particles) from the current population to the next generation without modification. These elite solutions are considered as benchmarks of high quality and contribute to maintaining diversity. Elitism guarantees that the best solutions are retained during optimization. MOPSO's diversity and exploration capabilities make it robust to noisy or uncertain data. It can discover solutions that are less sensitive to noise and outliers, leading to more reliable results.

MOPSO can be parallelized for more efficient exploration of the search space [36]. The population of particles can be divided into subsets, with each subset being subjected to individual processing by a separate computing unit. Each subset evolves independently, with its own set of particles, and periodically, these subsets exchange information to share promising solutions. The parallelization of MOPSO is feasible across multiple computing nodes in a distributed computing environment [37], such as a cluster or cloud infrastructure. This enables the algorithm to harness significant computational power, making it suitable for large-scale optimization problems.

MOPSO is capable of finding global optimal or near-optimal solutions, provided sufficient exploration is allowed. This makes it suitable for finding the best solutions in complex and high-dimensional spaces. Near-optimal solutions often represent practical and implementable solutions that are closer to real-world feasibility. This makes them valuable for applications in various spheres including engineering, design, and operations.

4.3.2 Limitations

MOPSO requires the tuning of many parameters, including the inertia weight, acceleration coefficients, and population size. The presence of very sensitive parameters can result in variable performance throughout multiple iterations of the algorithm, making it challenging to obtain consistently high-quality solutions. MOPSO algorithm for feature selection typically encompasses several hyperparameters, such as the number of particles, the maximum iteration count, and coefficients associated with the particle motion equation. Tuning these parameters for optimal performance can be challenging and may require substantial trial and error.

Another issue lies in achieving a balance between exploration (searching for new solutions) and exploitation (refining promising solutions). MOPSO may struggle achieving an optimal trade-off, and possibly leading to premature convergence or slow exploration. Evidently, feature selection problems often have non-convex search spaces with multiple local optima. Balancing exploration and exploitation in such spaces requires careful consideration of how to escape local optima while converging to global optima.

MOPSO is inherently designed for continuous optimization problems and may not adequately handle discrete or categorical features. Categorical attributes are not amenable to the continuous movement mechanisms used in MOPSO. This incompatibility can lead to difficulties in representing and updating discrete features. Adapting MOPSO to accommodate discrete features can be complex and may require additional discretization techniques. Additionally, the conversion of features into binary variables might result in the loss of significant information regarding their true influence on the objective functions. This can provide a greater difficulty for the algorithm to distinguish between subtle differences in feature importance.

4.3.3 Case Study: Multi-objective Particle Swarm Optimization (MOPSO) for Feature Selection in Weather Predictions

Background:

Weather prediction is a complex and critical task that relies on various meteorological data sources and models to forecast weather conditions accurately. Feature selection is of utmost importance in enhancing the precision and effectiveness of weather prediction models. This is achieved by finding the meteorological variables that are most pertinent and informative. The aim of this case study is to improve the precision of weather forecasts by accurately choosing the most pertinent meteorological characteristics while simultaneously minimizing the quantity of features employed [38, 39].

Dataset and Features:

The case study utilizes a historical weather dataset containing various meteorological variables such as temperature, humidity, wind speed, atmospheric pressure, and precipitation, collected over several years. These attributes have continuous numeric values.

Multi-Objective Feature Selection:

1. Objective Definition: The objectives can be formulated as follows:

   Objective 1 (Accuracy): Maximize prediction accuracy (e.g., minimizing mean squared error).

   Objective 2 (Dimensionality): Minimize the number of selected features to improve model simplicity and reduce computational costs.

   Objective 3 (Interpretability): Maximize model interpretability by selecting meteorological features that align with known weather patterns and meteorological principles.
2. Initialization: A population of potential feature subsets is generated. Every subset is a binary vector that indicates selection (1) or non-selection (0) of meteorological features. Velocity vectors are also initialized for each subset, which determine the direction and magnitude of movement in the feature space.
3. Evaluation: The quality of each feature subset is evaluated using a forecasting model. This model is trained and tested on historical weather data.
4. Pareto Dominance: Pareto dominance is used to compare solutions. One solution is considered better than another if it performs better or equivalent in all objectives and strictly better in at least one. Over multiple iterations (generations), the algorithm aims to converge toward a set of non-dominated solutions along the Pareto front.
5. Updating Velocity and Positions: The updates of velocity and position are guided by the trade-offs between objectives. The primary aim of this algorithm is to balance improving forecast accuracy while reducing the number of selected features and ensuring interpretability.
6. Termination: The algorithm runs for a predefined number of generations or until convergence criteria are met. After termination, the array of non-dominated solutions along the Pareto front is analyzed to select the most suitable feature subset(s) based on the problem's specific objectives and priorities.
7. Interpretability Analysis: To ensure that the selected features are interpretable, they are subjected to further analysis to evaluate their alignment with known meteorological principles and their contribution to weather patterns. Features that do not provide meaningful insights or are not relevant to weather prediction can be filtered out.

Challenges [40]:

1. Dynamic Data: Weather data is inherently dynamic, with temporal dependencies and varying patterns over time. MOPSO may need adaptations to account for time-dependent feature selection, which is essential for accurate weather forecasting.
2. Real-Time Application: In operational weather forecasting, decisions need to be made in real time. MOPSO-based feature selection should be efficient enough to work within the time constraints of real-time forecasting systems.

3. Limited Domain Knowledge: Effective feature selection often requires domain knowledge. Integrating meteorological expertise into the optimization process can be challenging, especially when domain experts are not familiar with optimization techniques.

4.4 Other Hybrid Approaches

4.4.1 Multi-objective Feature Selection with Metaheuristics

Metaheuristics are a class of algorithms that are characterized by their iterative nature and reliance on heuristics. These algorithms are designed to effectively navigate through extensive solution spaces in order to identify high-quality solutions for complex optimization problems [41]. When employed in context to multi-objective feature selection (MOFS), they aim to discover feature subsets that balance multiple objectives effectively. Several commonly used metaheuristics for MOFS include Simulated Annealing, Ant Colony Optimization (ACO), and Differential Evolution, among others [42]. The selection of algorithm is reliant upon the requirements of the problem at hand as well as the availability of computational resources.

A fitness function is defined to quantify the performance of a feature subset in relation to various objectives. The fitness function combines the values of these objectives into a single scalar value, often using weighted sums or Pareto dominance. This function guides the metaheuristic algorithm in its exploration of the solution space. The metaheuristic algorithm iteratively explores the solution space by creating new candidate solutions through operations such as perturbations. Upon the termination of the metaheuristic algorithm, a set of feature subsets that are Pareto-optimal will be obtained. These subsets offer different trade-offs between objectives.

Metaheuristic algorithms are well-suited for handling high-dimensional feature spaces, where traditional methods may struggle due to computational complexity. Moreover, these methods are generally data-driven [43] and do not require prior domain knowledge or manual feature engineering. They can discover relevant features automatically. In addition, many metaheuristic algorithms can be parallelized, allowing for efficient exploration of the solution space across multi-core processors or distributed computing environments.

However, metaheuristic algorithms, particularly those that involve a large number of iterations, can be computationally intensive and time-consuming for complex problems. Furthermore, properly configuring metaheuristic algorithms and tuning their corresponding hyperparameters can be challenging and may require extensive experimentation. While metaheuristics aim to find good solutions, they do not guarantee finding the global optima. The quality of solutions can be sensitive to algorithmic configurations and initial conditions.

4.4.2 Multi-objective Feature Selection with Deep Learning

In the hybrid approach of multi-objective feature selection with deep learning [16], integration with deep neural networks (DNNs) is vital to simultaneously optimize multiple objectives. This approach is advantageous, especially when dealing with high-dimensional and complex data. Deep learning architecture commonly includes Convolutional Neural Networks (CNNs) for image data, Recurrent Neural Networks (RNNs) for sequential data, and various layers of feedforward neural networks (including fully connected and deep learning models like deep autoencoders) for structured data. Selection of architecture depends on the nature of data available, and the problem being addressed.

A feature selection layer is integrated into the deep learning model [44]. This layer is responsible for selecting a subset of input features that will be used for training the subsequent layers. Feature selection layer can be added as an initial layer in the neural network. Subsequently, during training, the model learns to extract relevant features from the input data to optimize the defined objectives. Optimization of the model's hyperparameters is necessary in order to identify the optimal configuration that aligns with the desired objectives. Upon training the deep learning model and optimizing its hyperparameters, it is possible to obtain many configurations that constitute the Pareto front. This front illustrates the trade-offs between different objectives.

The primary advantage of this methodology is in its incorporation of multi-objective optimization alongside deep learning models, which facilitate autonomous acquisition of pertinent features from unprocessed data, eliminating the need for manual feature engineering and domain expertise. Deep neural networks can capture complex and hierarchical patterns in data, making them effective for feature selection in high-dimensional and unstructured data. Deep learning models can model nonlinear relationships between features and targets, allowing them to discover intricate feature interactions.

However, deep learning typically requires large volumes of labeled data for training, which may be tough to elicit, especially in domains with limited data. Training deep learning models can severely augment time complexity, especially for complex architectures and large datasets. Deep learning often relies on specialized hardware (e.g., GPUs or TPUs) for efficient training, which can be costly and may not be available to all users. Deep learning models are often deemed as black-box models, which can restrict their utility in certain domains.

4.4.3 Multi-objective Feature Selection with Ensemble Methods

Multi-objective feature selection with embedded methods entails the utilization of machine learning algorithms, typically integrated as part of the model-building process, to concurrently optimize various objectives associated to feature selection [22]. It allows for automated and data-driven feature selection, making it particularly useful when dealing with high-dimensional data and complex feature spaces.

Frequently employed machine learning techniques include decision tree-based algorithms (e.g., Random Forests, Gradient Boosting), regularized linear models (e.g., Lasso regression), and support vector machines (SVMs) [45]. The choice of algorithm depends on the type of data and the task at hand. The embedded algorithm or its objective function is modified to incorporate multiple objectives. Typically, the process requires modification of the loss function or regularization terms in order to incorporate the objectives associated with feature selection.

One of the notable advantages of ensemble methods is their ability to combine the decisions of multiple models, reducing the impact of individual model errors and enhancing robustness. Moreover, they can provide estimates of feature importance based on their contributions to model performance, thereby aiding feature selection. Feature importance scores are computed to facilitate the interpretation of the relevance of features in the model's predictions.

The drawback to these methods is that they require diverse and representative data to perform well. Ensembles may exhibit reduced effectiveness in situations when datasets are uneven or biased. Achieving a balance between diversity and accuracy among base models in the ensemble can be challenging. Overly similar models may not contribute significantly to feature selection. Ensemble methods are prone to overfitting, particularly if the ensemble is too complex or if individual models are not properly regularized.

5 Comparison of Multi-objective Feature Selection with Single-Objective Approaches

See Table 1.

6 Challenges and Limitations of Multi-objective Feature Selection

Defining appropriate and meaningful objective functions for MOFS (Multi-Objective Feature Selection) can be challenging. Combing multiple objectives, especially when they are conflicting, requires careful consideration. The selection of objectives and the determination of their respective weights can possess a certain degree of subjectivity [13]. The selection of distinct subsets of attributes may vary depending on how objectives are defined and weighted. This subjectivity can introduce bias into the feature selection process. MOFS often requires domain expertise to design meaningful objective functions for real-world problems and interpret the results effectively. Complex, and often conflicting, objectives may require advanced mathematical modeling or domain-specific knowledge. Without domain knowledge, it may be challenging to make informed decisions.

Table 1 Multi-objective feature selection versus single-objective feature selection

Parameter	Multi-objective feature selection	Single-objective feature selection
Objective	It considers trade-offs between multiple objectives such as accuracy, simplicity, and interpretability	It focuses on optimizing one objective function, often accuracy or precision; it does not consider other criteria concurrently
Solution(s)	A set of Pareto-optimal solutions is produced; final selection is made in accordance with the problem's requirements	Feature subset selected may be more biased toward that objective, potentially neglecting other important factors
Exploration of the solution space	It explores a broader solution space by considering a diverse set of non-dominated solutions along the Pareto front	Single-objective approaches explore a narrower solution space, typically focused on optimizing a single criterion
Robustness	Non-dominant solutions generated may offer better generalization to unseen data	It is more susceptible to overfitting and sensitive to noisy data as only one criterion is optimized
Computational intensity	Exploring the entire Pareto front requires evaluating many feature subsets. Dealing with high-dimensional data or complex objective functions requires high computational power	It is more computationally efficient and hence more appropriate for large datasets
Application	It is well-suited for complex decision-making scenarios where multiple criteria need to be considered, as discussed above	It more suitable for scenarios where a single criterion is of paramount importance, and other factors are less critical

When dealing with imbalanced datasets, MOFS may inadvertently assign higher priority to features belonging to the majority class, hence impacting the performance of models trained on the selected subsets [13]. In such cases, considering metrics like accuracy can be misleading because a model that is biased toward predicting the majority class can still achieve high accuracy, even if it performs inadequately on the minority class.

In high-dimensional datasets, the search space for feature subsets becomes exponentially large, making it computationally expensive and potentially impractical to explore exhaustively. Datasets characterized by a large crowding distance exhibit sparsity, which can give rise to challenges such as overfitting. This occurs when models encounter difficulties in identifying significant patterns within sparsely populated areas of the feature space. Additionally, MOFS may struggle with large datasets or those with many features due to computational and resource constraints. Storing high-dimensional datasets and intermediate results during MOFS can consume substantial memory, especially when dealing with large-scale data. Memory limitations can hinder the execution of MOFS algorithms on standard computing platforms. Scaling up MOFS for big data applications can be difficult.

MOFS algorithms can sometimes get stuck in local optima or fail to explore the entire Pareto front effectively. This can restrict the potential improvements in objectives for feature selection. Improving the exploration–exploitation balance is a popular research topic. Limited exploration often results in a lack of diversity among the solutions along the Pareto front. This implies that the selected features exhibit significant similarities to one another, presenting identical trade-offs between objectives, rather than providing a diverse set of options for decision-makers.

7 Future Scope

This chapter has explored complex domains of data science, optimization, and machine learning that intersect to address the issue of identifying the most informative characteristics from intricate datasets. Throughout this chapter, a wide array of topics has been covered, ranging from the fundamental concepts of multi-objective optimization to the practical implementation of MOFS in various fields. Extensive discussions have been conducted regarding advantages and limitations associated with algorithms that are most commonly employed.

Given the continuous expansion of data, it is imperative to ensure the scalability of multi-objective optimization methods in the context of feature selection. Researchers are continuously developing innovative approaches to enhance scalability and experimenting with various combinations of multi-objective feature selection strategies to establish hybrid methodologies. These studies aim to contribute to increasing the robustness of the selected features. For all the aforementioned methodologies, domain knowledge and demand human intervention are requisite. Techniques to automate feature selection without any human intervention could also be explored.

References

1. Gunantara N (2018) A review of multi-objective optimization: methods and its applications. Cogent Eng 5(1):1–16. https://doi.org/10.1080/23311916.2018.1502242
2. Kennedy MC, Ford ED, Singleton P, Finney M, Agee JK (2007) Informed multi-objective decision-making in environmental management using Pareto optimality. J Appl Ecol 45(1):181–192. https://doi.org/10.1111/j.1365-2664.2007.01367.x
3. Kohavi R, John GH, Wrappers for feature subset selection. [Online]. Available: http://robotics.stanford.edu/
4. Guyon I, De AM (2003) An introduction to variable and feature selection André Elisseeff
5. Brownlee J (2011) Clever algorithms: nature-inspired programming recipes, 1st ed. Lulu.com

6. Saeys Y, Inza I, Larrañaga P (2007) A review of feature selection techniques in bioinformatics. Bioinformatics 23(19). Oxford University Press, pp 2507–2517. https://doi.org/10.1093/bioinformatics/btm344.
7. Goldberg DE, Holland JH (1988) Genetic Algorithms and Machine Learning. Mach Learn 3(2):95–99. https://doi.org/10.1023/A:1022602019183
8. Deb K, Pratap A, Agarwal S, Meyarivan T (2002) A fast and elitist multiobjective genetic algorithm: NSGA-II. IEEE Trans Evol Comput 6(2):182–197. https://doi.org/10.1109/4235.996017
9. Coello CAC, Pulido GT, Lechuga MS (2004) Handling multiple objectives with particle swarm optimization. IEEE Trans Evol Comput 8(3):256–279. https://doi.org/10.1109/TEVC.2004.826067
10. Zhang Q, Li H (2007) MOEA/D: a multiobjective evolutionary algorithm based on decomposition. IEEE Trans Evol Comput 11(6):712–731. https://doi.org/10.1109/TEVC.2007.892759
11. Kennedy J, Eberhart R (1995) Particle swarm optimization. In: Proceedings of ICNN'95—international conference on neural networks, pp 1942–1948 vol 4. https://doi.org/10.1109/ICNN.1995.488968
12. Deb K, Roy PC, Hussein R (2021) Surrogate modeling approaches for multiobjective optimization: methods, taxonomy, and results. Mathem Comput Appl 26(1), https://doi.org/10.3390/mca26010005
13. Al-Tashi Q, Abdulkadir SJ, Rais HM, Mirjalili S, Alhussian H (2020) Approaches to multi-objective feature selection: a systematic literature review. IEEE Access 8:125076–125096. https://doi.org/10.1109/ACCESS.2020.3007291
14. Li L et al, Multi-label feature selection via information gain
15. Yuan W, Guan D, Shen L, Pan H (2014) An empirical study of filter-based feature selection algorithms using noisy training data. In: 2014 4th IEEE international conference on information science and technology, IEEE, pp 209–212. https://doi.org/10.1109/ICIST.2014.6920367
16. Zhou Z, Wong J, Yu K, Li G, Chen S (2023) Feature selection on deep learning models: an interactive visualization approach. [Online]. Available: https://api.semanticscholar.org/CorpusID:260140100
17. Kohavi R, Sommerfield D (1995) Feature subset selection using the wrapper method: overfitting and dynamic search space topology. In: Knowledge discovery and data mining, [Online]. Available: https://api.semanticscholar.org/CorpusID:5147685
18. Bolón-Canedo V, Rego-Fernández D, Peteiro-Barral D, Alonso-Betanzos A, Guijarro-Berdiñas B, Sánchez-Maroño N (2018) On the scalability of feature selection methods on high-dimensional data. Knowl Inf Syst 56(2):395–442. https://doi.org/10.1007/s10115-017-1140-3
19. Dong H, Sun J, Sun X, Ding R (2020) A many-objective feature selection for multi-label classification. Knowl Based Syst 208:106456. https://doi.org/10.1016/j.knosys.2020.106456
20. Ahmad IS, Bakar AA, Yaakub MR (2019) A review of feature selection in sentiment analysis using information gain and domain specific ontology. Int J Adv Comput Res 9(44):283–292. https://doi.org/10.19101/IJACR.PID90
21. Jumelet J, Zuidema W (2023) Feature interactions reveal linguistic structure in language models, [Online]. Available: http://arxiv.org/abs/2306.12181
22. Nouri-Moghaddam B, Ghazanfari M, Fathian M (2021) A novel multi-objective forest optimization algorithm for wrapper feature selection. Expert Syst Appl 175, https://doi.org/10.1016/j.eswa.2021.114737
23. Njoku UF, Abelló A, Bilalli B, Bontempi G (2023) Wrapper methods for multi-objective feature selection, https://doi.org/10.48786/edbt.2023.58
24. Simumba N, Okami S, Kodaka A, Kohtake N (2021) Comparison of profit-based multi-objective approaches for feature selection in credit scoring. Algorithms 14(9):260. https://doi.org/10.3390/a14090260
25. Simumba N, Okami S, Kodaka A, Kohtake N (2022) Multiple objective metaheuristics for feature selection based on stakeholder requirements in credit scoring. Decis Support Syst 155, https://doi.org/10.1016/j.dss.2021.113714

26. Hamdani TM, Won J-M, Alimi AM, Karray F (2007) Multi-objective feature selection with NSGA II. In: International conference on adaptive and natural computing algorithms, [Online]. Available: https://api.semanticscholar.org/CorpusID:26841901
27. Onay C, Öztürk E (2018) A review of credit scoring research in the age of big data. J Financ Regul Compliance 26(3):382–405. https://doi.org/10.1108/JFRC-06-2017-0054
28. Saroj and Jyoti (2014) Multi-objective genetic algorithm approach to feature subset optimization. In: 2014 IEEE international advance computing conference (IACC), pp 544–548. https://doi.org/10.1109/IAdCC.2014.6779383
29. Imani V, Sevilla-Salcedo C, Fortino V, Tohka J (2023) Multi-objective genetic algorithm for multi-view feature selection. [Online]. Available: http://arxiv.org/abs/2305.18352
30. Khan A, Baig AR (2015) Multi-objective feature subset selection using non-dominated sorting genetic algorithm. J Appl Res Technol 13(1):145–159. https://doi.org/10.1016/S1665-6423(15)30013-4
31. Hajimani E, Ruano MG, Ruano AE (2015) MOGA design for neural networks based system for automatic diagnosis of Cerebral Vascular Accidents. In: 2015 IEEE 9th international symposium on intelligent signal processing (WISP) proceedings, IEEE, pp 1–6. https://doi.org/10.1109/WISP.2015.713917.
32. Abdullah TAA, Zahid MSM, Ali W (2021) A review of interpretable ML in healthcare: taxonomy, applications, challenges, and future directions. Symmetry (Basel) 13(12):2439. https://doi.org/10.3390/sym13122439
33. Nguyen HB, Xue B, Zhang M, Similarity based multi-objective particle swarm optimisation for feature selection in classification
34. Raquel CR, Naval PC (2005) An effective use of crowding distance in multiobjective particle swarm optimization. In: Proceedings of the 7th annual conference on Genetic and evolutionary computation, New York, NY, USA: ACM, pp 257–264. https://doi.org/10.1145/1068009.1068047
35. Amoozegar M, Minaei-Bidgoli B (2018) Optimizing multi-objective PSO based feature selection method using a feature elitism mechanism. Expert Syst Appl 113:499–514. https://doi.org/10.1016/j.eswa.2018.07.013
36. Li J-Z, Chen W-N, Zhang J, Zhan Z-H (2015) A parallel implementation of multiobjective particle swarm optimization algorithm based on decomposition. In: 2015 IEEE symposium series on computational intelligence, IEEE, pp 1310–1317. https://doi.org/10.1109/SSCI.2015.187
37. Arun JP, Mishra M, Subramaniam SV (2011) Parallel implementation of MOPSO on GPU using OpenCL and CUDA. In: 2011 18th international conference on high performance computing, IEEE, pp 1–10. https://doi.org/10.1109/HiPC.2011.6152719
38. Sarangi S, Dash PK, Bisoi R (2023) Probabilistic prediction of wind speed using an integrated deep belief network optimized by a hybrid multi-objective particle swarm algorithm. Eng Appl Artif Intell 126, https://doi.org/10.1016/j.engappai.2023.107034
39. Lv SX, Wang L (2022) Deep learning combined wind speed forecasting with hybrid time series decomposition and multi-objective parameter optimization. Appl Energy 311, https://doi.org/10.1016/j.apenergy.2022.118674
40. Ahmad F, Tarik M, Ahmad M, Ansari MZ (2023) Weather forecasting using deep learning algorithms. In: 2023 international conference on recent advances in electrical, electronics & digital healthcare technologies (REEDCON), IEEE, pp 498–502. https://doi.org/10.1109/REEDCON57544.2023.10150439
41. Amarnath B, Appavu alias Balamurugan S (2016) Metaheuristic approach for efficient feature selection: a data classification perspective. Indian J Sci Technol 9(4), https://doi.org/10.17485/ijst/2016/v9i4/87039
42. Agrawal P, Abutarboush HF, Ganesh T, Mohamed AW (2021) Metaheuristic algorithms on feature selection: a survey of one decade of research (2009–2019). IEEE Access 9:26766–26791. https://doi.org/10.1109/ACCESS.2021.3056407
43. Gopalakrishnan A, Vadivel V (2023) Multi-objective metaheuristic optimization algorithms for wrapper-based feature selection: a literature survey. Bulletin Electr Eng Inform 12(5):3061–3066. https://doi.org/10.11591/eei.v12i5.4757

44. Zou Q, Ni L, Zhang T, Wang Q (2015) Deep learning based feature selection for remote sensing scene classification. IEEE Geosci Remote Sens Lett 12(11):2321–2325. https://doi.org/10.1109/LGRS.2015.2475299
45. Bolón-Canedo V, Alonso-Betanzos A (2019) Ensembles for feature selection: a review and future trends. Inform Fusion 52:1–12. https://doi.org/10.1016/j.inffus.2018.11.008